DIFFERENTIAL EQUATIONS
with Introduction to
LAPLACE TRANSFORM

DIFFERENTIAL EQUATIONS
with Introduction to
LAPLACE TRANSFORM

A N Das MSc (Gold Med.), PhD
Associate Professor in Mathematics
Alipurduar College, Jalpaiguri, West Bengal
Ex-Lecturer in Mathematics, St. Joseph's College
Darjeeling, West Bengal

New Central Book Agency (P) Ltd
LONDON
HYDERABAD ERNAKULAM BHUBANESWAR
NEW DELHI KOLKATA PUNE GUWAHATI

NCBA

REGD OFFICE
8/1 Chintamoni Das Lane, Kolkata 700 009, India .
email: ncbapvtltd@ncbapvtltd.com

OVERSEAS
NCBA (UK) Ltd, 149 Park Avenue North, Northampton, NN3 2HY, UK
email: ncbauk@yahoo.com

EXPORT
NCBA Exports Pvt. Ltd, 212 Shahpur Jat, New Delhi 110 049
email: ncbaexports@ncbapvtltd.com

BRANCHES

New Central Book Agency (P) Ltd
212 Shahpur Jat, **New Delhi** 110 049
email: ncbadel@ncbapvtltd.com

Shop Nos. 3 & 4, Vinayak Towers
681/B Budhwar Peth, Appa Balwant Chowk
Pune 411 002
email: ncbapune@ncbapvtltd.com

Shop No. 15, Bharati Towers, Ground Floor
Forest Park
Bhubaneswar 751 001
email: ncbabub@gmail.com

House No. 3-1-315, 1st Floor
Nimboliadda, Kachiguda, **Hyderabad** 500 027
email: ncbahydb@ncbapvtltd.com

GSS Shopping Complex, 1st Floor
Opposite College Post Office, Convent Road
Ernakulam 682 035
0484-4021054 / 53

Radhamayee Complex [Second Floor]
Opp. Axis Bank, Panbazar, Jaswanta Road
Guwahati 781 001
email: ncbaguwahati@ncbapvtltd.com

DIFFERENTIAL EQUATIONS WITH INTRODUCTION TO LAPLACE TRANSFORM ● A N Das

First Published: January 2012
Reprinted: February 2014, **January 2022**

PUBLISHER AND TYPESETTER
New Central Book Agency (P) Ltd
8/1 Chintamoni Das Lane, Kolkata 700 009

COVER DESIGNER
Soumen Paul

TECHNICAL EDITOR
Dipan Roy

PROJECT SUPERVISOR
Abhijit Sen

PROJECT TEAM
Suman Dutta, Koushik Samanta and Prabhat Jas

ISBN: 978 81 7381 663 5

Price ₹ 485.00

CONTENTS

LAPLACE TRANSFORM

PREFACE

This book on "Differential Equations with Laplace Transform" has been written as a textbook for the students of Honours courses of Indian universities and also for the students of engineering stream. Both the parts of the book have been presented in a simple way for easy understanding.

Several problems and theorems are given in different chapters of both sections so that the book can be considered a useful supplement to all current standard texts of those fields. A large number of supplementary problems—added at the end of each chapter—will serve as a complete review of the materials of each chapter.

My expectation will be fulfilled if the book serves its purpose. Any suggestion from the readers for the improvement of the book will be accepted thankfully.

Finally, I am thankful to Sri Amitabha Sen, Director, New Central Book Agency (P) Ltd, Kolkata, for his sincere care in publishing the book in time.

30 November 2011 **A. N. Das**
Alipurduar, West Bengal

Differential Equations

Chapter 1

ELEMENTARY CONCEPTS

1.1 Introduction

The governing equation of most of the problems of science and engineering are differential equations involving one or more unknown functions with certain prescribed conditions on the function and its derivatives. For example:

(i) One dimensional heat conduction equation,

(ii) One dimensional wave equation,

(iii) Longitudinal vibration of a beam,

(iv) Heat conduction in a cylinder, etc.

So to solve these problems means to solve the differential equation with given conditions associated with the problems.

Several analytical methods have been developed so far to solve different types of differential equations to get the exact solutions. In spite of that, these methods are not adequate to solve all types of differential equation and in such cases numerical methods are adopted to get an approximate solution.

1.2 Definitions

Differential equation is an equation involving an unknown function, its derivatives and independent variable(s).

Taking x as an independent variable and y as a dependent variable, we give here a few examples of differential equation:

(i) $y' + y = 0$, (ii) $y'' + y'^2 + 2y = 0$,

(iii) $\sqrt{y'^2 + y} = \sin x$, (iv) $\frac{\partial y}{\partial x} = c\frac{\partial y}{\partial t'}$,

(v) $\frac{\partial^2 z}{\partial x^2} + \frac{\partial^2 z}{\partial y^2} = 0$.

Differential equations are divided into two categories viz., *ordinary* and *partial* differential equations.

The differential equations which involve differential coefficients with respect to only one independent variable are called *ordinary differential equations* and the differential equations which involve partial differential coefficients with respect to more than one independent variable are called *partial differential equations.*

Examples (i)-(iii) are of ordinary differential equations and examples (iv), (v) are of partial differential equations.

The *order of differential equation* is the order of the highest ordered derivative present in the equation and the *degree of a differential equation* is the degree of the highest ordered derivative (after rationalizing it) present in the equation. In fact, the term 'degree' is used with reference to those differential equations which are expressed as a polynomial in the unknown function and its derivatives.

Order of the examples (i), (iii), (iv) is 1 and the rest are of order 2. Degree of all the examples except (iii) is 1 and degree of that example is 2.

1.3 Linear and Non-linear Differential Equations

If the coefficients of the dependent variable y and its derivatives of a first degree differential equation are functions of x only and the degree of y and all its derivatives is one, then the differential equation is called *linear* differential equation, otherwise it is said to be a *non-linear* differential equation.

The general form of a linear differential equation is

$$a_n(x)\frac{d^n y}{dx^n}+a_{n-1}(x)\frac{d^{n-1}y}{dx^{n-1}}+\cdots+a_2(x)\frac{d^2y}{dx^2}+a_1(x)\frac{dy}{dx}+a_0(x)y=f(x),$$

where $a_k(x), k=0,1,2,\ldots,n$ and $f(x)$ are functions of x only.

If $a_k(x)=c_k x^k, k=0,1,2,\cdots,n$, where c_k's are arbitrary constants, then the above form of linear differential equation is called *homogeneous linear differential equation.*

Examples: (i) $x\dfrac{dy}{dx}+y=0,$

(ii) $(3+2x)^2\dfrac{d^2y}{dx^2}+(3+2x)\dfrac{dy}{dx}+y=e^{2x},$

(iii) $x^2\dfrac{d^3y}{dx^3}+\cos x\dfrac{dy}{dx}+y=\sin x,$

(iv) $\dfrac{dy}{dx}+y^{3/2}=\cos x,$

(v) $x^2\dfrac{d^3y}{dx^3}+\dfrac{dy}{dx}+\sin y=0,$

(vi) $x^2\dfrac{d^3y}{dx^3}+y\dfrac{dy}{dx}+5y=0.$

In the above examples, equations given by (i)-(iii) are of linear differential equation, but in (iv) degree of $y^{3/2}$ is not one, in (v) sin y present, which is a transcendental function, in (vi) the term $y\frac{dy}{dx}$ is present, so examples (iv)-(vi) are of non-linear differential equations.

It is to be noted that every linear equation is of first degree, but the converse is not true as can be seen from the exampe (iv).

1.4 Formulation of Differential Equations

Origin of differential equations are basically different (i) geometrical problems, (ii) physical problems and (iii) elimination of arbitrary constants occurring in the relation between the independent variable x and the dependent variable y as can be seen from the following examples:

EXAMPLE 1

A plane curve is defined by the condition that at each point on it, the slope is proportional to the product of the coordinates of that point. Express the condition by means of a differential equation.

Let the plane curve be represented by the equation $y = f(x)$. Then at the point (x, y) on it, the slopc is given by $\frac{dy}{dx}$. So, using the given condition, we get

$$\frac{dy}{dx} = kxy, \; k \text{ being a non - zero real number.}$$

This is the differential equation representing the condition.

EXAMPLE 2

A body of mass m is dropped from some height under gravity in a medium which offers a resistance proportional to its velocity. Express the law according to which the velocity varies as it falls.

If at any time, t, v is the velocity of the body, then the applied force on the body is $mg - kv$, k is a non-zero real number and effective force is $m\frac{dv}{dt}$. Hence the equation of motion gives

$$m\frac{dv}{dt} = mg - kv.$$

This is the differential equation representing the law according to which the velocity varies as it falls.

EXAMPLE 3

Eliminate the arbitrary constants A, B from the relations $y = Ae^{mx} + Be^{-mx}$.

Differentiating the given relations two times (as the number of arbitrary constants are two) we get

$$y = Ae^{mx} + Be^{-mx}$$

$$y' = mAe^{mx} - mBe^{-mx}$$

$$y'' = m^2 Ae^{mx} + m^2 Be^{-mx}$$

$$= m^2\left(Ae^{mx} + Be^{-mx}\right) = m^2 y.$$

Thus the differential equation obtained after elimination of the arbitrary constants is $y'' = m^2 y$. Consider the equation

$$f(x, y, c_1, c_2, \ldots, c_n) = 0 \tag{1}$$

in, x, y and n arbitrary constants $c_1, c_2, \cdots, c_n$. To form a differential equation satisfied by the above relation between x, y for different values of the constants, we differentiate the relation with respect to x successively and get

$$\frac{\partial f}{\partial x} + \frac{\partial f}{\partial y} \cdot \frac{dy}{dx} = 0,$$

$$\frac{\partial^2 f}{\partial x^2} + 2\frac{\partial^2 f}{\partial x \partial y} \cdot \frac{dy}{dx} + \frac{\partial^2 f}{\partial y^2} \cdot \left(\frac{dy}{dx}\right)^2 + \frac{\partial f}{\partial y} \cdot \frac{d^2 y}{dx^2} = 0,$$

....

....

$$\frac{\partial^n f}{\partial x^n} + \cdots + \frac{\partial f}{\partial y} \cdot \frac{d^n y}{dx^n} = 0.$$

From these n equations and the given relation (1) we get a differential equation of the form

$$F\left(x, y, \frac{dy}{dx}, \frac{d^2 y}{dx^2}, \cdots, \frac{d^n y}{dx^n}\right) = 0 \tag{2}$$

after eliminating the constants $c_1, c_2, \ldots, c_n$. Clearly, (1) is the primitive of (2).

Conversely, if an n-th order differential equation is given, then integrating it we get a differential equation of order $n - 1$ and an arbitrary constant. The reduced form of the differential equation is called first integral of the given differential equation. Again integrating this first integral we get second integral of the given differential equation containing two arbitrary constants. Proceeding in this way we get, after nth step, the primitive of the given differential equation containing n arbitrary constants.

Thus we state the following theorem (without proof), which is called *fundamental theorem*,

An nth ordered differential equation has n, and cannot have more than n independent first integrals and so its general solution cannot have more than n arbitrary independent constants.

1.5 Solution of a Differential Equation

The *solution* or *integral* of a differential equation is a relation between the dependent and independent variables which, when put into the equation, converts it into an identity.

General Solution: A solution of a differential equation containing as many arbitrary independent constants as the order of the differential equation is called the *general solution* or *complete primitive.*

EXAMPLE

The general solution of the differential equation $y'' = m^2 y$ is $y = Ae^{mx} + Be^{-mx}$ as it contains two arbitrary independent constants and the order of the equation is two.

Particular Solution: A *particular solution* of a differential equation is obtained from the general solution for particular values of the constants occurring in the general solution.

EXAMPLE

In above example of general solution if we take $A = 1$, $B = 0$, then $y = e^{mx}$ is a particular solution of the differential equation $y'' = m^2 y$.

Singular Solution: The general solutions do not give all possible solutions of a differential equation, i.e., apart from the above solutions, there exist solutions which are neither general solutions nor can be obtained from the general solution by any manipulation of the arbitrary constants. Such solutions are called *singular solutions.*

EXAMPLE

Consider the equation

$$y = px + a\sqrt{p^2 + 1}, \text{ where } p = \frac{dy}{dx}. \tag{1}$$

Differentiating both sides with respect to x, we get $\frac{dp}{dx}\left\{x + \frac{ap}{\sqrt{p^2 + 1}}\right\} = 0$. Thus we have

$$y = px + a\sqrt{p^2 + 1} \tag{2}$$

$$\frac{dp}{dx} = 0, \text{ i.e., } p = C \tag{3}$$

and

$$x + \frac{ap}{\sqrt{p^2 + 1}} = 0. \tag{4}$$

Eliminating p from (2), (3), we get $y = Cx + a\sqrt{C^2 + 1}$, which is the general solution of the differential equation (1). Again, eliminating p from (2), (4), we get $x^2 + y^2 = a^2$, which is also a solution of the differential equation (1). But this one is neither a general solution nor a particular solution. This solution is the singular solution of the differential equation (1). It is to be noted that the general solutions of the differential equation (1) are tangents to the singular solution of it.

Geometrical Interpretations: Geometrically a differential equation represents a family of curves having some common property, i.e., the general solution of a differential equation represents a family of curves and particular solution represents one of this of curves, whereas singular solution of a differential equation represents the envelope of the family of the straight lines represented by the general solution of that differential equation.

WORKED-OUT EXAMPLES

1. Determine the degree and order of the following differential equation:

(i) $dy + (xy - 1)\,dx = 0$, (ii) $\sqrt[3]{\left[1+\left(\dfrac{dy}{dx}\right)^4\right]} = \dfrac{d^2y}{dx^2}$,

(iii) $\dfrac{d^2y}{dx^2}\dfrac{dy}{dx}+\left(\dfrac{dy}{dx}\right)^2 + y = 0$, (iv) $e^{y''} - xy' + y = 0$.

Solution

(i) Rewriting the given equation as $\dfrac{dy}{dx} + xy = 1$, we see that both order and degree of this equation are one.

(ii) Rewriting the given equation as $1+\left(\dfrac{dy}{dx}\right)^4 = \left(\dfrac{d^2y}{dx^2}\right)^3$, we see that the order of the equation is 2 and degree of the equation is 3.

(iii) In this equation the highest order derivative of y is 2 and its power is 1, so the equation is of order 2 and of degree 1.

(iv) In this equation the highest order derivative of y is 2 but $e^{y''}$ has no degree. So the equation is of order 2 but of no degree.

2. Check the solutions:

(i) $xy = e^{ax}$ of the equation $x\dfrac{dy}{dx} + y = y\ln(xy)$,

(ii) $y = a\tan^{-1}x + b$ of the equation $(1+x^2)\dfrac{d^2y}{dx^2} + 2x\dfrac{dy}{dx} = 0$.

Solution

(i) Differentiating $xy = e^{ax}$ with respect to x, we get

$$x\frac{dy}{dx} + y = ae^{ax} = axy = y\ln(xy).$$

Hence, $xy = e^{ax}$ is a solution of the equation $x\dfrac{dy}{dx} + y = y\ln(xy)$.

(ii) Differentiating $y = a\tan^{-1}x + b$ with respect to x, we get

$$\frac{dy}{dx} = \frac{a}{1+x^2}$$

or, $$(1+x^2)\frac{dy}{dx} = a.$$

Again differentiating it with respect to x, we get

$$\left(1+x^2\right)\frac{d^2y}{dx^2}+2x\frac{dy}{dx}=0.$$

Hence, $y=a\tan^{-1}x+b$ is a solution of the equation $\left(1+x^2\right)\frac{d^2y}{dx^2}+2x\frac{dy}{dx}=0.$

3. Find the differential equation of the following:

(i) $y=a\cos x+b\sin x+x\sin x,$ (ii) $y=e^x(a\cos x+b\sin x).$

Solution

(i) Differentiating $y=a\cos x+b\sin x+x\sin x$ with respect to x, we get

$$y'=-a\sin x+b\cos x+\sin x+x\cos x.$$

Again differentiating it with respect to x, we get

$$y''=-a\cos x-b\sin x+\cos x+\cos x-x\sin x$$

or, $y''=2\cos x-y$

or, $y''+y=2\cos x.$

(ii) Differentiating $y=e^x(a\cos x+b\sin x)$ with respect to x, we get

$$y'=e^x(a\cos x+b\sin x)+e^x(-a\sin x+b\cos x)$$

or, $y'=y+e^x\,(-a\sin x+b\cos x).$

Again differentiating it with respect to x, we get

$$y''=y'+e^x\,(-a\sin x+b\cos x)-e^x\,(a\cos x+b\sin x)$$

or, $y''=y'+y'-y-y$

or, $y''-2y'+2y=0.$

4. Find the differential equation of the curves for different values of the constants:

(i) $xy=Ae^x+Be^{-x}+x^2,$ (ii) $y=a+be^{2x}+ce^{3x}.$

Solution

(i) Differentiating $xy=Ae^x+Be^{-x}+x^2$ with respect to x twice, we get

$$xy'+y=Ae^x-Be^{-x}+2x$$

and $xy''+2y'=Ae^x+Be^{-x}+2$

$$=xy-x^2+2$$

or, $xy''+2y'-xy+x^2-2=0.$

(ii) Differentiating $y=a+be^{2x}+ce^{3x}$ with respect to x thrice, we get

$$y'=2be^{2x}+3ce^{3x},$$

$$y''=4be^{2x}+9ce^{3x},$$

and $y'''=8be^{2x}+27ce^{3x}.$

Solving the first two of these three equations for be^{2x} and ce^{3x}, we get

$$be^{2x}=\frac{1}{2}(3y'-y'')\text{ and }ce^{3x}=\frac{1}{3}(y''-2y').$$

Using these in the last equation, we get

$$y''' = 4(3y' - y'') + 9(y'' - 2y')$$

$$\text{or,} \quad y''' - 5y'' + 6y' = 0.$$

5. Show that the differential equation of the system of recangular hyperbola $xy = c^2$ is $xdy + ydx = 0$, and interpret the result geometrically.

Solution

Taking differential of the equation $xy = c^2$, we get $d(xy) = d(c^2)$, or, $xdy + ydx = 0$. This is the differential equation of the system of rectangular hyperbola $xy = c^2$, From $xdy + ydx = 0$, we get

$$\frac{dy}{dx} = -\frac{y}{x} = \tan\alpha \text{ (say).} \tag{1}$$

Thus the tangent to the rectangular hyperbola $xy = c^2$ makes an angle α with the positive direction of x-axis.

Again, equation of a radius vector, i.e., of a straight line joining origin to any point on the rectangular hyperbola $xy = c^2$ is $y = mx$, m being a parameter. Differentiating it with respect to x, we get

$$\frac{dy}{dx} = m = \frac{y}{x} = \tan(\pi - \alpha), \tag{2}$$

i.e., the radius vector is inclined to the positive direction of x-axis at an angle $\pi - \alpha$. Thus the radius vector of any point on the rectangular hyperbola and the tangent to the rectangular hyperbola at that point are equally inclined to the x-axis.

6. Find the differential equation of the circles passing through origin and having their centres on x-axis.

Solution

General equation of the circles through the origin is

$$x^2 + y^2 + 2gx + 2fy = 0.$$

If its centre $(-g, -f)$ lies on x- axis, then we have $f = 0$. So the differential equation of the circles passing through origin and having their centres on x-axis is

$$x^2 + y^2 + 2gx = 0, \text{ where } g \text{ is the only parameter.}$$

Differentiating with respect to x, we get

$$2xdx + 2ydy + 2gdx = 0$$

$$\text{or,} \quad g = -x - y\frac{dy}{dx}.$$

Eliminating g, we get

$$x^2 + y^2 - 2\left(x + y\frac{dy}{dx}\right)x = 0$$

$$\text{or,} \quad x^2 - y^2 + 2xy\frac{dy}{dx} = 0.$$

This is the differential equation of the circle passing through origin and having their centres on x-axis.

7. Find the differential equation of the circles having their centres on y-axis.

Solution

General equation of the circles is

$$x^2 + y^2 + 2gx + 2fy + c = 0.$$

If it's centre $(-g, -f)$ lies on y-axis, then we have $g = 0$. So the differential equation of the circles having their centres on y-axis is

$$x^2 + y^2 + 2fy + c = 0, \text{ where } f, c \text{ are the parameters.}$$

Differentiating with respect to x, we get

$$2xdx + 2ydy + 2fdy = 0$$

$$\text{or, } \quad \frac{x + yy'}{y'} = -f.$$

Again differentiating with respect to x, we get

$$\frac{y'\left(1 + y'^2 + yy''\right) - (x + yy')\, y''}{y'^2} = 0$$

$$\text{or, } \quad xy'' - y'^3 - y' = 0, \text{ where } y' = \frac{dy}{dx}, \text{ etc.}$$

This is the differential equation of the circles having their centres on y- axis.

8. Find the differential equation of the curves through the intersection of the circle $x^2 + y^2 = 1$ and the straight line $y = x$.

Solution

Equation of the curves through the intersection of the circle $x^2 + y^2 = 1$ and the straight line $y = x$ is

$$x^2 + y^2 - 1 + k\,(x - y) = 0, \text{ where } k \text{ is the only parameter.}$$

Differentiating with respect to x, we get

$$2x + 2yy' + k\,(1 - y') = 0, \text{ where } y' = \frac{dy}{dx}$$

$$\text{or, } \quad k = \frac{2\,(x + yy')}{y' - 1}.$$

Eliminating k, we get

$$x^2 + y^2 - 1 + \frac{2\,(x + yy')}{y' - 1}(x - y) = 0$$

$$\text{or, } \quad \left(x^2 + y^2 - 1\right)(y' - 1) + 2\,(x + yy')(x - y) = 0$$

$$\text{or, } \quad \left(x^2 - 2xy - y^2 + 1\right) dx + \left(x^2 + 2xy - y^2 - 1\right) dy = 0.$$

This is the differential equation of the curves through the intersection of the circle $x^2 + y^2 = 1$ and the straight line $y = x$.

9. Find the differential equation of all parabolas, with foci at the origin and axes along the x-axes.

Solution

Equation of all parabolas, with foci at the origin and axis along the x-axis is

$$y^2 = 4a(x+a), \text{ where } a \text{ is the only parameter.}$$

Differentiating it with respect to x, we get

$$2yy' = 4a \Rightarrow a = \frac{yy'}{2}.$$

Eliminating a, we get

$$y^2 = 4\frac{yy'}{2}\left(x + \frac{yy'}{2}\right) = yy'(2x + yy')$$

or, $\quad y = 2xy' + yy'^2$

or, $\quad yy'^2 + 2xy' - y = 0.$

10. Find the differential equation of all conics whose axes coincide with the coordinate axes.

Solution

General equation of a conic is

$$ax^2 + 2hxy + by^2 + 2gx + 2fy + c = 0.$$

Since axes of the conics are parallel to the coordinate axes, we must have $h = 0$. Again, since the centre of the conics is origin, we must have $g = 0 = f$. So, equation of all conics whose axes coincide with the coordinate axes is

$$ax^2 + by^2 + c = 0, \text{ where } a, b \text{ are the parameters.}$$

Differentiating it with respect to x, we get

$$2ax + 2byy' = 0 \quad \text{or,} \quad \frac{yy'}{x} = -\frac{a}{b}.$$

Differentiating with respect to x, we get

$$\frac{x\left(y'^2 + yy''\right) - yy'}{x^2} = 0$$

or, $\quad xyy'' + xy'^2 - yy' = 0.$

11. Find the differential equation of a system of con-focal conics $\dfrac{x^2}{a^2+\alpha} + \dfrac{y^2}{b^2+\alpha} = 1.$ where α is a parameter and a, b are constants.

Solution

Differentiating the given equation with respect to x, we get

$$\frac{x}{a^2+\alpha} + \frac{yy'}{b^2+\alpha} = 0, \text{ where } y' = \frac{dy}{dx}.$$

or, $\quad x\left(b^2+\alpha\right) + yy'\left(a^2+\alpha\right) = 0$

or, $\alpha(x+yy') = -(xb^2 + yy'a^2)$

or, $\alpha = -\dfrac{xb^2 + yy'a^2}{x+yy'}$.

So, $$a^2 + \alpha = \frac{a^2(x+yy') - xb^2 - yy'a^2}{x+yy'} = \frac{x(a^2-b^2)}{x+yy'}$$

and $$b^2 + \alpha = \frac{b^2(x+yy') - xb^2 - yy'a^2}{x+yy'} = \frac{yy'(b^2-a^2)}{x+yy'}.$$

So eliminating α from the given equation with the aid of the above results, we get

$$\frac{x^2(x+yy')}{x(a^2-b^2)} + \frac{y^2(x+yy')}{yy'(b^2-a^2)} = 1$$

or, $$x^2 + xyy' - y^2 - \frac{xy}{y'} = a^2 - b^2$$

or, $$x^2 - y^2 + xy\left(y' - \frac{1}{y'}\right) = a^2 - b^2.$$

It is interesting to note that the above equation remains unchanged if replace y' by $-\dfrac{1}{y'}$.

12. Find the differential equation of the parabolas touching the coordinate axes.

Solution

Equation of a parabola touching coordinate axes at a distance a, b from the origin is given by

$$\sqrt{\frac{x}{a}} + \sqrt{\frac{y}{b}} = 1.$$

Differentiating with respect to x, we get

$$\frac{1}{\sqrt{xa}} + \frac{1}{\sqrt{yb}}y' = 0, \text{ where } y' = \frac{dy}{dx}.$$

or, $$\sqrt{\frac{b}{a}} + \sqrt{\frac{x}{y}}y' = 0$$

Again differentiating with respect to x, we get

$$\frac{\left(\frac{1}{2\sqrt{x}}y' + \sqrt{x}y''\right)\sqrt{y} - \frac{\sqrt{x}y'^2}{2\sqrt{y}}}{y} = 0$$

or, $y(y' + 2xy'') - xy'^2 = 0$ or, $2xyy'' - xy'^2 + yy' = 0.$

13. Find the differential equation of the family of circles $x^2+y^2+2gx+2fy+c=0$.

Solution

Differentiating $x^2+y^2+2gx+2fy+c=0$ with respect to x twice, we get

$$x+yy'+g+fy'=0$$

and $\quad 1+y'^2+yy''+fy''=0$ or, $\dfrac{1+y'^2+yy''}{y''}=-f.$

Again differentiating, we get

$$\frac{(2y'\,y''+y'\,y''+yy''')\,y''-(1+y'^2+yy'')\,y'''}{y''^2}=0$$

or, $\quad (1+y'^2)\,y'''-3y'\,y''^2=0.$

14. Show that the substitution $z=\ln(1+2x)$ transform the equation

$$(1+2x)^2\,y''-6(1+2x)\,y'+16y=8(1+2x)^2, \text{ where } y'=\frac{dy}{dx}, \text{ etc.,}$$

into $\quad \dfrac{d^2y}{dz^2}-4\dfrac{dy}{dz}+4y=2e^{2z}.$

Solution

From $z=\ln(1+2x)$, we get $\dfrac{dy}{dz}=\dfrac{dy}{dx}\dfrac{dx}{dz}=\dfrac{1+2x}{2}\dfrac{dy}{dx}$, i.e., $(1+2x)\,y'=2\dfrac{dy}{dz}$ and

$$\frac{d^2y}{dz^2}=\frac{d}{dx}\left(\frac{dy}{dz}\right)\frac{dx}{dz}$$

$$=\frac{1+2x}{2}\frac{d}{dx}\left(\frac{1+2x}{2}\frac{dy}{dx}\right)=\frac{1+2x}{2}\left(\frac{dy}{dx}+\frac{1+2x}{2}\frac{d^2y}{dx^2}\right),$$

i.e., $\quad (1+2x)^2\,y''=4\dfrac{d^2y}{dz^2}-2(1+2x)\dfrac{dy}{dx}=4\dfrac{d^2y}{dz^2}-4\dfrac{dy}{dz}.$

Using these results in the given equation we see that the given equation, under the transformation $z=\ln(1+2x)$, reduces to

$$4\frac{d^2y}{dz^2}-4\frac{dy}{dz}-12\frac{dy}{dz}+16y=8e^{2z} \text{ or, } \frac{d^2y}{dz^2}-4\frac{dy}{dz}+4y=2e^{2z}.$$

15. Find the order and degree of the differential equation of the family of curves $y^2=2c\left(x+\sqrt{c}\right)$.

Solution

Differentiating, $y^2=2c\left(x+\sqrt{c}\right)$ with respect to x, we get

$$yy'=c.$$

Eliminating the parameter c from the given equation, we get

$$y=2y'\left(x+\sqrt{yy'}\right) \text{ or, } y^2-4xyy'+4x^2y'^2=4yy'^3.$$

This is a differential equation of order one and of degree three.

EXERCISE

1. Determine the order and degree of the following differential equation:

(i) $x\frac{dy}{dx}+y=5$ (ii) $y'''^2+2y''^3+y'=\sinh x$

(iii) $y''+3y'^3+y=\sin x$ (iv) $\sqrt{y''+y}=e^{2x}$

(v) $y'+x=\left(y-\frac{x}{y'}\right)^3$ (vi) $\sqrt[3]{1+\left(\frac{dy}{dx}\right)^4}=\frac{d^2y}{dx^2}$

(vii) $y=x\frac{dy}{dx}+k\frac{dx}{dy}$ (viii) $y=x\frac{dy}{dx}+a\sqrt{1+\left(\frac{dy}{dx}\right)^2}$

(ix) $\left(\frac{d^2y}{dx^2}\right)^{3/4}+\left(\frac{d^2y}{dx^2}\right)^{4/3}=0$ (x) $\frac{d^2}{dx^2}\left(\frac{d^2y}{dx^2}\right)^{-\frac{3}{2}}=0.$

2. Check the solutions:

(i) $xy=a\cos kx+b\sin kx$ of the equation $xy''+2y'+k^2xy=0$

(ii) $y=a\ln x+b$ of the equation $xy''+y'=0$,

(iii) $y=\frac{a}{x}+b$ of the equation $xy''+2y'=0$,

(iv) $y=a\sin\left(\sqrt{k}x+b\right)$ of the equation $y''+ky=0$,

(v) $y=a+b\cos x$ of the equation $y''=y'\cot x$.

3. Show that the differential equation of a system of concentric circles is $xdx+ydy=0$. Interpret the result geometrically.

[*Hints*: Equation of the family of concentric circles is $x^2+y^2=a^2$, a being the parameter. Taking differential of both sides we get the result. Further we see that, any radius of the circles is represented by $y=mx$. From this we get $\frac{dy}{dx}=\frac{y}{x}$ and from the equation we get $\frac{dy}{dx}=-\frac{x}{y}$, i.e., product of the slopes is –1]

4. Find the differential equation of all tangents to the parabola $x^2=y$.

5. Find the differential equation from each of the following relations:

(i) $y=ax+bx^2$, (ii) $xy=ae^x+be^{-x}+x^2$,

(iii) $y=a+be^{5x}$, (iv) $y=a+be^{5x}+ce^{-5x}$,

(v) $y=a+b\cos x$, (vi) $y=a\cosh x+b\sinh x$,

(vii) $y=a\cosh x+b\sinh x+c\sin x+d\cos x$,

(viii) $y=c\cosh\left(\frac{x}{c}\right)$.

6. Find the differential equation of the family of circles of constant radius c.

[*Hints*: Equation of the family of circles of constant radius c is

$$(x-a)^2+(y-b)^2=c^2.$$

Differentiating it with respect to x, we get

$$x - a + (y - b)\,y' = 0, \text{ where } y' = \frac{dy}{dx}. \tag{1}$$

Again differentiating, we get

$$1 + y'^2 + (y - b)\,y'' = 0. \tag{2}$$

Solving (2) we get the value of y and using it in (1) we get the value x. Finally, using theses values of x, y in the equation of the circle we get the required differential equation.]

7. (a) Find the differential equation of the family of circles touching x-axis at the origin.

[*Hints*: By symmetry, centre of a circle touching x-axis at the origin will lie on y-axis. So, equation of the family of circles touching x-axis at the origin is

$$x^2 + y^2 - 2ky = 0,\ k \text{ being the parameter.}]$$

(b) Find the differential equation of the family of circles touching y-axis at the origin.

8. Find the differential equation of the circles having centres on the line $y + x = 0$ and passing through the origin.

[*Hints*: Equation of the circles having centres on the line $y + x = 0$ and passing through the origin is

$$x^2 + y^2 - 2k(x - y) = 0,\ k \text{ being the parameter.}$$

Differentiating it with respect to x, we get

$$x + yy' - k(1 - y') = 0,$$

or, $$k = \frac{x + yy'}{1 - y'}.$$

Using this value of k in the equation of the circle, we get

$$x^2 + y^2 - 2\frac{x + yy'}{1 - y'}(x - y) = 0$$

or, $$\left(x^2 - 2xy - y^2\right)dx + \left(x^2 + 2xy - y^2\right)dy = 0.\ ]$$

9. Find the differential equation of the family of parabolas having their axes parallel to x-axis.

[*Hints*: Equation of the family of parabolas having their axes parallel to x-axis is

$$x = ay^2 + by + c, \text{ where } a,\ b,\ c \text{ are the parameters.}$$

Differentiating it with respect to x twice, we get

$$1 = 2ayy' + by',$$

and $$0 = 2a\left(y'^2 + yy''\right) + by''$$

or, $$\frac{y'^2 + yy''}{y''} = -\frac{b}{2a}.$$

Again differentiating it with respect to x, and simplifying, we get

$$y'y''' = 3y''^2.]$$

10. Find the differential equation of the family of parabolas having their axes parallel to y-axis.

[Hints: Equation of the family of parabolas having their axes parallel to y- axis is $y = ax^2 + bx + c$, where a, b, c are the parameters.]

11. Find the differential equation of the family of curves $y = a(1 + \cos x)$.

12. Find the differential equation of the system of circles of radius a and having centres on the x-axis. Obtain the equation if a is allowed to vary.

[*Hints*: Equation of the system of circles of radius a and having centres on the x-axis is

$$(x - k)^2 + y^2 = a^2, \text{ where } k \text{ is the only parameter.}$$

Differentiating it with respect to x, we get

$$x - k + yy' = 0.$$

Eliminating k, we get

$$y^2(1 + y'^2) = a^2.$$

If a is allowed to vary, then differentiating the above with respect to x, we get

$$yy'(1+y'^2) + y^2y'y'' = 0 \quad \text{or,} \quad yy''+y'^2+1 = 0\,]$$

13. Find the differential equation of the system of circles of radius a and having centres on the y-axis. Obtain the equation, if a is allowed to vary.

14. Find the differential equation of the system of parabolas $(x - \alpha)^2 = 4a(y - \beta)$, α, β are parameters and a is a constant.

15. Find the differential equation of the family of curves $c^2 + 2cy - x^2 + 1 = 0$, c being a parameter.

16. Show that the curves for which the length of the normal is equal to the radius vector are either circles or rectangular hyperbolas.

17. Find the differential equation of all straight lines in the xy-plane.

18. Show that the differential equation of the family of curves $y = c(x - c)^2$, where c is an arbitrary constant, is $y'^3 = 4y(xy'-2y)$.

19. Find the differential equation of the family of ellipses centered at the origin.

20. If $y = \dfrac{lx + m}{px + q}$, where l, m, p, q are arbitrary constants such that $l + q = 0$, then show that

$$(y - x)\,y'' = 2y'(1 + y').$$

[*Hints*: Putting $l = -q$, we get $y = \dfrac{x + a}{bx - 1}$, where $a = \dfrac{m}{l}, b = \dfrac{p}{l}$. Now differentiating two times the equation $(bx - 1)y = x + a$, we get

$$by + (bx - 1)\,y' = 1 \Rightarrow b(y + xy') = 1 + y'$$

and $$b(y'+xy''+y') = y'' \Rightarrow b(xy''+2y') = y''.$$

Now dividing, we get

$$\frac{y + xy'}{xy'' + 2y'} = \frac{1 + y'}{y''} \Rightarrow (y - x)\, y'' = 2y'(1 + y').]$$

21. Show that the substitution $z = \sin x$ transforms the equation

$$\frac{d^2y}{dx^2} + \frac{dy}{dx} \tan x + y \cos^2 x = 0$$

into
$$\frac{d^2y}{dz^2} + y = 0.$$

22. Show that the transformation $z = \sinh x$ transform the equation

$$(1 + x^2)\frac{d^2y}{dx^2} + x\frac{dy}{dx} - 4y = 0$$

into
$$\frac{d^2y}{dz^2} - 4y = 0.$$

ANSWERS

1. (i) Order = 1, Degree = 1, (ii) Order = 3, Degree = 2,
(iii) Order = 2, Degree = 1, (iv) Order = 2, Degree = 1,
(v) Order = 1, Degree = 4, (vi) Order = 2, Degree = 3,
(vii) Order = 1, Degree = 2, (viii) Order = 1, Degree = 2,
(ix) Order = 2, Degree = 16, (x) Order = 4, Degree = 1.

4. $(y')^2 + 4(y - xy') = 0,$

5. (i) $y = xy_1 - \frac{1}{2}x^2 y_2,$ (ii) $xy_2 + 2y_1 = xy - x^2 + 2,$
(iii) $y_2 - 5y_1 = 0,$ (iv) $y_3 - 25y_1 = 0,$
(v) $y_2 \sin x - y_1 \cos x = 0,$ (vi) $y_2 - y = 0,$
(vii) $y_4 - y = 0,$ (viii) $y \sinh^{-1} y_1 = x\sqrt{1 + y_1^2}$

10. $y_3 = 0,$ **11.** $(1 + \cos x)\, dy + y \sin x\, dx = 0,$

13. $x^2(1 + y_1^2) = a^2 y_1^2\ ;\ y_1^3 + y_1 = xy_2,$ **14.** $2ay'' = 1,$

15. $(1 - x^2)\, y'^2 + 2xy' + x^2 = 0.$ **17.** $\frac{d^2y}{dx^2} = 0.$

19. $xyy'' + xy'^2 - yy' = 0.$

Chapter 2

First Order and First Degree Differential Equations

2.1 Definition

A first order differential equation is of the form

$$F\left(x, y, \frac{dy}{dx}\right) = 0 \quad \text{or,} \quad \frac{dy}{dx} = f(x, y).$$

An equation of the form with prescribed condition may have

(i) unique solution,

(ii) infinite number of solutions.

For the existence and uniqueness of the solution of such equation, we have the following theorem:

Existence and Uniqueness Theorem: *If the function $f(x, y)$ and its partial derivative with respect to y are continuous in some domain D in xy-plane, containing the point (x_0, y_0), then there exists a unique solution $y = \varphi(x)$ of the differential equation*

$$\frac{dy}{dx} = f(x, y)$$

satisfying the condition $y(x_0) = y_0$.

The geometrical interpretation of the theorem is that there exists unique function $y = \varphi(x)$, which passes through the point (x_0, y_0) of the domain D.

The general solution of a first order differential equation is of the form $y = \varphi(x, C)$ where C is an arbitrary constant. For a definite value C_0 of C, $y = \varphi(x, C_0)$ is called a *particular solution of the first order differential equation.*

From geometrical point of view, the general solution is a family of curves represented by $y = \varphi(x, C)$ in xy-plane and a particular solution is associated with one of the members of such family that passes through a given point of the xy-plane.

A differential equation of the first order and first degree is of the form

$$M(x, y)\,dx + N(x, y)\,dy = 0, \tag{1}$$

where $M(x, y)$, $N(x, y)$ are functions of x, y or constants.

If this form of the differential equations is solvable, then these are classified according to the methods applied to solve these as follows:

i. Equation with separable variables,
ii. First order homogenenous equations,
iii. Equations reducible to homogeneous equations,
iv. Exact equations,
v. Linear equations,
vi. Equations reducible to linear equations.

2.1.1 Equations with separable variables

If the functions $M(x, y), N(x, y)$ can be expressed as $M(x, y) = f_1(x)\, g_1(y)$, $N(x, y) = f_2(x) g_2(y)$, then the equation (1) reduces to

$$\frac{f_1(x)}{f_2(x)} dx + \frac{g_2(y)}{g_1(y)} dy = 0,$$

provided neither $f_2(x)$ nor $g_1(y)$ vanish. Now it can be integrated to get the solution in terms of known functions. Sometimes, the variables x, y appear in an equation in such a way that separation of variables cannot be done and in such cases transformation of the dependent variable is needed.

EXAMPLE 1

Solve the Equation: $x^2(y-1)\,dx + y^2(x+1)\,dy = 0.$

Solution

Dividing the equation by $(y-1)(x+1)$, we get

$$\frac{x^2}{x+1} dx + \frac{y^2}{y-1} dy = 0$$

or, $$\frac{x^2-1+1}{x+1} dx + \frac{y^2-1+1}{y-1} dy = 0$$

or, $$\left(x-1+\frac{1}{x+1}\right)dx + \left(y+1+\frac{1}{y-1}\right)dy = 0.$$

Now integrating, we get

$$\frac{x^2}{2} - x + \ln(x+1) + \frac{y^2}{2} + y + \ln(y-1) = \frac{c}{2}$$

or, $$x^2 + y^2 - 2(x-y) + 2\ln(x+1)(y-1) = C,$$

where C is an arbitrary constant.

EXAMPLE 2

Solve the equation: $\dfrac{dy}{dx} = \sqrt{y-x}.$

Solution

We put $y - x = v^2$ so that $\frac{dy}{dx} - 1 = 2v\frac{dv}{dx}$. Then the given equation becomes

$$2v\frac{dv}{dx} + 1 = v$$

or, $$\frac{2vdv}{v-1} = dx$$

or, $$\frac{2v-2+2}{v-1}dv = dx$$

or, $$2dv + \frac{2dv}{v-1} = dx.$$

Integrating, we get

$$2v + 2\ln(v-1) = x + 2C$$

or, $$\sqrt{y-x} + \ln\left(\sqrt{y-x} - 1\right) = \frac{1}{2}x + C.$$

2.1.2 First order homogeneous equations

A function $f(x, y)$ is called a homogeneous function in x, y of degree n, if for any α the following identity is true:

$$f(\alpha x, \alpha y) = \alpha^n f(x, y).$$

A first order differential equation

$$\frac{dy}{dx} = f(x, y)$$

is called homogeneous in x, y, if the function $f(x, y)$ is a homogeneous function of degree zero in x, y. In other words, an equation of the form

$$M(x, y)\,dx + N(x, y)\,dy = 0$$

is said to be homogeneous, if and only if both $M(x, y), N(x, y)$ are homogeneous functions in x, y of same degree.

Thus a first order differential equation can be written as

$$\frac{dy}{dx} = f(x, y) = f(\alpha x, \alpha y) = f\left(1, \frac{y}{x}\right), \text{ taking } \alpha = \frac{1}{x}.$$

Taking $y = vx$, we get $\frac{dy}{dx} = v + x\frac{dv}{dx}$. So the above equation becomes

$$v + x\frac{dv}{dx} = f(1, v)$$

or, $$\frac{dv}{f(1, v) - v} = \frac{dx}{x}.$$

Now integrating above and then putting $v = \frac{y}{x}$ we get the required solution.

EXAMPLE 3

Solve the equation: $\left(1 + e^{\frac{x}{y}}\right)dx + e^{\frac{x}{y}}\left(1 - \frac{x}{y}\right)dy = 0.$

Solution

Rewriting the equation as

$$\frac{dx}{dy} = -\frac{e^{\frac{x}{y}}\left(1-\frac{x}{y}\right)}{\left(1+e^{\frac{x}{y}}\right)}.$$

This is a first order homogeneous differential equation. We put $x = vy$ and get $\frac{dx}{dy} = v + y\frac{dv}{dy}$. Using these in the above equation, we get

$$v + y\frac{dv}{dy} = \frac{e^v(1-v)}{1+e^v}$$

or, $$y\frac{dv}{dy} = -\frac{e^v(1-v)}{1+e^v} - v$$

or, $$y\frac{dv}{dy} = -\frac{e^v + v}{1+e^v}$$

or, $$\frac{1+e^v}{v+e^v}dv + \frac{dy}{y} = 0.$$

Integrating, we get

$$\ln\left(v+e^v\right) + \ln y = \ln C$$

or, $$y\left(v+e^v\right) = C$$

or, $$x + ye^{\frac{x}{y}} = C.$$

2.1.3 Equations reducible to homogeneous equations

Equations of the form

$$\frac{dy}{dx} = \frac{ax+by+c}{a_1x+b_1y+c_1},$$

where at least one of c, c_1 is non-zero, are reducible to homogeneous equations. To reduce it to the homogeneous form we put

$$x = x'+h,\ \ y = y'+k$$

in the above equation and get

$$\frac{dy'}{dx'} = \frac{ax+by'+ah+bk+c}{a_1x'+b_1y'+a_1h+b_1k+c_1}.$$

Now choose h, k in such a way that:

$$ah + bk + c = 0,$$

and $$a_1h + b_1k + c_1 = 0.$$

In this case the above equation reduces to

$$\frac{dy'}{dx'}=\frac{ax'+by'}{a_1x'+b_1y'}.$$

This can be solved using the method described in Art. 2.1.2, if $\begin{vmatrix} a & b \\ a_1 & b_1 \end{vmatrix} \neq 0$.

If $\begin{vmatrix} a & b \\ a_1 & b_1 \end{vmatrix} \neq 0$, then the given equation reduces to

$$\frac{dy}{dx}=\frac{ax+by+c}{k(ax+by)+c_1},$$ where k is any non-zero real number.

Now making the substitution,

$$z = ax + by,$$

$$\text{i.e.,}\quad \frac{dz}{dx}=a+b\frac{dy}{dx},$$

we get

$$\frac{dz}{dx}=a+b\frac{z+c}{kz+c_1}$$

or,
$$\frac{dz}{dx}=\frac{(ak+b)z+ac_1+bc}{kz+c_1}.$$

This is an equation with separated variables.

EXAMPLE 4

Solve the equation: $\dfrac{dy}{dx}=\dfrac{7x-3y-7}{7y-3x+3}$.

Solution

We put $x = x'+h,\ y = y'+k$ in the given equation and get

$$\frac{dy'}{dx'}=\frac{7x'-3y'+7h-3k-7}{7y'-3x'+7k-3h+3}.$$

We choose h, k in such a way that

$$7h-3k-7=0 \text{ and } 3h-7k-3=0.$$

Solving these we get

$$\frac{h}{-40}=\frac{k}{0}=\frac{1}{-40}, \text{ i.e., } h=1, k=0.$$

Hence, the transformed equation becomes

$$\frac{dy'}{dx'}=\frac{7x'-3y'}{7y'-3x'}.$$

Putting $y' = vx'$, we get $v+x'\dfrac{dv}{dx'}=\dfrac{7-3v}{7v-3}$

or,
$$x'\frac{dv}{dx'}=\frac{7-3v}{7v-3}-v=\frac{7(1-v^2)}{7v-3}$$

or, $$\frac{7v-3}{7(1-v^2)}dv = \frac{dx'}{x'}.$$

Integrating we get

$$-\frac{1}{2}\ln(1-v^2) - \frac{3}{14}\ln\frac{1+v}{1-v} = \ln x' + \ln C$$

or, $$-\frac{1}{2}\ln(x'^2 - y'^2) - \frac{3}{14}\ln\frac{x'+y'}{x'-y'} = \ln C$$

or, $$\ln\left[(x'+y')^{-\frac{5}{7}}(x'-y')^{-\frac{2}{7}}\right] = \ln C$$

or, $$(y+x-1)^5(x-y-1)^2 = K.$$

EXAMPLE 5

Solve the equation: $\dfrac{dy}{dx} = \dfrac{x-y-3}{2(x-y)+5}$.

Solution

We put $x - y = v$, then the given equation becomes

$$1 - \frac{dv}{dx} = \frac{v-3}{2v+5}$$

or, $$\frac{dv}{dx} = 1 - \frac{v-3}{2v+5} = \frac{v+8}{2v+5}$$

or, $$\frac{2v+5}{v+8}dv = dx$$

or, $$\frac{2v+16-11}{v+8}dv = dx$$

or, $$2dv - 11\frac{dv}{v+8} = dx.$$

Integrating, we get

$$2v - 11\ln(v+8) = x + C$$

or, $$2(x-y) - 11\ln(x-y+8) = x + C$$

or, $$x - 2y - 11\ln(x-y+8) = C.$$

EXAMPLE 6

Solve the equation: $(x^2+2y^2-3)xdx - (2x^2+y^2-3)ydy = 0$.

Solution

Putting $x^2 = u$, $y^2 = v$ in the given equation, we get

$$\frac{dv}{du} = \frac{u+2v-3}{2u+v-3}.$$

To make it a homogeneous equation, we use $u = u'+h,\ v = v'+k$ and get

$$\frac{dv'}{du'} = \frac{u'+2v'+h+2k-3}{2u'+v'+2h+k-3}.$$

We chose h, k in such a way that $h + 2k = 3$ and $2h + k = 3$. Solving we get $h = 1$, $k = 1$.

Next we put $v' = zu'$, i.e., $\frac{dv'}{du'} = z + u'\frac{dz}{du'}$ in the above equation and get

$$z + u'\frac{dz}{du'} = \frac{1+2z}{2+z}$$

or, $$u'\frac{dz}{du'} = \frac{1+2z}{2+z} - z = \frac{1-z^2}{2+z}$$

or, $$\frac{du'}{u'} = \frac{z+2}{1-z^2}dz.$$

Integrating, we get

$$\ln u' = -\frac{1}{2}\ln\left(1-z^2\right) + \ln\frac{1+z}{1-z} + \frac{1}{2}\ln C$$

or, $$u'^2 = C\left(\frac{1+z}{1-z}\right)^2 \frac{1}{1-z^2}$$

or, $$u'^2 = C\left(\frac{v'+u'}{u'-v'}\right)^2 \frac{u'^2}{u'^2-v'^2}$$

or, $$(u'-v')^3 = C(v'+u')$$

or, $$(u-v)^3 = C(v+u-2)$$

or, $$\left(x^2-y^2\right)^3 = C\left(y^2+x^2-2\right).$$

2.1.4 Exact equations

An equation of the form

$$M(x, y)\,dx + N(x, y)\,dy = 0$$

is said to be an exact differential equation, if it can be expressed as $du(x, y) = 0$ without multiplying it by any factor, $\mu(x, y)$, i.e., $M(x, y)\,dx + N(x, y)\,dy$ becomes a perfect differential.

THEOREM 2.1.1

An equation of the form

$$M(x, y)\,dx + N(x, y)\,dy = 0$$

is said to be an exact differential equation, if the differentialble functions $M(x, y)$ $N(x, y)$ are such that $\frac{\partial M}{\partial y}$, $\frac{\partial N}{\partial x}$ are continuous in some region and satisfy

$$\frac{\partial M}{\partial y} = \frac{\partial N}{\partial x}$$

and conversely.

Proof: If the equation $Mdx + Ndy = 0$ be exact, then there exists some functuion $u(x, y)$ such that

$$Mdx + Ndy = du = \frac{\partial u}{\partial x}dx + \frac{\partial u}{\partial y}dy.$$

Then

$$M = \frac{\partial u}{\partial x} \text{ and } N = \frac{\partial u}{\partial y}.$$

From these, we get

$$\frac{\partial M}{\partial y} = \frac{\partial^2 u}{\partial y \partial x} \text{ and } \frac{\partial N}{\partial x} = \frac{\partial^2 u}{\partial x \partial y}.$$

By Schwartz's theorem,

$$\frac{\partial M}{\partial y} = \frac{\partial N}{\partial x}.$$

Thus the condition $\frac{\partial M}{\partial y} = \frac{\partial N}{\partial x}$ is the necessary condition for the equation $Mdx + Ndy = 0$ to be exact.

Conversely, let $\frac{\partial M}{\partial y}, \frac{\partial N}{\partial x}$ be continuous in some region and satisfy

$$\frac{\partial M}{\partial y} = \frac{\partial N}{\partial x}.$$

Now if the function $V(x, y)$ be such that $\frac{\partial V}{\partial x} = M$, then

$$\frac{\partial N}{\partial x} = \frac{\partial M}{\partial y} = \frac{\partial^2 V}{\partial y \partial x} = \frac{\partial^2 V}{\partial x \partial y} = \frac{\partial}{\partial x}\left(\frac{\partial V}{\partial y}\right).$$

Integrating with respect to x, we get

$$N = \frac{\partial V}{\partial y} + f(y), \text{ where } f(y) \text{ is some function of } y.$$

So,

$$\begin{aligned} Mdx + Ndy &= \frac{\partial V}{\partial x}dx + \frac{\partial V}{\partial y}dy + f(y)\,dy \\ &= dV + d[g(y)], \text{ where } g'(y) = f(y). \\ &= du, \text{ where } u(x, y) = V(x, y) + g(y). \end{aligned}$$

Hence, $Mdx + Ndy = 0$ is an exact differential equation.

The method of solving an exact equation of the form $Mdx + Ndy = 0$ is as follows:

Integrate the terms in M(x, y) with respect to x, keeping y constant, then integrate the terms in N(x, y) with respect to y, keeping x constant. In these results at least one term will be same. Now, equate the sum of these two results, after writing the repeated terms once, to a constant to get the solution.

THEOREM 2.1.2

If the equation, $M(x, y)\,dx + N(x, y)\,dy = 0$ be an exact differential equation where $M(x, y)$, $N(x, y)$ are both homogeneous functions of x, y of same degree ($\neq -1$), then the solution is given by

$$M(x, y)\,x + N(x, y)\,y = C.$$

Proof: Let $M(x, y)$, $N(x, y)$ be homogeneous functions of degree $n(\neq -1)$ and $\phi = Mx + Ny$, then differentiating $\phi(x, y)$ partially with respect to x, we get

$$\frac{\partial \phi}{\partial x} = M + x\frac{\partial M}{\partial x} + y\frac{\partial N}{\partial x}$$

$$= M + x\frac{\partial M}{\partial x} + y\frac{\partial M}{\partial y}, \text{ since the equation is exact,}$$

$$= M + nM, \text{ by Euler's theorem on homogeneous functions.}$$

$$= (n+1)\,M.$$

Similarly, $$\frac{\partial \phi}{\partial y} = (n+1)\,N.$$

Now $$d\phi = \frac{\partial \phi}{\partial x}dx + \frac{\partial \phi}{\partial y}dy$$

$$= (n+1)\left[Mdx + Ndy\right] = 0$$

Integrating, we get $\phi = Mx + Ny = \text{constant}$.

EXAMPLE 7

Solve the equation: $\left(1 + 4xy + 2y^2\right)dx + \left(1 + 4xy + 2x^2\right)dy = 0$.

Solution

Here, $M = 1 + 4xy + 2y^2$ and $N = 1 + 4xy + 2x^2$. So,

$$\frac{\partial M}{\partial y} = 4(x+y) = \frac{\partial N}{\partial x}.$$

Hence the given equation is exact,

Now, $$\int\left(1 + 4xy + 2y^2\right)dx = x + 2x^2y + 2y^2x$$

and $$\int\left(1 + 4xy + 2x^2\right)dy = y + 2x^2y + 2y^2x.$$

So the solution is $x + 2x^2y + 2y^2x + y = C$, where C is an arbitrary constant.

EXAMPLE 8

Solve the equation $\left(xy^2 + x^3\right)dx + \left(x^2y + y^3\right)dy = 0$.

Solution

Here, $M = (x, y) = xy^2 + x^3$, $N = (x, y) = x^2y + y^3$. So,

$$\frac{\partial M}{\partial y} = 2xy = \frac{\partial N}{\partial x}.$$

Thus the equation is exact in which both $M(x, y), N(x, y)$ are homogeneous functions of degree three. Therefore, the solution is

$$(xy^2 + x^3)x + (x^2y + x^3)y = C$$

$$\text{or,} \quad x^4 + 2x^2y^2 + y^4 = C.$$

2.1.5 Linear equation

An equation which is linear in the unknown function y and its derivative $\frac{dy}{dx}$, is called a ***first order linear equation***. It is of the form

$$\frac{dy}{dx} + P(x)y = Q(x),$$

where $P(x), Q(x)$ are continuous functions of x only.

To find the solution of this equation we assume

$$y(x) = u(x)v(x),$$

where one of these may be arbitrary and other is to be determined from the associated homogeneous equation. Differentiating the above expression with respect to x, we get

$$\frac{dy}{dx} = u\frac{dv}{dx} + v\frac{du}{dx}.$$

Using this in the given equation, we get

$$u\frac{dv}{dx} + v\frac{du}{dx} + Puv = Q \quad \text{or,} \quad u\left(\frac{dv}{dx} + Pv\right) + v\frac{du}{dx} = Q.$$

If we chose v such that

$$\frac{dv}{dx} + Pv = 0,$$

i.e., $v(x)$ is the solution of the associated homogeneous equation of the first order linear differential equation, then

$$\frac{dv}{v} + Pdx = 0.$$

Solving we get $v = Ce^{-\int pdx}$.

So, the above equation reduces to

$$\frac{du}{dx} = \frac{Q}{v}.$$

Integrating we get

$$u = \int \frac{Q}{v}\,dx + C'.$$

So the solution of the given linear differential equation is

$$y(x) = v(x)\left[\int \frac{Q(x)}{v(x)}\,dx + C'\right] \quad \text{or,} \quad y(x) = e^{-\int p(x)\,dx}\left[\int Q(x)\,e^{\int P(x)\,dx} + K\right],$$

where the constant K is to be determined using some prescribed condition.

The function $e^{\int P(x)\,dx}$ is called integrating factor (see chapter 3 for details) of the given first order linear differential equation.

If $Q \neq 0$, the method of finding the solution of a linear differential equation, described above, is known as the method of variation of parameter to solve a first order linear equation.

EXAMPLE 9

Solve the equation: $\dfrac{dy}{dx} + \dfrac{1-2x}{x^2}\,y = 1.$

Solution

Here, $\displaystyle\int \frac{1-2x}{x^2}\,dx = \int \frac{1}{x^2}\,dx - 2\int \frac{1}{x}\,dx = -\frac{1}{x} - 2\ln x.$

So the integrating factor is $e^{-\frac{1}{x}-2\ln x} = \dfrac{1}{x^2}e^{-\frac{1}{x}}$. Multiplying the given equation by this factor, we get

$$\frac{1}{x^2}e^{-\frac{1}{x}}\frac{dy}{dx} + \frac{1}{x^2}e^{-\frac{1}{x}}\frac{1-2x}{x^2}\,y = \frac{1}{x^2}e^{-\frac{1}{x}} \quad \text{or,} \quad \frac{d}{dx}\left[\frac{y}{x^2}e^{-\frac{1}{x}}\right] = \frac{1}{x^2}e^{-\frac{1}{x}}.$$

Integrating, we get

$$\frac{y}{x^2}e^{-\frac{1}{x}} = \int \frac{1}{x^2}e^{-\frac{1}{x}}dx + C = e^{-\frac{1}{x}} + C$$

$$\text{or,} \quad y = x^2\left[1 + Ce^{\frac{1}{x}}\right], \text{ where } C \text{ is an arbitrary constant.}$$

2.1.6 Equation reducible to linear equation (Bernoulli's Equation)

An equation of the form

$$\frac{dy}{dx} + P(x)\,y = Q(x)\,y^n,$$

where $P(x), Q(x)$ are continuous functions of x only and $n \neq 0,1$ (as these values reduce it to a linear differential equation) is known as *Bernoulli's equation.*

This can be reduced to a linear differential equation as follows:

Dividing the equation by y^n, we get

$$\frac{1}{y^n}\frac{dy}{dx}+P(x)\frac{1}{y^{n-1}}=Q(x).$$

Putting

$$\frac{1}{y^{n-1}}=z,$$

we get
$$\frac{dz}{dx}=\frac{1-n}{y^n}\frac{dy}{dx}.$$

So the above equation becomes

$$\frac{dz}{dx}+(1-n)P(x)z=(1-n)Q(x).$$

This is a linear differential equation and can be solved using the method described in the previous article.

EXAMPLE 10

Solve the equation: $\dfrac{dy}{dx}+\dfrac{\sin 2y}{x}=x^3\cos^2 y.$

Solution

Dividing the equation by $\cos^2 y$, we get

$$\sec^2 y\frac{dy}{dx}+\frac{2}{x}\tan y=x^3,$$

Putting $z=\tan y$, we get

$$\frac{dz}{dx}+\frac{2}{x}z=x^3.$$

This is a linear differential equation whose integrating factor is $e^{\int \frac{2}{x}dx}=e^{2\log x}=x^2$.

Multiplying the above equation by x^2, we get

$$x^2\frac{dz}{dx}+2xz=x^5 \text{ or, } \frac{d}{dx}\left[zx^2\right]=x^5.$$

Integrating, we get

$$zx^2=\frac{x^6}{6}+C \text{ or, } z=\frac{x^4}{6}+Cx^{-2}$$

$$\text{or, } \tan y=\frac{x^4}{6}+Cx^{-2}.$$

WORKED-OUT EXAMPLES

1. Solve the equation: $y - x\frac{dy}{dx} = a\left(y^2 + \frac{dy}{dx}\right)$.

Solution

First we separate the variables of the given equation as follows:

$$y - x\frac{dy}{dx} = a\left(y^2 + \frac{dy}{dx}\right)$$

or, $$y(1-ay) = (a+x)\frac{dy}{dx}$$

or, $$\frac{dx}{a+x} = \frac{dy}{y(1-ay)}.$$

Integrating, we get

$$\int \frac{dx}{a+x} = \int \frac{dy}{y(1-ay)} = \int \frac{dy}{y} + a\int \frac{dy}{1-ay}$$

or, $$\ln(a+x) = \ln y - \ln(1-ay) + \ln C$$

or, $$\ln(a+x) = \ln \frac{Cy}{1-ay}$$

or, $$(a+x)(1-ay) = Cy.$$

2. Solve the equation: $\frac{dy}{dx} + 2xy = x^2 + y^2$.

Solution

We rewrite the given equation as:

$$\frac{dy}{dx} = (x-y)^2.$$

To separate the variables, we put

$$x - y = v, \text{ i.e., } 1 - \frac{dy}{dx} = \frac{dv}{dx}.$$

Using these in the above equation, we get

$$1 - \frac{dv}{dx} = v^2 \text{ or, } \frac{dv}{1-v^2} = dx.$$

Integrating we get

$$\frac{1}{2}\ln\left|\frac{1+v}{1-v}\right| = x + \frac{C}{2}$$

or, $$\ln\left|\frac{1+x-y}{1-x+y}\right| = 2x + C.$$

3. Solve the equation: $\dfrac{dy}{dx} = 1 + e^{x+y}$.

Solution

To separate the variables, we put

$$y - x = v, \text{ i.e., } \frac{dy}{dx} - 1 = \frac{dv}{dx}.$$

Using these in the given equation, we get

$$\frac{dv}{dx} = e^{2x+v} \quad \text{or, } e^{-v}dv = e^{2x}dx.$$

Integrating, we get

$$\int e^{-v}dv = \int e^{2x}dx$$

$$\text{or, } -e^{-v} = \frac{1}{2}e^{2x} - C$$

$$\text{or, } -e^{x-y} = \frac{1}{2}e^{2x} - C$$

$$\text{or, } e^{-y} = -\frac{1}{2}e^{x} + Ce^{-x}.$$

4. Solve the equation: $aye^{\frac{x}{y}}dx + \left(y - axe^{\frac{x}{y}}\right)dy = 0$.

Solution

To separate the variables we put $x = vy$, i.e., $\dfrac{dx}{dy} = v + y\dfrac{dv}{dy}$. Using these in the given equation, after rewriting the equation as

$$\frac{dx}{dy} = \frac{axe^{x/y} - y}{aye^{x/y}},$$

we get

$$v + y\frac{dv}{dy} = \frac{ave^{v} - 1}{ae^{v}}$$

$$\text{or, } y\frac{dv}{dy} = \frac{ave^{v} - 1}{ae^{v}} - v = -\frac{1}{ae^{v}}$$

$$\text{or, } ae^{v}dv + \frac{dy}{y} = 0.$$

Integrating, we get

$$ae^{v} + \ln y = C \quad \text{or, } ae^{\frac{x}{y}} + \ln y = C.$$

5. Solve the equation: $xy\dfrac{dy}{dx} - y^2 = (x+y)^2 e^{-\frac{y}{x}}$.

Solution

We rewrite the equation as

$$\frac{dy}{dx} = \frac{y^2 + (x+y)^2 e^{-\frac{y}{x}}}{xy}.$$

Now we put $y = vx$, i.e., $\frac{dy}{dx} = v + x\frac{dv}{dx}$ in the above equation and get

$$v + x\frac{dv}{dx} = \frac{v^2 + (1+v)^2 e^{-v}}{v}$$

or, $$x\frac{dv}{dx} = \frac{(1+v)^2 e^{-v}}{v}$$

or, $$\frac{ve^v dv}{(1+v)^2} = \frac{dx}{x}$$

or, $$\frac{e^v dv}{1+v} - \frac{e^v dv}{(1+v)^2} = \frac{dx}{x}.$$

Integrating, we get

$$\int \frac{e^v dv}{1+v} - \int \frac{e^v dv}{(1+v)^2} = \int \frac{dx}{x}$$

or, $$\frac{e^v}{1+v} + \int \frac{e^v dv}{(1+v)^2} - \int \frac{e^v dv}{(1+v)^2} = \ln x + C$$

or, $$\frac{xe^{\frac{y}{x}}}{x+y} = \ln x + C.$$

6. Solve the equation: $\frac{dy}{dx} = \frac{6x - 2y - 7}{3y + 2x - 6}.$

Solution

We put $x = x' + h,\ y = y' + k$ in the given equation and get

$$\frac{dy'}{dx'} = \frac{6x' - 2y' + 6h - 2k - 7}{3y' + 2x' + 3k + 2h - 6}.$$

We choose h, k in such a way that:

$$6h - 2k - 7 = 0 \text{ and } 3k + 2h - 6 = 0.$$

Solving these we get

$$\frac{h}{3} = \frac{k}{2} = \frac{1}{2}, \text{ i.e., } h = \frac{3}{2}, k = 1.$$

Hence, the transformed equation becomes

$$\frac{dy'}{dx'} = \frac{6x' - 2y'}{3y' + 2x'}.$$

Putting $y' = vx'$, we get

$$v + x'\frac{dv}{dx'} = \frac{6 - 2v}{3v + 2}$$

or, $$x'\frac{dv}{dx'} = \frac{6 - 2v}{3v + 2} - v = \frac{6 - 4v - 3v^2}{3v + 2}$$

or, $$\frac{3v + 2}{6 - 4v - 3v^2}dv = \frac{dx'}{x'}.$$

Integrating, we get

$$-\frac{1}{2}\ln\left(6-4v-3v^2\right)=\ln x'+\ln C$$

or, $$-\frac{1}{2}\ln\left(6x'^2-4x'y'-3y'^2\right)=\ln C$$

or, $$\ln\left[6\left(x-\frac{3}{2}\right)^2-4\left(x-\frac{3}{2}\right)(y-1)-3(y-1)^2\right]=\ln K$$

or, $$6x^2-4xy-3y^2-14x+12y=K'$$

7. Solve the equation: $\dfrac{dy}{dx}=\dfrac{x+y+1}{2x+2y+1}$.

Solution

We put $x+y=v$, then the given equation becomes

$$\frac{dv}{dx}-1=\frac{v+1}{2v+1}$$

or, $$\frac{dv}{dx}=\frac{v+1}{2v+1}+1=\frac{3v+2}{2v+1}$$

or, $$\frac{2v+1}{3v+2}dv=dx$$

or, $$\frac{2}{3}dv-\frac{1}{3}\frac{1}{3v+2}dv=dx.$$

Integrating, we get

$$\frac{2}{3}v-\frac{1}{9}\ln(3v+2)=x+C$$

or, $$\frac{2}{3}(x+y)-\frac{1}{9}\ln(3x+3y+2)=x+C$$

or, $$6x+6y-\ln(3x+3y+2)=9x+9C$$

or, $$6y-3x-\ln(3x+3y+2)=C'.$$

8. Solve the equation: $\left(e^{2y}-y\cos xy\right)dx+\left(2xe^{2y}-x\cos xy+2y\right)dy=0.$

Solution

Taking $M=e^{2y}-y\cos xy$ and $N=2xe^{2y}-x\cos xy+2y,$ we see that

$$\frac{\partial M}{\partial y}=2e^{2y}-\cos xy+xy\sin xy \quad\text{and}\quad \frac{\partial N}{\partial x}=2e^{2y}-\cos xy+xy\sin xy.$$

Thus the equation is exact. Now

$$\int\left(e^{2y}-y\cos xy\right)dx=xe^{2y}-\sin xy \quad \text{(considering } y \text{ as constant)}$$

and $$\int\left(2xe^{2y}-x\cos xy+2y\right)dy=xe^{2y}-\sin xy+y^2 \quad \text{(considering } x \text{ as constant).}$$

The term $xe^{2y}-\sin xy$ is common to both the integral, so the general solution is

$$xe^{2y}-\sin xy+y^2=C.$$

9. Solve the equation: $\left(1+e^{\frac{x}{y}}\right)dx + e^{\frac{x}{y}}\left(1-\frac{x}{y}\right)dy = 0.$

Solution

Taking $M = 1 + e^{\frac{x}{y}}$ and $N = e^{\frac{x}{y}}\left(1-\frac{x}{y}\right)$, we see that

$$\frac{\partial M}{\partial y} = -\frac{xe^{\frac{x}{y}}}{y^2} \text{ and } \frac{\partial N}{\partial x} = \frac{1}{y}e^{\frac{x}{y}} - \frac{1}{y}e^{\frac{x}{y}} - \frac{xe^{\frac{x}{y}}}{y^2} = -\frac{xe^{\frac{x}{y}}}{y^2}.$$

Thus the equation is exact. Now

$$\int\left(1+e^{x/y}\right)dx = x + ye^{x/y} \text{ (considering } y \text{ as constant).}$$

and $$\int e^{x/y}\left(1-\frac{x}{y}\right)dy = -\int \frac{e^{zx}(1-zx)}{z^2}dz, \text{ where } z = \frac{1}{y}$$

$$= x\int\frac{e^{zx}}{z}\,dz - \int\frac{e^{zx}}{z^2}dz \quad = \frac{e^{zx}}{z} + \int\frac{e^{zx}}{z^2}dz - \int\frac{e^{zx}}{z^2}dz$$

$$= ye^{x/y} \text{ (considering } x \text{ as constant)}$$

The term $ye^{x/y}$ is common to both the integral, so the general solution is

$$x + ye^{x/y} = C.$$

10. Solve the equation: $y^2dx + \left(x - \frac{1}{y}\right)dy = 0.$

Solution

We rewrite the equation as

$$\frac{dx}{dy} + \frac{x}{y^2} = \frac{1}{y^3}.$$

This is a linear differential equation, whose integrating factor is

$$e^{\int\frac{dy}{y^2}} = e^{-\frac{1}{y}}.$$

Multiplying the equation by $e^{-\frac{1}{y}}$, we get

$$\left(\frac{dx}{dy} + \frac{x}{y^2}\right)e^{-\frac{1}{y}} = \frac{1}{y^3}e^{-\frac{1}{y}}$$

or, $$\frac{d}{dy}\left(xe^{-\frac{1}{y}}\right) = \frac{1}{y^3}e^{-\frac{1}{y}}.$$

Integrating both sides, we get

$$xe^{-\frac{1}{y}} = \int \frac{1}{y^3} e^{-\frac{1}{y}} dy + C = -\int ze^z dz + C, \text{ where } z = -\frac{1}{y}$$

$$= -(z-1)e^z + C = \left(\frac{1}{y}+1\right) e^{-\frac{1}{y}} + C$$

or, $\quad x = \left(\frac{1}{y}+1\right) + Ce^{\frac{1}{y}}.$

11. Solve the equation: $(1+y^2)\,dx - (\tan^{-1} y - x)\,dy = 0.$

Solution

We rewrite the equation as

$$\frac{dx}{dy} + \frac{x}{1+y^2} = \frac{\tan^{-1} y}{1+y^2}.$$

This is a linear differential equation, whose integrating factor is

$$e^{\int \frac{dy}{1+y^2}} = e^{\tan^{-1} y}.$$

Multiplying the equation by $e^{\tan^{-1} y}$, we get

$$\left(\frac{dx}{dy} + \frac{x}{1+y^2}\right) e^{\tan^{-1} y} = \frac{\tan^{-1} y}{1+y^2} e^{\tan^{-1} y} \quad \text{or,} \quad \frac{d}{dy}\left(xe^{\tan^{-1} y}\right) = \frac{\tan^{-1} y}{1+y^2} e^{\tan^{-1} y}.$$

Integrating both sides, we get

$$xe^{\tan^{-1} y} = \int \frac{\tan^{-1} y}{1+y^2} e^{\tan^{-1} y} dy = \int ze^z dz, \text{ where } z = \tan^{-1} y$$

$$= (z-1)\,e^z + C = \left(\tan^{-1} y - 1\right) e^{\tan^{-1} y} + C$$

or, $x = \tan^{-1} y - 1 + Ce^{-\tan^{-1} y}.$

12. Solve the equation: $\dfrac{dy}{dx} + \dfrac{y}{\sqrt{(1-x^2)^3}} = \dfrac{x+\sqrt{1-x^2}}{(1-x^2)^2}.$

Solution

The given equation is a linear differential equation whose integrating factor is

$$e^{\int \frac{dx}{\sqrt{(1-x^2)^3}}} = e^{\int \sec^2 \theta d\theta}, \text{ where } x = \sin\theta$$

$$= e^{\tan\theta} = e^{\frac{x}{\sqrt{1-x^2}}}.$$

Multiplying the given equation by $e^{\frac{x}{\sqrt{1-x^2}}}$, we get

$$\left(\frac{dy}{dx}+\frac{y}{\sqrt{(1-x^2)^3}}\right)e^{\frac{x}{\sqrt{1-x^2}}}=\frac{x+\sqrt{1-x^2}}{(1-x^2)^2}e^{\frac{x}{\sqrt{1-x^2}}}$$

or,
$$\frac{d}{dx}\left(ye^{\frac{x}{\sqrt{1-x^2}}}\right)=\frac{x+\sqrt{1-x^2}}{(1-x^2)^2}e^{\frac{x}{\sqrt{1-x^2}}}.$$

Integrating, we get

$$ye^{\frac{x}{\sqrt{1-x^2}}}=\int\frac{x+\sqrt{1-x^2}}{(1-x^2)^2}e^{\frac{x}{\sqrt{1-x^2}}}dx$$

$$=\int\frac{\sin\theta+\cos\theta}{\cos^3\theta}e^{\tan\theta}d\theta,\ \text{where } x=\tan\theta$$

$$=\int(1+z)e^z dz,\ \text{where } z=\tan\theta$$

$$=ze^z+C=\tan\theta\, e^{\tan\theta}+C$$

$$=\frac{x}{\sqrt{1-x^2}}e^{\frac{x}{\sqrt{1-x^2}}}+C$$

or,
$$y=\frac{x}{\sqrt{1-x^2}}+Ce^{-\frac{x}{\sqrt{1-x^2}}}.$$

13. Solve the equation: $\frac{dy}{dx}+(2x\tan^{-1}y-x^3)(1+y^2)=0.$

Solution

We rewrite the equation as

$$\frac{dy}{dx}+2x\tan^{-1}y(1+y^2)=x^3(1+y^2).$$

Dividing $1+y^2$, we get

$$\frac{1}{1+y^2}\frac{dy}{dx}+2x\tan^{-1}y=x^3.$$

Putting $z=\tan^{-1}y$, we get

$$\frac{dz}{dx}+2xz=x^3.$$

This is a linear differential equation whose integrating factor is

$$e^{\int 2xdx}=e^{x^2}.$$

Multiplying the above equation by e^{x^2}, we get

$$\left(\frac{dz}{dx}+2xz\right)e^{x^2}=x^3e^{x^2} \quad \text{or,} \quad \frac{d}{dx}\left(ze^{x^2}\right)=x^3e^{x^2}.$$

Integrating both sides, we get

$$ze^{x^2}=\int x^3e^{x^2}\,dx+C$$

$$=\frac{1}{2}\int ue^u\,du+C, \text{ where } u=x^2$$

$$=\frac{1}{2}(u-1)e^u+C=\frac{1}{2}\left(x^2-1\right)e^{x^2}+C$$

or, $$\tan^{-1}y=\frac{1}{2}\left(x^2-1\right)+Ce^{-x^2}.$$

14. Solve the equation: $\dfrac{dy}{dx}=y^2+\dfrac{1}{2x^2}$.

Solution

$$\frac{dy}{dx}=y^2+\frac{1}{2x^2}$$

$$\frac{dy}{dx}=\frac{2x^2y^2+1}{2x^2} \qquad x\frac{dy}{dx}+y=\frac{2x^2y^2+1}{2x}+y=\frac{2x^2y^2+2xy+1}{2x}$$

$$\Rightarrow d(xy)=\frac{1+(1+2xy)^2}{4x}dx \Rightarrow d(2xy+1)=\frac{1+(1+2xy)^2}{2x}dx$$

$$\frac{d(2xy+1)}{1+(1+2xy)^2}=\frac{dx}{2x}.$$

Integrating, we get

$$\tan^{-1}(1+2xy)=\frac{1}{2}\ln x+C.$$

15. Show that in an equation of the form $yf(xy)\,dx+xg(xy)\,dy=0$ the variables can be separated by the substitution $xy=v$. Hence find the solution of the equation

$$\left(x^3y^3+x^2y^2+xy+1\right)ydx+\left(x^3y^3-x^2y^2-xy+1\right)xdy=0.$$

Solution

Putting $xy=v$, i.e., $x\dfrac{dy}{dx}+y=\dfrac{dv}{dx}$ in the given equation, we get

$$yf(v)+g(v)\left[\frac{dv}{dx}-y\right]=0$$

$$\text{or,} \quad y[f(v)-g(v)]+g(v)\frac{dv}{dx}=0 \quad \text{or,} \quad \frac{g(v)\,dv}{v[f(v)-g(v)]}+\frac{dx}{x}=0.$$

Thus the variables x, v are separated and now the above expression can be integrated.

Thus under the substitution $xy = v$, the equation

$$\left(x^3y^3 + x^2y^2 + xy + 1\right) ydx + \left(x^3y^3 - x^2y^2 - xy + 1\right) xdy = 0$$

becomes

$$\frac{v^3 - v^2 - v + 1}{v\left(v^3 + v^2 + v + 1 - v^3 + v^2 + v - 1\right)} dv + \frac{dx}{x} = 0$$

or, $$\frac{(v-1)^2}{v^2} dv = -\frac{2dx}{x}$$

or, $$\left(1 - \frac{2}{v} + \frac{1}{v^2}\right) dv = -2\frac{dx}{x}.$$

Integrating, we get

$$v - 2\ln v - \frac{1}{v} = -2\ln x + C$$

or, $$xy - 2\ln y - \frac{1}{xy} = C.$$

EXERCISE

❑ Solve the following differential equations:

1. $x^2(y-1)\,dx + y^2(x-1)\,dy = 0,$
2. $xydy + \sqrt{\left(x^2-1\right)\left(y^2-1\right)}\, dx = 0,$
3. $x\sqrt{y}dx + (1+y)\sqrt{1+x}dy = 0,$
4. $\left(1-x^2\right)(1-y)\,dx = xy(1+y)\,dy,$
5. $(x+y)\,dy + (x-y)\,dx = 0,$
6. $\dfrac{dy}{dx} = \sin(x+y) + \cos(x+y),$
7. $\ln\left(\dfrac{dx}{dy}\right) = ax + by,$
8. $\dfrac{dy}{dx} = \sqrt{y-x},$
9. $\dfrac{dy}{dx} = \cos(x-y),$
10. $(x-y)^2\dfrac{dy}{dx} = a^2,$
11. $\dfrac{dy}{dx} = (x+y+1)^2,$
12. $\dfrac{dy}{dx} + 2xy = x^2 + y^2,$
13. $\left(x^4 - 2xy^3\right)\dfrac{dy}{dx} + y^4 - 2x^3y = 0,$
14. $x\dfrac{dy}{dx} = y + \sqrt{x^2+y^2},$
15. $x^3\dfrac{dy}{dx} = y^3 + y^2\sqrt{y^2-x^2},$
16. $\dfrac{dy}{dx} = \dfrac{y}{x} + \cos\dfrac{y}{x},$
17. $\left(1 + 3e^{\frac{x}{y}}\right)dx + 3e^{\frac{x}{y}}\left(1 - \dfrac{x}{y}\right)dy = 0,$ [*Hints*: Put $x = vy$.]
18. $\left(x + y\cos\dfrac{y}{x}\right)dx = x\cos\dfrac{y}{x}dy,$
19. $x\sin\dfrac{y}{x}dy = \left(y\sin\dfrac{y}{x} - x\right)dx,$

20. $x\cos\frac{y}{x}(y\,dx + x\,dy) = y\sin\frac{y}{x}(x\,dy - y\,dx)$,

[*Hints*: Rewrite the equation as $\frac{d(xy)}{xy} = \tan\frac{y}{x}\,d\left(\frac{y}{x}\right)$]

21. $(3y - 7x + 7)\,dx + (7y - 3x + 3)\,dy = 0$,

22. $(6x - 5y + 4)\,dy + (y - 2x - 1)\,dx = 0$,

23. $(3x + y - 5)\,dy = (2x + 2y - 2)\,dx$,

24. $(x + y + 1)\,dx - (2x + 2y + 1)\,dy = 0$,

25. $(x + y - a)(x + y + b)\frac{dy}{dx} = (x + y + a)(x + y - b)$,

26. $(2x - 2y + 5)\,dy - (x - y + 3)\,dx = 0$,

27. $(x^2 + 3y^2 - 7)\,xdx - (3x^2 + y^2 - 5)\,ydy = 0$,

28. $\left(x^2 + y^2 - 2a^2\right)xdx + \left(x^2 - y^2\right)ydy = 0$,

29. $\left(x^2 + y^2 - a^2\right)xdx + \left(x^2 + y^2 + a^2\right)ydy = 0$,

[*Hints*: In 27-29, put $x^2 = u,\ y^2 = v$]

30. $y(1 + xy)\,dx + x(1 - xy)\,dy = 0$,

31. $\left(x^2y^2 + xy\right)ydx + (x^2y^2 - 1)\,xdy = 0$,

32. $y(2xy + 1)\,dx + x(1 + xy)^2\,dy = 0$,

33. $\cos^2 x\frac{dy}{dx} + y = \tan x$,

34. $x\cos x\frac{dy}{dx} + y(x\sin x + \cos x) = 1$,

35. $\sqrt{a^2 + x^2}\frac{dy}{dx} + y = \sqrt{a^2 + x^2} - x$,

36. $\left(1 - x^2\right)\frac{dy}{dx} + 2xy = x\sqrt{1 - x^2}$,

37. $\frac{dy}{dx} + yf'(x) = f(x)\,f'(x)$, [*Hints*: Here I.F. $= e^{\int f'(x)dx} = e^{f(x)}$. So after multiplication, the given equation becomes

$$\frac{d}{dx}\left[ye^{f(x)}\right] = f(x)\,f'(x)\,e^{f(x)} \text{ or, } d\left[ye^{f(x)}\right] = z\,e^{z}dz, \text{ where } z = f(x).]$$

38. $\phi_1(x)\frac{dy}{dx} + y\phi_2(x) = 1$,

39. $(x + y + 1)\frac{dy}{dx} = 1$, [*Hints*: Rewrite the equation as $\frac{dx}{dy} - x = y + 1$]

40. $\left(1 + y^2\right)dx = \left(\tan^{-1} y - x\right)dy$,

[*Hints*: Rewrite the equation as $\frac{dx}{dy} + \frac{x}{1 + y^2} = \frac{\tan^{-1} y}{1 + y^2}$]

41. $\left(1 + y^2\right)dx = \left(e^{\tan^{-1} y} + x\right)dy$,

[*Hints*: Rewrite the equation as $\frac{dx}{dy} - \frac{x}{1 + y^2} = \frac{e^{\tan^{-1} y}}{1 + y^2}$]

42. $\left[1+x\left(1+y^2\right)\right]dy+y\left(1+y^2\right)dx=0,$

43. $y\ln y\,dx+(x-\ln y)\,dy=0,$

44. $\sin 2x\dfrac{dy}{dx}=y+\tan x,$

45. $dx+xdy=e^{-y}\sec^2 y\,dy,$

46. $\left(x^2y^3+2xy\right)dy=dx,$

47. $\left(x+2y^3\right)\dfrac{dy}{dx}=y,$

48. $\dfrac{dy}{dx}+\dfrac{1}{x}=\dfrac{e^y}{x^2},$

49. $\dfrac{dy}{dx}-\dfrac{\tan y}{1+x}=(1+x)e^x\sec y,$

50. $\dfrac{dy}{dx}+\dfrac{y\ln y}{x}=\dfrac{y(\ln y)^2}{x^2},$

51. $\dfrac{dy}{dx}+\dfrac{xy}{1-x^2}=x\sqrt{y},$

52. $\sin x\dfrac{dy}{dx}-y\cos x+y^2=0,$

53. $xy-\dfrac{dy}{dx}=y^3e^{-x^2},$

54. $\dfrac{dz}{dx}+\dfrac{z}{x}\ln z=\dfrac{z(\ln z)^2}{x^2},$

55. $x\dfrac{dy}{dx}+3y=x^3y^2,$

56. $\left(xy+x^3y^3\right)\dfrac{dy}{dx}=1,$

[*Hints*: Rewrite the equation as $\dfrac{1}{x^3}\dfrac{dx}{dy}-\dfrac{y}{x^2}=y^3$ and put $z=\dfrac{1}{x^2}$]

57. $\left(x^2+y^2+2y\right)dy+2xdx=0,$

58. $y\left(2xy+e^x\right)dx-e^xdy=0,$

59. Show that the general solution of the equation

$$\frac{dy}{dx}+P(x)y=Q(x)$$

can be written in the form $y=k(u-v)+v,$ where k is a constant and u, v are two particular solutions of the equation.

[*Hints*: If u v are two particular solutions of the equation, then any linear combination of these, say $y = au + bv$, where a, b are arbitrary constants, will be a general solution of the given equation. Choosing $a = k$, $b = 1 - k$, we see that

$$y=ku+(1-k)v=k(u-v)+v$$

is a general solution of the given equation.]

60. Show that the solution of the equation

$$\frac{dy}{dx}+P(x)y=Q(x)$$

can be put in the form

$$y=\frac{Q}{P}-e^{-\int Pdx}\left[\int e^{\int Pdx}d\left(\frac{Q}{P}\right)+C\right].$$

[*Hints*: Let $z=y-\dfrac{Q}{P}$, then $\dfrac{dz}{dx}=\dfrac{dy}{dx}-\dfrac{d}{dx}\left(\dfrac{Q}{P}\right)$.

Now, $\frac{dy}{dx}+Py=Q \quad \frac{d}{dx}\left(z+\frac{Q}{P}\right)+P\left(z+\frac{Q}{P}\right)=Q \quad \frac{dz}{dx}+Pz=-\frac{d}{dx}\left(\frac{Q}{P}\right)$. This is a linear differential equation whose integrating factor is $e^{\int Pdx}$. So the solution is

$$z=-e^{-\int Pdx}\left[\int e^{\int Pdx}\frac{d}{dx}\left(\frac{Q}{P}\right)dx+C\right] \quad \text{or,} \quad y=\frac{Q}{P}-e^{-\int Pdx}\left[\int e^{\int Pdx}d\left(\frac{Q}{P}\right)+C\right].$$

61. If u, v are two solutions of the equation

$$\frac{dy}{dx}+P(x)y=Q(x)$$

and if $v=uz$, then $z=1+Ce^{-\int \frac{Q}{u}dx}$, where C is an arbitrary constant.

[*Hints*: Given that $\frac{du}{dx}+Pu=Q$ and $\frac{dv}{dx}+Pv=Q$. Putting $v=uz$ in the second equation we get

$$u\frac{dz}{dx}+z\left(\frac{du}{dx}+Pu\right)=Q \text{ or, } \frac{d(z-1)}{dx}+\frac{(z-1)Q}{u}=0.$$

This is a linear differential equation whose integrating factor is $e^{\int \frac{Q}{u}dx}$. So the solution is

$$(z-1)e^{\int \frac{Q}{u}dx}=C \text{ or, } z=1+Ce^{-\int \frac{Q}{u}dx},$$ where C is an arbitrary constant.]

62. Show that the equation

$$(P+Qx)\frac{dy}{dx}=R+Qy,$$

where P, R are homogeneous functions in x, y of degree n and Q is a homogeneous function in x, y of degree m can be solved by the substitution $y=vx$.

ANSWERS

1. $x^2+y^2+2x+2y+2\ln(x-1)(y-1)=C$, **2.** $\sqrt{x^2-1}+\sqrt{y^2-1}-\sec^{-1}x=C$,

3. $(x-2)\sqrt{x+1}+(y+3)\sqrt{y}=C$, **4.** $\ln\left[x\left(1-y^2\right)\right]=\frac{x^2-y^2}{2}-2y+C$,

5. $\frac{1}{2}\ln\left(x^2+y^2\right)+\tan^{-1}\left(\frac{y}{x}\right)=C$, **6.** $\ln\left(1+\tan\frac{x+y}{2}\right)=x+C$,

7. $ae^{by}+be^{-ax}=C$,

8. $2\sqrt{y-x}+2\ln\left(\sqrt{y-x}-1\right)=x+C$, **9.** $\cot(x-y)+\text{cosec}(x-y)+x=C$,

10. $\frac{a}{2}\ln\frac{x-y-a}{x-y+a}=y+C$, **11.** $x+y+1=\tan(x+C)$,

12. $\ln\left(\frac{1+x-y}{1-x+y}\right)=2x+C$, **13.** $x^3+y^3=Cxy$,

14. $y+\sqrt{x^2+y^2}=cx^2$,

15. $y+\sqrt{y^2-x^2}=cxy$,

16. $\sec\dfrac{y}{x}+\tan\dfrac{y}{x}=Cx$,

17. $x+3ye^{\frac{x}{y}}=C$,

18. $\ln x=\sin\left(\dfrac{y}{x}\right)+C$,

19. $\ln x=\cos\dfrac{y}{x}+C$,

20. $\sec\dfrac{y}{x}=Cxy$,

21. $(y-x+1)^2(y+x-1)^5=C$,

22. $(5y-2x-3)^4=C(4y-4x-3)$,

23. $(y-x+3)^4=C(y+2x-3)$,

24. $3x-6y+\ln(3x+3y+2)=C$,

25. $(b-a)\ln\left[(x+y)^2-ab\right]=2(x-y)+C$,

26. $2y-x-\ln(x-y+2)=C$,

27. $x^2+y^2-3=C\left(x^2-y^2-1\right)^2$,

28. $\tan^{-1}\left(\dfrac{y^2-a^2}{x^2-a^2}\right)-\dfrac{1}{2}\ln\left[\left(y^2-a^2\right)^2+\left(x^2-a^2\right)^2\right]=C$,

29. $x^4+y^4+2x^2y^2-2a^2\left(x^2-y^2\right)=C$,

30. $\ln\left(\dfrac{x}{y}\right)-\dfrac{1}{xy}=C$,

31. $xy-\ln y=C$,

32. $2x^2y^2\ln y-4xy-1=Cx^2y^2$,

33. $y=\tan x-1+Ce^{-\tan x}$,

34. $xy=\sin x+C\cos x$,

35. $\left(x+\sqrt{a^2+x^2}\right)y=a^2\ln\left(x+\sqrt{a^2+x^2}\right)+C$,

36. $y=\sqrt{1-x^2}+C\left(1-x^2\right)$,

37. $y=Ce^{-f(x)}+f(x)-1$,

38. $y=e^{-\int\phi(x)dx}\left[\int\dfrac{1}{\phi_1(x)}e^{\int\phi(x)dx}dx+C\right]$, where $\phi(x)=\dfrac{\phi_2(x)}{\phi_1(x)}$

39. $x+y+2=Ce^y$,

40. $x=\tan^{-1}y-1+Ce^{-\tan^{-1}y}$,

41. $xe^{-\tan^{-1}y}=\tan^{-1}y+C$

42. $xy+\tan^{-1}y=C$,

43. $x\ln y=\dfrac{1}{2}(\ln y)^2+C$,

44. $y=\tan x+C\sqrt{\tan x}$,

45. $xe^y=\tan y+C$,

46. $\dfrac{1}{x}=\dfrac{1}{2}\left(1-y^2\right)C+e^{-y^2}$,

47. $x=y\left(y^2+C\right)$,

48. $2x=e^y\left(1+Cx^2\right)$,

49. $\sin y=(1+x)\left(e^x+C\right)$,

50. $x=\left(Cx^2+\dfrac{1}{2}\right)\ln y$,

51. $\sqrt{y} = -\frac{1}{3}(1-x^2) + C\sqrt[4]{1-x^2}$,

52. $\sin x = y(x - C)$,

53. $e^{x^2} = (2x + C)y^2$,

54. $\frac{1}{x \ln z} = \frac{1}{2x^2} + C$,

55. $\frac{1}{yx^3} = C - \ln x$,

56. $\frac{1}{x^2} = 1 - y^2 + Ce^{-y^2}$

57. $(x^2 + y^2)e^y = C$,

58. $e^x = y(C - x^2)$.

Chapter 3

INTEGRATING FACTORS

3.1 Introduction

In the previous chapter we have discussed about the exact differential equation. Sometimes it is found that an equation of the form $Mdx + Ndy = 0$ is not exact, but there exist some suitable functions of x, y called multiplying factors, with the help of which the equation can be made exact one. This multiplying factors are called *integrating factors.*

Consider the first order linear differential equation

$$\frac{dy}{dx} + Py = Q,$$

where P, Q are functions of x only.

Rewriting the equation as $Pydx + dy = Qdx$

we see that the term Qdx is exact, but the terms in the left-hand side is not an exact differential. To make it so, we multiply the above equation by a function $R(x)$ and get

$$RPydx + Rdy = QRdx.$$

Using the condition of integrability, we get

$$\frac{\partial(RPy)}{\partial y} = \frac{\partial R}{\partial x}$$

$$\text{or,} \quad \frac{dR}{dx} = RP$$

$$\text{or,} \quad \frac{dR}{R} = Pdx$$

$$\text{or,} \quad R(x) = e^{\int Pdx}.$$

The function $R(x)$ is called integrating factor of the above linear differential equation.

Now the question is:

Is the integrating factor of an equation of the form $Mdx + Ndy = 0$ unique?

The following theorem will give the answer of the question.

Theorem 3.1.1: The differential equation $Mdx + Ndy = 0$ possesses an infinite number of integrating factors.

Proof: Let $f(x, y)$ be an integrating factor of the given differential equation, then $f(x, y)(Mdx + Ndy)$ will be an exact differential, i.e., there must be a function $u(x, y)$ such that

$$f(x, y)(Mdx + Ndy) = du.$$

Multiplying the above equation by a function $g(u)$, we get

$$g(u)\, f(Mdx + Ndy) = g(u)\, du = d[G(u)] \text{ (say)}.$$

Since the expression in the right-hand side is an exact differential, the expression in the left-hand side will also be an exact differential. This shows that $g(u) f(x, y)$ is also an integrating factor of the given equation. Since the function $g(u)$ is arbitrary, the given equation has infinite number of integrating factors.

3.2 Rules for Finding the Integrating Factors

An equation of the form $Mdx + Ndy = 0$ has always integrating factors. Here we shall discuss the rules for finding them.

3.2.1 By inspection

Integrating factors of the equation $Mdx + Ndy = 0$ can be found by rearranging its terms, or by dividing it by suitable function of x, y. Here we give a list of expressions of exact differentials which would help in solving an equation of the above form.

(i) $d\left(\dfrac{y}{x}\right) = \dfrac{xdy - ydx}{x^2}$, (ii) $d\left(\dfrac{x}{y}\right) = \dfrac{ydx - xdy}{y^2}$,

(iii) $d(xy) = xdy + ydx$, (iv) $\dfrac{1}{2} d(x^2 + y^2) = xdx + ydy$

(v) $d[\ln xy] = \dfrac{xdy + ydx}{xy}$, (vi) $d\left[\ln\left(\dfrac{x}{y}\right)\right] = \dfrac{ydx - xdy}{xy}$,

(vii) $d\left[\ln\left(\dfrac{y}{x}\right)\right] = \dfrac{xdy - ydx}{xy}$, (viii) $d\left[\dfrac{1}{2}\ln(x^2 + y^2)\right] = \dfrac{xdx + ydy}{x^2 + y^2}$,

(ix) $d\left(\tan^{-1}\dfrac{y}{x}\right) = \dfrac{xdy - ydx}{x^2 + y^2}$, (x) $d\left(\tan^{-1}\dfrac{x}{y}\right) = \dfrac{ydx - xdy}{x^2 + y^2}$

(xi) $d[\sin^{-1} xy] = \dfrac{xdy + ydx}{\sqrt{1 - x^2 y^2}}$ (xii) $d\left[\sin^{-1}\dfrac{y}{x}\right] = \dfrac{xdy - ydx}{x\sqrt{x^2 - y^2}}$.

ILLUSTRATIVE EXAMPLES

(i) Solve: $xdy + ydx = k(xdy - ydx)$.

Solution

Dividing by xy, we get

$$\frac{d(xy)}{xy} = k\left(\frac{xdy - ydx}{xy}\right) \quad \text{or,} \quad d[\ln xy] = k\, d\left(\ln\frac{y}{x}\right)$$

Integrating, we get

$$\ln xy = k \ln \frac{y}{x} + \ln C \quad \text{or,} \quad xy = C\left(\frac{y}{x}\right)^k, \text{ where } C \text{ is an arbitrary constant.}$$

(ii) Solve: $(x^2 + y^2)(xdx + ydy) + xdy - ydx = 0.$

Solution

We rewrite the equation as $\frac{1}{2}d(x^2 + y^2) + \frac{xdy - ydx}{x^2 + y^2} = 0.$

Integrating, we get $x^2 + y^2 + 2\tan^{-1}\frac{y}{x} = C$, where C is an arbitrary constant.

(iii) Solve: $x^2 \frac{dy}{dx} + xy + 2\sqrt{1 - x^2y^2} = 0.$

Solution

We rewrite the equation as $x(xdy + ydx) + 2\sqrt{1 - x^2y^2}\,dx = 0$

$$\text{or,} \quad \frac{xdy + ydx}{\sqrt{1 - x^2y^2}} + 2\frac{dx}{x} = 0 \qquad \text{or,} \quad d\left[\sin^{-1} xy\right] + 2d[\ln x] = 0.$$

Integrating we get $\sin^{-1} xy + 2\ln x = C$, where C is an arbitrary constant.

(iv) Solve: $xdx + ydy + \left(1 + \frac{y^2}{x^2}\right)(ydx - xdy) = 0.$

Solution

We rewrite the equation as $\frac{xdx + ydy}{x^2 + y^2} + \frac{ydx - xdy}{x^2} = 0$

$$\text{or,} \quad d\left[\frac{1}{2}\ln(x^2 + y^2)\right] - d\left(\frac{y}{x}\right) = 0.$$

Integrating we get $\ln(x^2 + y^2) = 2\frac{y}{x} + C$, where C is an arbitrary constant.

(v) Solve: $xdy - ydx = 2\sqrt{x^2 - y^2}\,dx.$

Solution

We rewrite the equation as $\frac{xdy - ydx}{x\sqrt{x^2 - y^2}} = \frac{2}{x}dx$ or, $d\left[\sin^{-1}\frac{y}{x}\right] = 2d(\ln x)$

Integrating we get $\sin^{-1}\frac{y}{x} = 2\ln x + C$, where C is an arbitrary constant.

3.2.2 If the equation $Mdx + Ndy = 0$ is a homogeneous equation of degree n and $Mx + Ny \neq 0$, then $\frac{1}{Mx + Ny}$ will be an integrating factor.

Proof: The condition of integrability of the equation $\frac{Mdx + Ndy}{Mx + Ny} = 0$

is $\frac{\partial}{\partial y}\left(\frac{M}{Mx + Ny}\right) = \frac{\partial}{\partial x}\left(\frac{N}{Mx + Ny}\right).$

Now, $\dfrac{\partial}{\partial y}\left(\dfrac{M}{Mx+Ny}\right)$

$$= \frac{(Mx+Ny)\dfrac{\partial M}{\partial y} - M\left(x\dfrac{\partial M}{\partial y} + y\dfrac{\partial N}{\partial y} + N\right)}{(Mx+Ny)^2} = \frac{Ny\dfrac{\partial M}{\partial y} - M\left(y\dfrac{\partial N}{\partial y} + N\right)}{(Mx+Ny)^2}$$

$$= \frac{N\left(x\dfrac{\partial M}{\partial x} + y\dfrac{\partial M}{\partial y}\right) - M\left(y\dfrac{\partial N}{\partial y} + N\right) - Nx\dfrac{\partial M}{\partial x}}{(Mx+Ny)^2}$$

$$= \frac{(n-1)\,MN - My\dfrac{\partial N}{\partial y} - Nx\dfrac{\partial M}{\partial x}}{(Mx+Ny)^2}$$

and $\dfrac{\partial}{\partial x}\left(\dfrac{N}{Mx+Ny}\right)$

$$= \frac{(Mx+Ny)\dfrac{\partial N}{\partial x} - N\left(x\dfrac{\partial M}{\partial x} + M + y\dfrac{\partial N}{\partial x}\right)}{(Mx+Ny)^2}$$

$$= \frac{Mx\dfrac{\partial N}{\partial x} - N\left(x\dfrac{\partial M}{\partial x} + M\right)}{(Mx+Ny)^2}$$

$$= \frac{M\left(x\dfrac{\partial N}{\partial x} + y\dfrac{\partial N}{\partial y}\right) - N\left(x\dfrac{\partial M}{\partial x} + M\right) - My\dfrac{\partial N}{\partial y}}{(Mx+Ny)^2}$$

$$= \frac{(n-1)\,MN - My\dfrac{\partial N}{\partial y} - Nx\dfrac{\partial M}{\partial x}}{(Mx+Ny)^2}.$$

Thus from (2) and (3) we see that the condition of integrability is satisfied, hence $\dfrac{1}{Mx+Ny}$ is an integrating factor of the given equation.

Alternative Method: We rewrite $Mdx + Ndy$ as

$$Mdx + Ndy = \frac{1}{2}\left[(Mx+Ny)\left(\frac{dx}{x} + \frac{dy}{y}\right) + (Mx-Ny)\left(\frac{dx}{x} - \frac{dy}{y}\right)\right].$$

Dividing by $Mx + Ny$ as $Mx + Ny \neq 0$, we get

$$\frac{Mdx + Ndy}{Mx+Ny} = \frac{1}{2}\left[d\,(\ln xy) + \frac{Mx-Ny}{Mx+Ny}\,d\left(\ln\frac{x}{y}\right)\right].$$

If $Mx + Ny$ be a homogeneous function, then $Mx - Ny$ is also so and hence we assume

$$\frac{Mx-Ny}{Mx+Ny} = f\left(\frac{x}{y}\right) = f\left(e^{\ln\frac{x}{y}}\right) = F\left(\ln\frac{x}{y}\right).$$

So $$\frac{Mdx + Ndy}{Mx + Ny} = \frac{1}{2}\left[d(\ln xy) + F\left(\ln\frac{x}{y}\right)d\left(\ln\frac{x}{y}\right)\right].$$

This shows that $\dfrac{Mdx + Ndy}{Mx + Ny}$ is integrable, so $\dfrac{1}{Mx + Ny}$ is an integrating factor of the equation

$$Mdx + Ndy = 0.$$

ILLUSTRATIVE EXAMPLES

(i) Solve: $x^2dy - (xy + 2y^2)\,dx = 0.$

Solution

The given equation is a homogeneous equation and $Mx + Ny = -x^2y - 2y^2x + x^2y$ $= -2y^2x \neq 0$, so $-\dfrac{1}{2y^2x}$ is an integrating factor of the equation. Multiplying the equation by $-\dfrac{1}{2y^2x}$, we get

$$\frac{x^2dy - (xy + 2y^2)\,dx}{y^2x} = 0 \quad \text{or,} \quad \frac{x(xdy - ydx)}{y^2x} - 2\frac{dx}{x} = 0$$

or, $$\frac{xdy - ydx}{y^2} - 2\frac{dx}{x} = 0 \quad \text{or,} \quad d\left(\frac{x}{y}\right) + 2d(\ln x) = 0.$$

Integrating, we get $\dfrac{x}{y} + 2\ln x = C$, where C is an arbitrary constant.

(ii) Solve: $(x^2y - 2xy^2)\,dx - (x^3 - 3x^2y)\,dy = 0.$

Solution

The given equation is a homogeneous equation and

$$Mx + Ny = x^3y - 2x^2y^2 - x^3y + 3x^2y^2 = x^2y^2 \neq 0,$$

so $\dfrac{1}{x^2y^2}$ is an integrating factor of the equation. Multiplying the equation by $\dfrac{1}{x^2y^2}$, we get

$$\frac{(x^2y - 2xy^2)\,dx - (x^3 - 3x^2y)dy}{x^2y^2} = 0. \quad \text{or,} \quad \left(\frac{1}{y} - \frac{2}{x}\right)dx - \left(\frac{x}{y^2} - \frac{3}{y}\right)dy = 0$$

or, $$\frac{ydx - xdy}{y^2} - \frac{2}{x}dx + \frac{3}{y}dy = 0 \quad \text{or,} \quad d\left(\frac{x}{y}\right) - 2d(\ln x) + 3d(\ln y) = 0.$$

Integrating, we get

$$\frac{x}{y} - 2\ln x + 3\ln y = C, \text{ where } C \text{ is an arbitrary constant.}$$

(iii) Solve: $(y^3 - 2yx^2)dx + (2xy^2 - x^3)dy = 0.$

Solution

The given equation is a homogeneous equation and

$$Mx + Ny = y^3x - 2x^3y - x^3y + 2xy^3 = 3xy(y^2 - x^2) \neq 0,$$

so $\dfrac{1}{xy(y^2 - x^2)}$ is an integrating factor of the equation. Multiplying the equation by $\dfrac{1}{xy(y^2 - x^2)}$, we get

$$\frac{(y^3 - 2yx^2)\,dx + (2xy^2 - x^3)\,dy}{xy(y^2 - x^2)} = 0$$

or,
$$\frac{(y^3 - yx^2)\,dx + (xy^2 - x^3)\,dx + xy^2dy - yx^2dx}{xy(y^2 - x^2)} = 0$$

or,
$$\frac{dx}{x} + \frac{dy}{y} + \frac{1}{2}\frac{d(y^2 - x^2)}{(y^2 - x^2)} = 0.$$

Integrating, we get $\ln x^2 + \ln y^2 + \ln(y^2 - x^2) = \ln C$

or, $\quad x^2y^2(y^2 - x^2) = C,$ where C is an arbitrary constant.

3.2.3: If $\dfrac{1}{N}\left(\dfrac{\partial M}{\partial y} - \dfrac{\partial N}{\partial x}\right)$ is a function of x alone, say $f(x)$, then $e^{\int f(x)dx}$ is an integrating factor of the equation $Mdx + Ndy = 0$.

Proof: If $\mu(x)$ be an integrating factor of the equation $Mdx + Ndy = 0$, then

$$\mu Mdx + \mu Ndy = 0$$

is an exact equation, i.e., we must have

$$\mu\frac{\partial}{\partial y}M = \frac{\partial}{\partial x}(\mu N).$$

This gives
$$N\frac{d\mu}{dx} = \mu\left(\frac{\partial M}{\partial y} - \frac{\partial N}{\partial x}\right)$$

or,
$$\frac{d\mu}{\mu} = \frac{1}{N}\left(\frac{\partial M}{\partial y} - \frac{\partial N}{\partial x}\right)dx = f(x)\,dx.$$

Integrating, we get

$$\ln\mu = \int f(x)\,dx \quad \text{i.e.,} \quad \mu(x) = e^{\int f(x)\,dx}.$$

Alternative Method: Given $\dfrac{1}{N}\left(\dfrac{\partial M}{\partial y} - \dfrac{\partial N}{\partial x}\right) = f(x)$

$$\text{i.e.,} \quad \frac{\partial M}{\partial y} = \frac{\partial N}{\partial x} + Nf(x)dx.$$

Multiplying both sides by $e^{\int f(x)\,dx}$, we get

$$\frac{\partial M}{\partial y} e^{\int f(x)\,dx} = \frac{\partial N}{\partial x} e^{\int f(x)\,dx} + N f(x) e^{\int f(x)\,dx} dx$$

$$\text{or,}\quad \frac{\partial}{\partial y}\left(M e^{\int f(x)\,dx}\right) = \frac{\partial}{\partial x}\left(N e^{\int f(x)\,dx}\right).$$

This proves exactness of the equation

$$(Mdx + Ndy)\, e^{\int f(x)\,dx} = 0.$$

ILLUSTRATIVE EXAMPLES

(i) Solve: $\left(xy^2 - e^{\frac{1}{x^3}}\right) dx - x^2 y dy = 0.$

Solution

Here $M = xy^2 - e^{\frac{1}{x^3}}$, $N = -x^2 y$ and $\frac{1}{N}\left(\frac{\partial M}{\partial y} - \frac{\partial N}{\partial x}\right) = \frac{2xy + 2xy}{-x^2 y} = -\frac{4}{x}$. So, $e^{-\int \frac{4}{x} dx} = e^{-4\ln x} = \frac{1}{x^4}$ is an integrating factor of the equation. Multiplying the equation by $\frac{1}{x^4}$, we get

$$\frac{\left(xy^2 - e^{1/x^3}\right) dx - x^2 y dy}{x^4} = 0$$

$$\text{or,}\quad \frac{xy\,(ydx - xdy)}{x^4} - \frac{1}{x^4} e^{1/x^3} dx = 0$$

$$\text{or,}\quad -\frac{y}{x} d\left(\frac{y}{x}\right) + \frac{1}{3} e^{\frac{1}{x^3}} d\left(\frac{1}{x^3}\right) = 0.$$

Integrating, we get $-\frac{y^2}{2x^2} + \frac{1}{3} e^{\frac{1}{x^3}} = C$, where C is an integrating factor.

(ii) Solve: $xdy - \left(4y + x^6 e^x\right) dx = 0.$

Solution

Here $M = -x^6 e^x - 4y$, $N = x$ and $\frac{1}{N}\left(\frac{\partial M}{\partial y} - \frac{\partial N}{\partial x}\right) = \frac{-4-1}{x} = -\frac{5}{x}$. So, $e^{-\int \frac{5}{x} dx} = e^{-5\ln x} = \frac{1}{x^5}$ is an integrating factor of the equation. Multiplying the equation by $\frac{1}{x^5}$, we get

$$\frac{xdy - (4y + x^6 e^x)dx}{x^5} = 0 \quad \text{or,}\quad \frac{xdy - 4ydx}{x^5} - xe^x dx = 0$$

$$\text{or,}\quad \frac{x^4 dy - 4x^3 y dx}{x^8} - xe^x dx = 0 \quad \text{or,}\quad d\left(\frac{y}{x^4}\right) - xe^x dx = 0.$$

Integrating, we get $yx^{-4} - (x-1)e^x = C$, where C is an arbitrary constant.

(iii) Solve: $(x^3 + xy^4)dx + 2y^3dy = 0.$

Solution

Here $M = x^3 + xy^4$, $N = 2y^3$ and $\frac{1}{N}\left(\frac{\partial M}{\partial y} - \frac{\partial N}{\partial x}\right) = \frac{4xy^3 - 0}{2y^3} = 2x.$

So, $e^{\int 2x\,dx} = e^{x^2}$ is an integrating factor of the equation. Multiplying the equation by e^{x^2} we get

$$(x^3 + xy^4)e^{x^2}dx + 2y^3e^{x^2}dy = 0 \quad \text{or,} \quad x^2e^{x^2}d(x^2) + d\left(y^4e^{x^2}\right) = 0.$$

Integrating, we get

$$(x^2 - 1)e^{x^2} + y^4e^{x^2} = C \text{ or, } y^4 + x^2 = 1 + Ce^{-x^2}, \text{ where } C \text{ is an arbitrary constant.}$$

3.2.4: If $\frac{1}{M}\left(\frac{\partial M}{\partial y} - \frac{\partial N}{\partial x}\right)$ is a function of y alone, say $f(y)$, then $e^{-\int f(y)\,dy}$ is an integrating factor of the equation $Mdx + Ndy = 0$.

Proof: If $\mu(y)$ be an integrating factor of the equation $Mdx + Ndy = 0$, then

$$\mu M dx + \mu N dy = 0$$

is an exact equation, i.e., we must have

$$\frac{\partial}{\partial y}(\mu M) = \mu \frac{\partial N}{\partial x}.$$

This gives
$$M\frac{d\mu}{dy} = -\mu\left(\frac{\partial M}{\partial y} - \frac{\partial N}{\partial x}\right)$$

or,
$$\frac{d\mu}{\mu} = -\frac{1}{M}\left(\frac{\partial M}{\partial y} - \frac{\partial N}{\partial x}\right)dy = -f(y)\,dy$$

Integrating, we get

$$\ln \mu = -\int f(y)\,dy \text{ i.e., } \mu(y) = e^{-\int f(y)\,dy}.$$

ILLUSTRATIVE EXAMPLES

(i) Solve: $xydx + (2x^2 + 3y^2 - 20)\,dy = 0.$

Solution

Here $M = xy$, $N = 2x^2 + 3y^2 - 20$, so $\frac{1}{M}\left(\frac{\partial M}{\partial y} - \frac{\partial N}{\partial x}\right) = \frac{1}{xy}(x - 4x) = -\frac{3}{y}.$

Hence, $e^{\int \frac{3}{y}dy} = e^{3\ln y} = y^3$ is an integrating factor of the equation. Multiplying the equation by y^3, we get

$$xy^4dx + (2x^2 + 3y^2 - 20)y^3dy = 0 \text{ or, } \frac{1}{2}d(x^2y^4) + 3y^5dy - 20y^3dy = 0.$$

Integrating, we get

$$x^2y^4 + y^6 - 10y^4 = C, \text{ where } C \text{ is an arbitrary constant.}$$

(ii) Solve: $(xy^3 + y)dx + 2(x^2y^2 + x + y^4)dy = 0.$

Solution

Here, $M = xy^3 + y, N = 2x^2y^2 + 2x + 2y^4,$ so

$\frac{1}{M}\left(\frac{\partial M}{\partial y} - \frac{\partial N}{\partial x}\right) = \frac{1}{xy^3 + y}(3xy^2 + 1 - 4xy^2 - 2) = -\frac{1}{y}$. Hence, $e^{\int \frac{dy}{y}} = e^{\ln y} = y$ is an integrating factor of the equation. Multiplying the equation by y, we get

$$(xy^3 + y)ydx + 2(x^2y^2 + x + y^4)ydy = 0$$

or, $$xy^4dx + 2x^2y^3dy + y^2dx + 2xydy + y^5dy = 0$$

or, $$\frac{1}{2}d(x^2y^4) + d(xy^2) + y^5dy = 0.$$

Integrating, we get

$$3x^2y^4 + 6xy^2 + y^6 = C, \text{ where } C \text{ is an arbitrary constant.}$$

(iii) Solve: $(2xy^4e^y + 2xy^3 + y)dx + (x^2y^4e^y - x^2y^2 - 3x)dy = 0.$

Solution

Here, $M = 2xy^4e^y + 2xy^3 + y, N = x^2y^4e^y - x^2y^2 - 3x.$

So, $$\frac{1}{M}\left(\frac{\partial M}{\partial y} - \frac{\partial N}{\partial x}\right)$$

$$= \frac{1}{2xy^4e^y + 2xy^3 + y}(8xy^3e^y + 2xy^4e^y + 6xy^2 + 1 - 2xy^4e^y + 2xy^2 + 3)$$

$$= \frac{8xy^3e^y + 8xy^2 + 4}{2xy^4e^y + 2xy^3 + y} = \frac{4}{y}.$$

Hence, $e^{-\int \frac{4}{y}dy} = e^{-4\ln y} = \frac{1}{y^4}$ is an integrating factor of the equation. Multiplying the equation by $\frac{1}{y^4}$, we get

$$\frac{1}{y^4}(2xy^4e^y + 2xy^3 + y)dx + \frac{1}{y^4}(x^2y^4e^y - x^2y^2 - 3x)dy = 0$$

or, $$d(x^2e^y) + \frac{2xydx - x^2dy}{y^2} + \frac{y^3dx - 3xy^2dy}{y^6} = 0$$

or, $$d\left(x^2e^y + \frac{x^2}{y} + \frac{x}{y^3}\right) = 0.$$

Integrating, we get

$$x^2e^y + \frac{x^2}{y} + \frac{x}{y^3} = C, \text{ where } C \text{ is an integrating constant.}$$

3.2.5: If $Mx - Ny \neq 0$, then $\dfrac{1}{Mx - Ny}$ is an integrating factor of the equation $Mdx + Ndy = 0$, where $M = yf(xy)$, $N = xg(xy)$.

Proof: If $\dfrac{1}{Mx - Ny}$ is an integrating factor of the equation $Mdx + Ndy = 0$, then

$$\frac{Mdx + Ndy}{Mx - Ny} = \frac{yf(xy)\,dx + xg(xy)\,dy}{xy(f(xy) - g(xy))} = \frac{yfdx + xgdy}{xy(f - g)} = \frac{f}{x(f - g)}dx + \frac{g}{y(f - g)}dy$$

will be exact.

Now $$\frac{\partial}{\partial y}\left(\frac{f}{x(f - g)}\right) = \frac{1}{x}\frac{(f - g)\dfrac{\partial f}{\partial y} - f\dfrac{\partial}{\partial y}(f - g)}{(f - g)^2} = \frac{1}{x}\frac{f\dfrac{\partial g}{\partial y} - g\dfrac{\partial f}{\partial y}}{(f - g)^2}$$ and

$$\frac{\partial}{\partial x}\left(\frac{g}{y(f - g)}\right) = \frac{1}{y}\frac{(f - g)\dfrac{\partial g}{\partial x} - g\dfrac{\partial}{\partial x}(f - g)}{(f - g)^2} = \frac{1}{y}\frac{f\dfrac{\partial g}{\partial x} - g\dfrac{\partial f}{\partial x}}{(f - g)^2}.$$

So, $$\frac{\partial}{\partial y}\left(\frac{f}{x(f - g)}\right) - \frac{\partial}{\partial x}\left(\frac{g}{y(f - g)}\right)$$

$$= \frac{1}{xy}\frac{y\left(f\dfrac{\partial g}{\partial y} - g\dfrac{\partial f}{\partial y}\right) - x\left(f\dfrac{\partial g}{\partial x} - g\dfrac{\partial f}{\partial x}\right)}{(f - g)^2}$$

$$= \frac{1}{xy}\frac{g\left(x\dfrac{\partial f}{\partial x} - y\dfrac{\partial f}{\partial y}\right) + f\left(y\dfrac{\partial g}{\partial y} - x\dfrac{\partial g}{\partial x}\right)}{(f - g)^2}$$

$= 0$, since $x\dfrac{\partial f}{\partial x} = xy\,f' = yx\,f' = y\dfrac{\partial f}{\partial y}$ and $y\dfrac{\partial g}{\partial y} = x\dfrac{\partial g}{\partial x}$.

Hence, the equation $\dfrac{f}{x(f - g)}dx + \dfrac{g}{y(f - g)}dy = 0$, i.e., $\dfrac{1}{Mx - Ny}(Mdx + Ndy) = 0$ is exact. In orther words, $\dfrac{1}{Mx - Ny}$ is an integrating factor of the equation $Mdx + Ndy = 0$.

Alternative Method: We rewrite $Mdx + Ndy$ as

$$Mdx + Ndy = \frac{1}{2}\left[(Mx + Ny)\left(\frac{dx}{x} + \frac{dy}{y}\right) + (Mx - Ny)\left(\frac{dx}{x} - \frac{dy}{y}\right)\right].$$

Dividing by $Mx - Ny$, as $Mx - Ny \neq 0$, we get

$$\frac{Mdx + Ndy}{Mx - Ny} = \frac{1}{2}\left[\frac{Mx + Ny}{Mx - Ny}d(\ln xy) + d\left(\ln\frac{x}{y}\right)\right]$$

$$= \frac{1}{2}\left[\frac{f + g}{f - g}d(\ln xy) + d\left(\ln\frac{x}{y}\right)\right]$$

$$= \frac{1}{2}\left[F(xy)d(\ln xy) + d\left(\ln\frac{x}{y}\right)\right]$$

$$= \frac{1}{2}\left[F(e^{\ln xy})\,d(\ln xy) + d\left(\ln\frac{x}{y}\right)\right].$$

This shows that $\frac{Mdx + Ndy}{Mx - Ny}$ is integrable, so $\frac{1}{Mx - Ny}$ is an integrating factor of the equation

$$Mdx + Ndy = 0.$$

ILLUSTRATIVE EXAMPLES

(i) Solve: $y(xy + 2x^2y^2)dx + x(xy - x^2y^2)dy = 0.$

Solution

The given equation is of the form $Mdx + Ndy = 0$, where $M = yf(xy)$ and $N = xg(xy)$. Now $Mx - Ny = 3x^3y^3 \neq 0$, so $\frac{1}{3x^3y^3}$ is an integrating factor of it.

Multiplying the equation by $\frac{1}{3x^3y^3}$, we get

$$\frac{y(xy + 2x^2y^2)dx + x(xy - x^2y^2)dy}{3x^3y^3} = 0$$

or, $$\frac{ydx + xdy + xy(2ydx - xdy)}{x^2y^2} = 0$$

or, $$\frac{d(xy)}{(xy)^2} + 2\frac{dx}{x} - \frac{dy}{y} = 0.$$

Integrating, we get

$$-\frac{1}{xy} + 2\ln x - \ln y = C, \text{ where } C \text{ is an arbitrary constant.}$$

(ii) Solve: $(x^2y^2 + xy + 1)\,ydx + (x^2y^2 - xy + 1)\,xdy = 0.$

Solution

The given equation is of the form $Mdx + Ndy = 0$, where $M = yf(xy)$ and $N = xg(xy)$. Now $Mx - Ny = 2x^2y^2 \neq 0$, so $\frac{1}{2x^2y^2}$ is an integrating factor of it. Multiplying the equation by $\frac{1}{2x^2y^2}$, we get

$$\frac{(x^2y^2 + xy + 1)\,ydx + (x^2y^2 - xy + 1)\,xdy}{x^2y^2} = 0$$

or, $$\left(1+\frac{1}{x^2y^2}\right)(ydx+xdy)+\frac{dx}{x}-\frac{dy}{y}=0$$

or, $$\left(1+\frac{1}{(xy)^2}\right)d(xy)+\frac{dx}{x}-\frac{dy}{y}=0.$$

Integrating, we get

$$xy-\frac{1}{xy}+\ln\left(\frac{x}{y}\right)=C, \text{ where } C \text{ is an arbitrary constant.}$$

(iii) Solve: $y(1+2xy)\,dx+x(1+2xy-x^3y^3)\,dy=0.$

Solution

The given equation is of the form $Mdx + Ndy = 0$, where $M = yf(xy)$ and $N = xg(xy)$. Now $Mx - Ny = x^4y^4 \neq 0$, so $\frac{1}{x^4y^4}$ is an integrating factor of it. Multiplying the equation by $\frac{1}{x^4y^4}$, we get

$$\frac{y(1+2xy)\,dx+x(1+2xy-x^3y^3)\,dy}{x^4y^4}=0$$

or, $$\left(\frac{1}{(xy)^4}+\frac{2}{(xy)^3}\right)d(xy)-\frac{dy}{y}=0.$$

Integrating, we get

$$\frac{1}{3x^3y^3}+\frac{1}{x^2y^2}+\ln y=C, \text{ where } C \text{ is an arbitrary constant.}$$

3.2.6: If $\frac{1}{M-N}\left(\frac{\partial M}{\partial y}-\frac{\partial N}{\partial x}\right)$ is a function of the form $f(x+y)$, then the equation $Mdx + Ndy = 0$ has an integrating factor of the form $e^{-\int f(x+y)\,d(x+y)}$.

Proof: If $\mu(x+y)$ be an integrating factor of the differential equation $Mdx + Ndy = 0$, then

$$\mu Mdx + \mu Ndy = 0$$

is an exact equation, i.e., we must have

$$\frac{\partial}{\partial x}(\mu N)=\frac{\partial}{\partial y}(\mu M)$$

or, $$N\frac{\partial\mu}{\partial x}+\mu\frac{\partial N}{\partial x}=M\frac{\partial\mu}{\partial y}+\mu\frac{\partial M}{\partial y}$$

or, $$\mu\left(\frac{\partial N}{\partial x}-\frac{\partial M}{\partial y}\right)=(M-N)\,\mu'$$

or, $$\frac{d\mu}{\mu}=\frac{1}{M-N}\left(\frac{\partial N}{\partial x}-\frac{\partial M}{\partial y}\right)d(x+y)=-f(x+y)\,d(x+y)$$

or, $$\mu(x+y)=e^{-\int f(x+y)\,d(x+y)}.$$

ILLUSTRATIVE EXAMPLES

(i) Solve: $(y-1)dx-(x+1)dy=0$.

Solution

Here, $M=y-1, N=-x-1$, so $\dfrac{1}{M-N}\left(\dfrac{\partial N}{\partial x}-\dfrac{\partial M}{\partial y}\right)=\dfrac{-2}{x+y}$.

Hence, $e^{-\int\frac{2}{x+y}d(x+y)}=\dfrac{1}{(x+y)^2}$ is an integrating factor. Multiplying the equation by $\dfrac{1}{(x+y)^2}$, we get

$$\frac{(y-1)dx-(x+1)dy}{(x+y)^2}=0 \text{ or, } \frac{ydx-xdy-dx-dy}{(x+y)^2}=0$$

$$\text{or, } -\frac{xdy-ydx}{x^2}\frac{1}{\left(1+\frac{y}{x}\right)^2}-\frac{d(x+y)}{(x+y)^2}=0 \text{ or, } \frac{d\left(\frac{y}{x}\right)}{\left(1+\frac{y}{x}\right)^2}+\frac{d(x+y)}{(x+y)^2}=0.$$

Integrating, we get

$$-\frac{1}{1+\frac{y}{x}}-\frac{1}{x+y}=-C \text{ or, } x+1=C(x+y),$$ where C is an arbitrary constant.

(ii) Solve: $(x^2+2)dy+(y^2+2)dx=0$.

Solution

Here $M=y^2+2, N=x^2+2$, so $\dfrac{1}{M-N}\left(\dfrac{\partial N}{\partial x}-\dfrac{\partial M}{\partial y}\right)=\dfrac{-2(y-x)}{(y-x)(y+x)}=\dfrac{-2}{x+y}$.

Hence, $e^{-\int\frac{2}{x+y}d(x+y)}=\dfrac{1}{(x+y)^2}$ is an integrating factor. Multiplying the equation by $\dfrac{1}{(x+y)^2}$, we get

$$\frac{(x^2+2)dy+(y^2+2)dx}{(x+y)^2}=0$$

$$\text{or, } \frac{x^2dy+y^2dx}{(x+y)^2}+2\frac{d(x+y)}{(x+y)^2}=0$$

$$\text{or, } \frac{x(x+y)dy+y(x+y)dx-xyd(x+y)}{(x+y)^2}+2\frac{d(x+y)}{(x+y)^2}=0$$

$$\text{or, } \frac{(x+y)(xdy+ydx)-xyd(x+y)}{(x+y)^2}+2\frac{d(x+y)}{(x+y)^2}=0$$

or, $$\frac{(x+y)d(xy)-xyd(x+y)}{(x+y)^2}+2\frac{d(x+y)}{(x+y)^2}=0$$

or, $$d\left(\frac{xy}{x+y}\right)-2d\left(\frac{1}{x+y}\right)=0.$$

Integrating, we get

$$\frac{xy-2}{x+y}=C,\text{ where } C \text{ is an integrating constant.}$$

(iii) Solve: $(2xy-x^2-x)dy+(2xy-y^2-y)dx=0.$

Solution

Here, $M=2xy-y^2-y$, $N=2xy-x^2-x$, so $\dfrac{1}{M-N}\left(\dfrac{\partial N}{\partial x}-\dfrac{\partial M}{\partial y}\right)=\dfrac{-4}{x+y+1}.$

Hence $e^{-\int\frac{4}{x+y+1}d(x+y)}=\dfrac{1}{(x+y+1)^4}$ is an integrating factor. Multiplying the equation by $\dfrac{1}{(x+y+1)^4}$, we get

$$\frac{(2xy-x^2-x)dy+(2xy-y^2-y)dx}{(x+y+1)^4}=0$$

or, $$\frac{3xy(dx+dy)-(x+y+1)xdy-(x+y+1)ydx}{(x+y+1)^4}=0$$

or, $$\frac{3xyd(x+y+1)-(x+y+1)(xdy+ydx)}{(x+y+1)^4}=0$$

or, $$\frac{3xy(x+y+1)^2d(x+y+1)-(x+y+1)^3d(xy)}{(x+y+1)^6}=0$$

or, $$d\left(\frac{xy}{(x+y+1)^3}\right)=0.$$

Integrating, we get

$$\frac{xy}{(x+y+1)^3}=C,\text{ where } C \text{ is an integrating constant.}$$

3.2.7: If $\dfrac{1}{xM-yN}\left(\dfrac{\partial M}{\partial y}-\dfrac{\partial N}{\partial x}\right)$ is a function of the form $f(xy)$, then the equation $Mdx + Ndy = 0$ has an integrating factor of the form $e^{-\int f(xy)d(xy)}$.

Proof: If $\mu(xy)$ be an integrating factor of the differential eqnation $Mdx + Ndy = 0$, then

$$\mu Mdx+\mu Ndy=0$$

is an exact equation, i.e., we must have

$$\frac{\partial}{\partial x}(\mu N)=\frac{\partial}{\partial y}(\mu M) \text{ or, } N\frac{\partial \mu}{\partial x}+\mu\frac{\partial N}{\partial x}=M\frac{\partial \mu}{\partial y}+\mu\frac{\partial M}{\partial y}$$

$$\text{or, } Ny\mu'+\mu\frac{\partial N}{\partial x}=Mx\mu'+\mu\frac{\partial M}{\partial y} \text{ or, } \mu\left(\frac{\partial N}{\partial x}-\frac{\partial M}{\partial y}\right)=(xM-yN)\,\mu'$$

$$\text{or, } \frac{d\mu}{\mu}=\frac{1}{xM-yN}\left(\frac{\partial N}{\partial x}-\frac{\partial M}{\partial y}\right)d(x+y)=-f(xy)\,d(xy) \text{ or, } \mu(xy)=e^{-\int f(xy)\,d(xy)}.$$

ILLUSTRATIVE EXAMPLES

(i) Solve: $y(1+xy)\,dx+x(1-xy)\,dy=0.$

Solution

Here $M=y(1+xy)$, $N=x(1-xy)$, so $\dfrac{1}{yN-xM}\left(\dfrac{\partial M}{\partial y}-\dfrac{\partial N}{\partial x}\right)=-\dfrac{2}{xy}.$

Hence, $e^{-\int\frac{2}{xy}d(xy)}=\dfrac{1}{x^2y^2}$ is an integrating factor of the equation. Multiplying the equation by $\dfrac{1}{x^2y^2}$, we get

$$\frac{y(1+xy)\,dx+x(1-xy)\,dy}{x^2y^2}=0$$

$$\text{or, } \frac{ydx+xdy}{x^2y^2}+\frac{dx}{x}-\frac{dy}{y}=0$$

$$\text{or, } \frac{d(xy)}{(xy)^2}+\frac{dx}{x}-\frac{dy}{y}=0.$$

Integrating we get

$$\ln\frac{x}{y}-\frac{1}{xy}=C, \text{ where } C \text{ is an arbitrary constant.}$$

(ii) Solve: $(x^4y^2-y)\,dx+(x^2y^4-x)\,dy=0.$

Solution

Here $M=x^4y^2-y$, $N=x^2y^4-x$, so $\dfrac{1}{yN-xM}\left(\dfrac{\partial M}{\partial y}-\dfrac{\partial N}{\partial x}\right)=-\dfrac{2}{xy}.$

Hence, $e^{-\int\frac{2}{xy}d(xy)}=\dfrac{1}{x^2y^2}$ is an integrating factor of the equation. Multiplying the equation by $\dfrac{1}{x^2y^2}$, we get

$$\frac{(x^4y^2-y)\,dx+(x^2y^4-x)\,dy}{x^2y^2}=0$$

or, $$x^2dx + y^2dy - \frac{xdy + ydx}{x^2y^2} = 0$$

or, $$d(x^3) + d(y^3) - 3\frac{d(xy)}{(xy)^2} = 0.$$

Integrating, we get

$$x^3 + y^3 + \frac{3}{xy} = C, \text{ where } C \text{ is an arbitrary constant.}$$

(iii) Solve: $(xy^2 + 2x^2y^3)dx + (x^2y - x^3y^2)dy = 0.$

Solution

Here $M = xy^2 + 2x^2y^3$, $N = x^2y - x^3y^2$, so $\frac{1}{yN - xM}\left(\frac{\partial M}{\partial y} - \frac{\partial N}{\partial x}\right) = -\frac{3}{xy}$.

Hence, $e^{-\int \frac{3}{xy} d(xy)} = \frac{1}{x^3y^3}$ is an integrating factor of the equation. Multiplying the equation by $\frac{1}{x^3y^3}$, we get

$$\frac{(xy^2 + 2x^2y^3)dx + (x^2y - x^3y^2)dy}{x^3y^3} = 0$$

or, $$\frac{ydx + xdy}{x^2y^2} + 2\frac{dx}{x} - \frac{dy}{y} = 0$$

or, $$\frac{d(xy)}{(xy)^2} + 2d(\ln x) - d(\ln y) = 0.$$

Integrating, we get

$$\ln\frac{x^2}{y} - \frac{1}{xy} = C, \text{ where } C \text{ is an arbitrary constant.}$$

3.2.8: If $\frac{y^2}{xM + yN}\left(\frac{\partial M}{\partial y} - \frac{\partial N}{\partial x}\right)$ is a function of the form $f\left(\frac{x}{y}\right)$, then the equation $Mdx + Ndy = 0$ has an integrating factor of the form $e^{\int f\left(\frac{x}{y}\right)d\left(\frac{x}{y}\right)}$.

Proof: If $\mu\left(\frac{x}{y}\right)$ be an integrating factor of the differential equation $Mdx + Ndy = 0$, then

$$\mu Mdx + \mu Ndy = 0$$

is an exact equation, i.e., we must have

$$\frac{\partial}{\partial x}(\mu N) = \frac{\partial}{\partial y}(\mu M)$$

or, $$N\frac{\partial \mu}{\partial x} + \mu\frac{\partial N}{\partial x} = M\frac{\partial \mu}{\partial y} + \mu\frac{\partial M}{\partial y}$$

or, $$\frac{N\mu'}{y} + \mu\frac{\partial N}{\partial x} = -\frac{M\mu' x}{y^2} + \mu\frac{\partial M}{\partial y}$$

$$\text{or,}\quad \mu\left(\frac{\partial N}{\partial x}-\frac{\partial M}{\partial y}\right)=-\frac{(xM+yN)\,\mu'}{y^2}$$

$$\text{or,}\quad \frac{d\mu}{\mu}=-\frac{y^2}{xM+yN}\left(\frac{\partial N}{\partial x}-\frac{\partial M}{\partial y}\right)d\left(\frac{x}{y}\right)=f\left(\frac{x}{y}\right)d\left(\frac{x}{y}\right)$$

$$\text{or,}\quad \mu\left(\frac{x}{y}\right)=e^{\int f\left(\frac{x}{y}\right)d\left(\frac{x}{y}\right)}.$$

ILLUSTRATIVE EXAMPLES

(i) Solve: $\left(\frac{2x^2}{y}+\frac{x}{y}\right)dx+2xdy=0.$

Solution

Here $M=\frac{2x^2}{y}+\frac{x}{y},\ N=2x.$

$$\text{So}\ \frac{y^2}{xM+yN}\left(\frac{\partial M}{\partial y}-\frac{\partial N}{\partial x}\right)=\frac{y^3}{x^2(2x+1)+2xy^2}\left(-\frac{x(2x+1)}{y^2}-2\right)=-\frac{y}{x}.$$

So, $e^{-\int\frac{y}{x}d\left(\frac{x}{y}\right)}=e^{-\ln\frac{x}{y}}=\frac{y}{x}$ is an integrating factor of the equation. Multiplying the equation by $\frac{y}{x}$, we get

$$\left(\frac{2x^2}{y}+\frac{x}{y}\right)\frac{y}{x}dx+2x\frac{y}{x}dy=0$$

$$\text{or,}\quad (2x+1)\,dx+2ydy=0.$$

Integrating we get

$$x^2+y^2+x=C,\ \text{where } C \text{ is an arbitrary constant.}$$

(ii) Solve: $\left(2y+\frac{y}{x}\right)dx+\frac{2y^2}{x}dy=0.$

Solution

Here $M=2y+\frac{y}{x},\ N=\frac{2y^2}{x}.$

$$\text{So}\ \frac{y^2}{xM+yN}\left(\frac{\partial M}{\partial y}-\frac{\partial N}{\partial x}\right)=\frac{xy^2}{(2xy+y)x+2y^3}\left(2+\frac{1}{x}+\frac{2y^2}{x^2}\right)=\frac{y}{x}.$$

So, $e^{\int\frac{y}{x}d\left(\frac{x}{y}\right)}=e^{\ln\frac{x}{y}}=\frac{x}{y}$ is an integrating factor of the equation. Multiplying the equation by $\frac{x}{y}$, we get

$$\left(2y+\frac{y}{x}\right)\frac{x}{y}dx+\frac{2y^2}{x}\frac{x}{y}dy=0\quad\text{or,}\ (2x+1)\,dx+2ydy=0.$$

Integrating we get

$$x^2 + y^2 + x = C, \text{ where } C \text{ is an arbitrary constant.}$$

3.2.9: If $\dfrac{x^2}{xM + yN}\left(\dfrac{\partial M}{\partial y} - \dfrac{\partial N}{\partial x}\right)$ is a function of the form $f\left(\dfrac{y}{x}\right)$, then the equation $Mdx + Ndy = 0$ has an integrating factor of the form $e^{-\int f\left(\frac{y}{x}\right)d\left(\frac{y}{x}\right)}$.

Proof: If $\mu\left(\dfrac{y}{x}\right)$ be an integrating factor of the differential equation $Mdx + Ndy = 0$, then

$$\mu Mdx + \mu Ndy = 0$$

is an exact equation, i.e., we must have

$$\frac{\partial}{\partial x}(\mu N) = \frac{\partial}{\partial y}(\mu M)$$

$$\text{or,}\quad N\frac{\partial \mu}{\partial x} + \mu\frac{\partial N}{\partial x} = M\frac{\partial \mu}{\partial y} + \mu\frac{\partial M}{\partial y}$$

$$\text{or,}\quad \frac{-N\mu' y}{x^2} + \mu\frac{\partial N}{\partial x} = \frac{M\mu'}{x} + \mu\frac{\partial M}{\partial y}$$

$$\text{or,}\quad \mu\left(\frac{\partial N}{\partial x} - \frac{\partial M}{\partial y}\right) = \frac{(xM + yN)\mu'}{x^2}$$

$$\text{or,}\quad \frac{d\mu}{\mu} = \frac{x^2}{xM + yN}\left(\frac{\partial N}{\partial x} - \frac{\partial M}{\partial y}\right)d\left(\frac{y}{x}\right) = -f\left(\frac{y}{x}\right)d\left(\frac{y}{x}\right).$$

$$\text{or,}\quad \mu\left(\frac{y}{x}\right) = e^{-\int f\left(\frac{y}{x}\right)d\left(\frac{y}{x}\right)}.$$

ILLUSTRATIVE EXAMPLES

(i) Solve: $\left(\dfrac{2x^2}{y} + \dfrac{x}{y}\right)dx + 2xdy = 0.$

Solution

Here $M = \dfrac{2x^2}{y} + \dfrac{x}{y}$, $N = 2x$.

So $$\frac{x^2}{xM + yN}\left(\frac{\partial M}{\partial y} - \frac{\partial N}{\partial x}\right) = \frac{yx^2}{x^2(2x+1) + 2xy^2}\left(-\frac{x(2x+1)}{y^2} - 2\right) = -\frac{x}{y}.$$

So, $e^{\int \frac{x}{y} d\left(\frac{y}{x}\right)} = e^{\ln\frac{y}{x}} = \dfrac{y}{x}$ is an integrating factor of the equation. Multiplying the equation by $\dfrac{y}{x}$, we get

$$\left(\frac{2x^2}{y} + \frac{x}{y}\right)\frac{y}{x}dx + 2x\frac{y}{x}dy = 0 \quad\text{or,}\quad (2x+1)dx + 2ydy = 0.$$

Integrating, we get

$$x^2 + y^2 + x = C, \text{ where } C \text{ is an arbitrary constant.}$$

(ii) Solve: $\left(2y+\frac{y}{x}\right)dx+\frac{2y^2}{x}dy=0.$

Solution

Here $M=2y+\frac{y}{x}, N=\frac{2y^2}{x}.$

$$\text{So } \frac{x^2}{xM+yN}\left(\frac{\partial M}{\partial y}-\frac{\partial N}{\partial x}\right)=\frac{x^3}{(2xy+y)x+2y^3}\left(2+\frac{1}{x}+\frac{2y^2}{x^2}\right)=\frac{x}{y}.$$

So, $e^{-\int\frac{x}{y}d\left(\frac{y}{x}\right)}=e^{-\ln\frac{y}{x}}=\frac{x}{y}$ is an integrating factor of the equation. Multiplying the equation by $\frac{y}{x}$, we get

$$\left(2y+\frac{y}{x}\right)\frac{x}{y}dx+\frac{2y^2}{x}\frac{x}{y}dy=0 \text{ or, } (2x+1)dx+2ydy=0.$$

Integrating we get

$$x^2+y^2+x=C, \text{ where } C \text{ is an arbitrary constant.}$$

3.2.10: For the equation $x^a y^b(mydx+nxdy)=0, x^{m-1-a}y^{n-1-b}\varphi(x^m y^n)$, is an integrating factor where $\varphi(x^m y^n)$ is an arbitrary function.

Proof: Multiplying the equation $x^a y^b(mydx+nxdy)=0$ by, $x^{m-1-a}y^{n-1-b}\varphi(x^m y^n)$, we get

$$x^{m-1}y^{n-1}\varphi(x^m y^n)(mydx+nxdy)=0$$

$$\text{or, } \varphi(x^m y^n)\left[y^n d(x^m)+x^m d(y^n)\right]=0$$

$$\text{or, } \varphi(x^m y^n)d(x^m y^n)=0$$

$$\text{or, } d\left[f(x^m y^n)\right]=0, \text{ where } f'(x^m y^n)=\varphi(x^m y^n).$$

This proves the exactness of the equation

$$x^{m-1}y^{n-1}\varphi(x^m y^n)(mydx+nxdy)=0.$$

Putting $a=0=b$ in the above results we see that $x^{m-1}y^{n-1}\varphi(x^m y^n)$ is an integrating factor of the equation

$$mydx+nxdy=0.$$

ILLUSTRATIVE EXAMPLES

(i) Solve: $x^2y^3(5ydx+3xdy)=0.$

Solution

Here $a=2, b=3, m=5, n=3$, so $x^{m-1-a}y^{n-1-b}\varphi(x^m y^n)$, i.e., $x^2y^{-1}\varphi(x^5y^3)$ is an integrating factor of the equation. Multiplying the equation by $x^2y^{-1}\varphi(x^5y^3)$, we get

$$x^4y^2(5ydx+3xdy)\varphi(x^5y^3)=0 \text{ or, } \varphi(x^5y^3)d(x^5y^3)=0.$$

Integrating, we get

$\int \varphi(x^5y^3)d(x^5y^3) = C$, where $\varphi(x^5y^3)$ is an arbitrary function of x^5y^3 and C is an arbitrary constant.

(ii) Solve: $x^2y^3(5ydx + 3xdy) = 0$.

Solution

Here $a = 2$, $b = 3$, $m = 5$, $n = 3$, so $x^{km-1-a}y^{kn-1-b}$, i.e., $x^{5k-3}y^{3k-4}$, where k has any value, is an integrating factor of the equation. Multiplying the equation by $x^{5k-3}y^{3k-4}$, we get

$$x^{5k-1}y^{3k-1}(5ydx + 3xdy) = 0 \text{ or, } d\left(x^{5k}y^{3k}\right) = 0.$$

Integrating, we get

$$x^{5k}y^{3k} = C, \text{ where } C \text{ is an arbitrary constant.}$$

3.2.11: For the equation $x^a y^b(mydx + nxdy) + x^c y^d(pydx + qxdy) = 0$, where $mq \neq np$, $x^h y^k$ will be an integrating factor, if

$$\frac{a+h+1}{m} = \frac{b+k+1}{n} \text{ and } \frac{c+h+1}{p} = \frac{d+k+1}{q}.$$

Proof: If $x^h y^k$ is an integrating factor of the equation

$$x^a y^b(mydx + nxdy) + x^c y^d(pydx + qxdy) = 0,$$

then $\quad x^{h+a}y^{k+b}(mydx + nxdy) + x^{h+c}y^{k+d}(pydx + qxdy)$

would be an exact differential. In other words, we must have

$$\frac{\partial}{\partial y}\left(mx^{h+a}y^{k+b+1} + px^{h+c}y^{k+d+1}\right) = \frac{\partial}{\partial x}\left(nx^{h+a+1}y^{k+b} + qx^{h+c+1}y^{k+d}\right)$$

or, $\quad m(k+b+1)x^{h+a}y^{k+b} + p(k+d+1)x^{h+c}y^{k+d}$

$$= n(h+a+1)x^{h+a}y^{k+b} + q(h+c+1)x^{h+c}y^{k+d}.$$

Comparing both sides we get

$$m(k+b+1) = n(h+a+1) \text{ and } p(k+d+1) = q(h+c+1)$$

or, $$\frac{a+h+1}{m} = \frac{b+k+1}{n} \text{ and } \frac{c+h+1}{p} = \frac{d+k+1}{q}.$$

These equations will give the values of h, k, i.e., the integrating factor $x^h y^k$. Now, the above equations in h, k are solvable, if the coefficient determinant does not vanish, i.e., if

$$\begin{vmatrix} n & -m \\ q & -p \end{vmatrix} \neq 0, \text{ i.e., if } mq \neq np.$$

ILLUSTRATIVE EXAMPLES

(i) Solve: $2ydx + 3xdy + 2xy(3ydx + 4xdy) = 0.$

Solution

Here $a = 0, b = 0, c = 1, d = 1, m = 2, n = 3, p = 6, q = 8$ and $mq - np = 16 - 18 \neq 0.$ Hence, $x^h y^k$ will be an integrating factor where h, k satisfy

$$\frac{h+1}{2} = \frac{k+1}{3} \text{ and } \frac{h+2}{6} = \frac{k+2}{8},$$

$$\text{or,} \quad 3h - 2k + 1 = 0 \text{ and } 4h - 3k + 2 = 0.$$

Solving, we get

$$\frac{h}{-4+3} = \frac{k}{4-6} = \frac{1}{-9+8}, \text{ i.e., } h = 1, k = 2.$$

Multiplying the equation by xy^2, we get

$$xy^2(2ydx + 3xdy) + 2x^2y^3(3ydx + 4xdy) = 0$$

$$\text{or,} \quad d(x^2y^3) + 2d(x^3y^4) = 0.$$

Integrating, we get

$$x^2y^3 + 2x^3y^4 = C, \text{ where } C \text{ is an arbitrary constant.}$$

(ii) Solve: $(y^2 + 2x^2y)dx + (2x^3 - xy)dy = 0.$

Solution

Rewriting the equation as $y(ydx - xdy) + 2x^2(ydx + xdy) = 0$ we see that, here

$$a = 0, b = 1, c = 2, d = 0, m = 1, n = -1, p = 1, q = 1$$

and $mq - np = 1 + 1 \neq 0$. Hence $x^h y^k$ will be an integrating factor where h, k satisfy

$$\frac{h+1}{1} = \frac{k+2}{-1} \text{ and } \frac{h+3}{1} = \frac{k+1}{1},$$

$$\text{or,} \quad h + k + 3 = 0 \text{ and } h - k + 2 = 0.$$

Solving, we get

$$h = -\frac{5}{2}, k = -\frac{1}{2}.$$

Multiplying the equation by $x^{-\frac{5}{2}}y^{-\frac{1}{2}}$, we get

$$x^{-\frac{5}{2}}y^{\frac{1}{2}}(ydx - xdy) + 2x^{-\frac{1}{2}}y^{-\frac{1}{2}}(ydx + xdy) = 0$$

$$\text{or,} \quad x^{-\frac{5}{2}}y^{\frac{3}{2}}dx - x^{-\frac{3}{2}}y^{\frac{1}{2}}dy + 2x^{-\frac{1}{2}}y^{\frac{1}{2}}dx + 2x^{\frac{1}{2}}y^{-\frac{1}{2}}dy = 0$$

$$\text{or,} \quad 4d\left(x^{\frac{1}{2}}y^{\frac{1}{2}}\right) - \frac{2}{3}d\left(x^{-\frac{3}{2}}y^{\frac{3}{2}}\right) = 0.$$

Integrating, we get

$$6x^{\frac{1}{2}}y^{\frac{1}{2}} - x^{-\frac{3}{2}}y^{\frac{3}{2}} = C, \text{ where } C \text{ is an arbitrary constant.}$$

EXERCISE

1. Solve the following equations:

(i) $xdy - ydx = (x^2 + y^2)dx$, (ii) $xdy - ydx = \sqrt{x^2 + y^2}\,dx$,

(iii) $xdy - ydx = x\sqrt{x^2 + y^2}\,dx$, (iv) $xdx + ydy = \sqrt{x^2 + y^2}\,dx$,

(v) $xdy - ydx = \cos\frac{1}{x}dx$,

[*Hints*: Multiply both sides by $\frac{1}{x^2}$ and put $z = \frac{1}{x}$.]

(vi) $xdy + ydx = y^2 \ln x\, dx$,

(vii) $(x + y)(dx - dy) = dx + dy$,

[*Hints*: Rewrite the equation as $dx - dy = \frac{d(x+y)}{x+y}$]

(viii) $(xdx + ydy)(x^2 + y^2) + xdy - ydx = 0$,

(ix) $xdy - ydx = 2\sqrt{x^2 - y^2}\,dx$,

(x) $\frac{x + yy'}{xy' - y} = \sqrt{\frac{1 - x^2 - y^2}{x^2 + y^2}}$, where $y' \frac{dy}{dx}$,

[*Hints*: Rewrite the equation as $\frac{d\left(\sqrt{x^2 + y^2}\right)}{\sqrt{1 - x^2 - y^2}} = \frac{d\left(\frac{y}{x}\right)}{1 + \left(\frac{y}{x}\right)^2}$.]

(xi) $xdx + ydy + a^2\frac{ydx - xdy}{x^2 + y^2} = 0$,

(xii) $xdx + ydy = m(xdy - ydx)$,

[*Hints*: Rewrite the equation as $\frac{d(x^2 + y^2)}{x^2 + y^2} = 2m\frac{d\left(\frac{y}{x}\right)}{1 + \left(\frac{y}{x}\right)^2}$.]

(xiii) $x\cos\frac{y}{x}(ydx + xdy) = y\sin\frac{y}{x}(xdy - ydx)$,

(xiv) $x^2dy - (xy + 2y^2)dx = 0$, (xv) $2xydy + (y^2 - 3x^2)dx = 0$,

(xvi) $(x^2 + y^2)dy - 2xydx = 0$, (xvii) $(x^3 + y^3)dx - xy^2dy = 0$,

(xviii) $x^2ydx - (x^3 + y^3)dy = 0$,

(xix) $y(xy + 2x^2y^2)dx + x(xy - x^2y^2)dy = 0$,

(xx) $y(xy + 1)dx + x(1 - xy)dy = 0$,

(xxi) $(1 + 4xy + 2y^2)dx + (1 + 4xy + 2x^2)dy = 0$,

(xxii) $3x^2ydx + (x^3 + y^3)dy = 0$,

(xxiii) $(1 + 3x^2 + 6xy^2)dx + (1 + 3y^2 + 6yx^2)dy = 0$,

(xxiv) $(x^3y^3 + x^2y^2 + xy + 1)ydx + (x^3y^3 - x^2y^2 - xy + 1)xdy = 0$,

(xxv) $(x^2 + y^2 + 1)dy - 2xydx = 0$,

(xxvi) $(x^2 + y^2 + 1)dx + x(x - 2y)dy = 0$,

(xxvii) $(x^2 + y^2 + x)dx + xydy = 0$,

(xxviii) $(x^3 - 2y^3)dx + 2xy^2dy = 0$,

(xxix) $(x^3 + xy^4)dx + 2y^3dy = 0$,

(xxx) $(3x^2y^4 + 2xy)dx + (2x^3y^3 - x^2)dy = 0$,

(xxxi) $(xy^3 + y)dx + 2(x^2y^2 + x + y^4)dy = 0$,

(xxxii) $(3x + 2y^2)ydx + 2x(2x + 3y^2)dy = 0$,

(xxxiii) $(3x^2y^3 + 2x)ydx + (2xy^3 - 1)x^2dy = 0$,

(xxxiv) $(2x^2 - 3y^3)ydx + (3x^2 + 2y^3)xdy = 0$,

(xxxv) $(y + 2x^2)ydx + (2x^2 - y)xdy = 0$,

(xxxvi) $x(3ydx + 2xdy) + 8y^4(ydx + 3xdy) = 0$,

(xxxvii) $x^2y(2ydx - xdy) + (ydx + 3xdy) = 0$,

(xxxviii) $(2ydx + 3xdy) + 2xy(3ydx + 4xdy) = 0$,

(xxxix) $x(4ydx + 2xdy) + y^3(3ydx + 5xdy) = 0$,

(xl) $3ydx + 5xdy + xy^3(ydx + 2xdy) = 0$,

(xli) $\left(\frac{3}{x} + 2y^4\right)dx + \left(\frac{3}{y} - xy^3\right)dy = 0$,

(xlii) $x^2(xdx + ydy) + 2y(xdy - ydx) = 0$,

[*Hints*: Rewrite the equation as $\dfrac{d(x^2 + y^2)}{\sqrt{x^2 + y^2}} + 4\dfrac{\frac{y}{x}d\left(\frac{y}{x}\right)}{\sqrt{1 + \left(\frac{y}{x}\right)^2}} = 0$.]

2. Show that e^{x^2} is an integrating factor of the equation $(x + y^4)xdx + 2y^3dy = 0$, and hence solve it.

3. Show that $\dfrac{1}{x(x^2 - y^2)}$ is an integrating factor of the equation $(x^2 + y^2)dx - 2xydy = 0$ and hence solve it.

4. Show that $\dfrac{1}{(x + y + 1)^4}$ is an integrating factor of the equation

$(2xy - y^2 - y)dx + (2xy - x^2 - x)dy = 0$. Hence solve it.

5. Show that $\frac{x}{x-1}$ is an integrating factor of the equation $x(x-1)\frac{dy}{dx}-y=[x(x-1)]^2$ and hence solve it.

[*Hints*: In the problem no 2-5, multiply the equation by the I.F. and then prove the exactness of the resulting equations.]

6. If $(x^2+y^2)^k$ be an integrating factor of the equation $(x+y)\,dx+(y-x)\,dy=0$, find k, and hence solve it.

7. If $x^a y^b$ be an integrating factor of the equation $\left(\frac{3}{x}+2y^4\right)dx+\left(\frac{3}{y}-xy^3\right)dy=0$, then find a, b.

8. If $x^a y^b$ be an integrating factor of the equation $(y^2+2x^2y)\,dx+(2x^3-xy)\,dy=0$, then find a, b.

9. If $x^a y^b$ be an integrating factor of the equation $(x-y^2)\,dx+2xy\,dy=0$, find a, b.

ANSWERS

1. (i) $\tan^{-1}\frac{y}{x}=x+C$, (ii) $y+\sqrt{x^2+y^2}=Cx^2$,

(iii) $y=x\sinh(x+C)$, (iv) $\sqrt{x^2+y^2}=x+C$,

(v) $\frac{y}{x}+\sin\frac{1}{x}=C$, (vi) $1+Cxy=y(1+\ln x)$,

(vii) $x-y=\ln(x+y)+C$, (viii) $x^2+y^2+2\tan^{-1}\frac{y}{x}=C$,

(ix) $y=x\sin(\ln Cx^2)$, (x) $x^2+y^2=\sin^2\left(\tan^{-1}\frac{y}{x}+C\right)$,

(xi) $x^2+y^2+2a^2\tan^{-1}\left(\frac{x}{y}\right)=C$, (xii) $\ln(x^2+y^2)=C-2m\tan^{-1}\left(\frac{x}{y}\right)$,

(xiii) $xy\cos\frac{y}{x}=C$, (xiv) $x+y\ln x^2=Cy$,

(xv) $x(x^2-y^2)=C$, (xvi) $x^2-y^2=Cy$,

(xvii) $y^3=3x^3\ln x+Cx^3$, (xviii) $y=ce^{\frac{x^3}{3y^3}}$,

(xix) $\ln\frac{x^2}{y}-\frac{1}{xy}=C$, (xx) $\ln\frac{x}{y}-\frac{1}{xy}=C$,

(xxi) $(x+y)(1+2xy)=C$, (xxii) $y(4x^3+y^3)=C$,

(xxiii) $x+y+x^3+y^3+3x^2y^2=C$, (xxiv) $x^2y^2-1=2xy(\ln y+C)$,

(xxv) $x^2-y^2=1+Cx$ (xxvi) $x(x+y)=1+Cx+y^2$,

(xxvii) $3x^4 + 6x^2y^2 + 4x^3 = C$, (xxviii) $x = ce^{\frac{-2y^3}{3x^3}}$,

(xxix) $e^{x^2}(x^2 - 1 + y^4) = C$, (xxx) $x^3y^3 + x^2 = Cy$,

(xxxi) $3x^2y^4 + 6xy^2 + y^6 = C$, (xxxii) $x^3y^4 + x^2y^6 = C$,

(xxxiii) $6\sqrt{xy} - \left(\frac{x}{y}\right)^{\frac{3}{2}} = C$, (xxxiv) $5x^{-\frac{36}{13}}y^{\frac{24}{13}} + 12x^{-\frac{10}{13}}y^{-\frac{15}{13}} = C$,

(xxxv) $6\sqrt{xy} - \left(\frac{y}{x}\right)^{\frac{3}{2}} = C$, (xxxvi) $x^3y^3 + 4x^2y^7 = C$,

(xxxvii) $4x^{\frac{10}{7}}y^{-\frac{5}{7}} - 5x^{-\frac{4}{7}}y^{-\frac{12}{7}} = C$, (xxxviii) $x^2y^3 + 2x^3y^4 = C$,

(xxxix) $x^3y^2(x + y^3) = C$, (xl) $x^3y^5(4 + xy^3) = C$,

(xli) $\frac{1}{x^3y^3} + \frac{y}{x^2} = C$, (xlii) $(x^2 + y^2)(x + 2)^2 = Cx^2$,

6. $k = -1, \ln(x^2 + y^2) - 2\tan^{-1}\frac{y}{x} = C$, **7.** $a = -4, b = -4$,

8. $h = -\frac{5}{2}, k = -\frac{1}{2}$, **9.** $a = -2, b = 0$.

Chapter 4

EQUATIONS OF FIRST ORDER BUT OF HIGHER DEGREE

4.1 Introduction

A first order differential equation is of the form $f(x, y, p) = 0$, where $p = y'$. If the degree of p in the equation is greater than one, say n (a positive integer), then it can be expressed as

$$P_0 p^n + P_1 p^{n-1} + P_2 p^{n-2} + \cdots + P_n = 0, \tag{4.1}$$

where $P_0, P_1, P_2, \cdots, P_n$ are functions of x, y and $P_0 \neq 0$.

There types of equations are called *first order non-linear equations* and can be solved if these are of the following types:

(i) Equations solvable for p,

(ii) Equations solvable for x,

(iii) Equations solvable for y,

(iv) Equations homogeneous in x, y.

4.2 Equations Solvable for p

If the equations (4.1) is solvable for p, then the expression in the left-hand side of (4.1) can be factorized into n linear factors. Thus the equations (4.1) reduces to

$$(p - f_1)(p - f_2)(p - f_3)\ldots(p - f_n) = 0,$$

where $f_k = f_k(x, y), k = 1, 2, 3, \ldots, n$ are functions of x, y. From the above equation we get

$$\frac{dy}{dx} = f_k(x, y), k = 1, 2, 3, \cdots, n.$$

Solving we get $g_k(x, y, C) = 0, k = 1, 2, 3, \cdots, n$.

Thus the general solution of the equation is

$$g_1(x, y, C)\, g_2(x, y, C)\, g_3(x, y, C) \ldots g_n(x, y, C) = 0.$$

ILLUSTRATIVE EXAMPLES

1. Solve: $p^3 - 6p^2 + 11p - 6 = 0.$

Solution

The given equation can be written as $(p-1)(p-2)(p-3) = 0.$

From $p - 1 = 0$. we get $dy - dx = 0$. Integrating we get $y - x + C = 0.$

From $p - 2 = 0$, we get $dy - 2dx = 0$. Integrating we get $y - 2x + C = 0.$

From $p - 3 = 0$, we get $dy - 3dx = 0$. Integrating we get $y - 3x + C = 0.$

Hence, the general solution is $(y - x + C)(y - 2x + C)(y - 3x + C) = 0.$

2. Solve: $p(p - y) = x(x + y).$

Solution

The equation can be written as

$$p^2 - x^2 - py - xy = 0$$

or, $$(p - x)(p + x) - y(p + x) = 0$$

or, $$(p - x - y)(p + x) = 0.$$

From $p + x = 0$, we get

$$dy + xdx = 0\,.$$

Integrating we get $$y + \frac{x^2}{2} + C = 0.$$

From $p - x - y = 0$, we get

$$\frac{dy}{dx} - y = x\,.$$

This is a linear differential equation whose integrating factor is $e^{-\int dx} = e^{-x}$. Multiplying the above equation by e^{-x}, we get

$$\frac{d}{dx}(ye^{-x}) = xe^{-x}.$$

Integrating, we get $$ye^{-x} = \int xe^{-x}dx = -(x+1)e^{-x} + C$$

or, $$y = -x - 1 + Ce^{x}.$$

Hence the general solution is

$$\left(y + \frac{x^2}{2} + C\right)(y + x + 1 - Ce^{x}) = 0.$$

3. Solve: $x + yp^2 = p(1 + xy).$

Solution

We rewrite the equation as

$$yp^2 - p - pxy + x = 0$$

or, $$(yp - 1)(p - x) = 0.$$

From $p-x=0$ we get $dy = xdx$. Integrating, we get

$$y=\frac{x^2}{2}+\frac{C}{2}.$$

From $yp - 1 = 0$, we get $ydy - dx = 0$. Integrating, we get

$$x=\frac{y^2}{2}+\frac{C}{2}.$$

Hence, the general solution is

$$(2y-x^2-C)(2x-y^2-C)=0.$$

4. Solve: $xy(p^2-1)=(x^2-y^2)p$.

Solution

We rewrite the equation as

$$xyp^2-x^2p+y^2p-xy=0$$

or, $$(yp-x)(xp+y)=0.$$

From $yp-x=0$, we get $xdx-ydy=0$. Integrating, we get

$$x^2-y^2-2C=0.$$

From $xp+y=0$, we get $xdy+ydx=0$. Integrating, we get

$$xy-C=0.$$

Hence the general solution is

$$(x^2-y^2-2C)(xy-C)=0.$$

5. Solve: $p^2+2py\cot x-y^2=0$.

Solution

We rewrite the equation as

$$p^2+2py\cot x+y^2\cot^2 x=y^2+y^2\cot^2 x$$

or, $$(p+y\cot x)^2=y^2\operatorname{cosec}^2 x$$

or, $$p=-y(\cot x\pm\operatorname{cosec} x).$$

From $p=-y(\cot x+\operatorname{cosec} x)$, we get

$$\frac{dy}{y}=-(\cot x+\operatorname{cosec} x)\,dx.$$

Integrating, we get

$$\ln y=-\ln\sin x-\ln(\operatorname{cosec} x-\cot x)+\ln C$$

or, $$\ln y=\ln C-\ln(1-\cos x)$$

or, $$y(1-\cos x)=C.$$

Similarly, from $p=-y(\cot x-\operatorname{cosec} x)$, we get

$$y(1+\cos x)=C.$$

So the general solution is

$$[y(1-\cos x)-C][y(1+\cos x)-C]=0.$$

6. Solve: $3p^2y^2-2xyp+4y^2=x^2$.

Solution

We rewrite the equation as

$$9p^2y^2-6xyp+x^2=4(x^2-3y^2)$$

or, $$(3py-x)^2=4(x^2-3y^2)$$

or, $$p=\frac{1}{3y}\left[x\pm2\sqrt{x^2-3y^2}\right]$$

or, $$3ydy-xdx=\pm2\sqrt{x^2-3y^2}\,dx$$

or, $$\tfrac{1}{2}d\left(x^2-3y^2\right)=\mp2\sqrt{x^2-3y^2}\,dx$$

or, $$\frac{d\left(x^2-3y^2\right)}{\sqrt{x^2-3y^2}}=\mp4dx.$$

Integrating, we get

$$\sqrt{x^2-3y^2}=\mp2x+3C.$$

Squaring both sides, we get

$$x^2-3y^2=4x^2\mp12xC+9C^2$$

or, $$3x^2+3y^2\mp12xC+9C^2=0$$

or, $$x^2+y^2\mp4xC+3C^2=0.$$

7. Solve: $x+\dfrac{p}{\sqrt{1+p^2}}=a.$

Solution

We rewrite the equation as

$$p=\frac{x-a}{\sqrt{1-(x-a)^2}}\quad\text{or,}\quad dy=\frac{x-a}{\sqrt{1-(x-a)^2}}dx.$$

Integrating, we get

$$-(y+C)=\sqrt{1-(x-a)^2}\quad\text{or,}\quad(x-a)^2+(y+C)^2=1.$$

4.3 Equations Solvable for x

In this case, equation (4.1) can be written as

$$x=f(y,p).$$

Differentiating both sides with respect to y, we get

$$\frac{1}{p}=\frac{dx}{dy}=\frac{\partial f}{\partial y}+\frac{\partial f}{\partial p}\frac{dp}{dy}=F\left(y,p,\frac{dp}{dy}\right).$$

This is an equation in y, p. If $y = g(p,c)$ be its solution, then using it in $x = f(y,p)$, we get

$$x = G(p, c),\ y = g(p, c).$$

These two give the parametric form of the general solution of the given equation. If p is eliminated from these two, we get the required solution in the form of a function in x, y.

If the equation does not contain y and is solvable for x, then it can be expressed as

$$x = f(p).$$

So the above method can be applied to solve it. On the other hand, if it can be expressed as

$$p = f(x).$$

Then the solution is $\quad y = \int f(x)\,dx + C.$

ILLUSTRATIVE EXAMPLES

1. Solve: $x = y + ap^2$.

Solution

Differentiating both sides with respect to y, we get

$$\frac{1}{p} = \frac{dx}{dy} = 1 + 2ap\frac{dp}{dy}$$

or, $$dy = \frac{2ap^2}{1-p}dp$$

or, $$dy = -2a(p+1)\,dp + 2a\frac{dp}{1-p}.$$

Integrating, we get $\quad y = -ap^2 - 2ap - 2a\ln(p-1) + C.$

Using this in the given equation, we get $x = -2ap - 2a\ln(p-1) + C.$

The p eliminant of these two will give the general solution.

2. Solve: $x = py - ap^2$.

Solution

Differentiating both sides with respect to y, we get

$$\frac{1}{p} = \frac{dx}{dy} = p + (y - 2ap)\frac{dp}{dy}$$

or, $$\frac{dy}{dp} = \frac{yp - 2ap^2}{1-p^2}$$

or, $$\frac{dy}{dp} + \frac{yp}{p^2-1} = \frac{2ap^2}{p^2-1}.$$

This is a linear differential equation, whose integrating factor is

$$e^{\int \frac{pdp}{p^2-1}} = e^{\frac{1}{2}\ln(p^2-1)} = \sqrt{p^2-1}.$$

Multiplying the above equation by $\sqrt{p^2-1}$, we get

$$\frac{d}{dp}\left\{y\sqrt{p^2-1}\right\} = \frac{2ap^2}{\sqrt{p^2-1}} = 2a\left\{\sqrt{p^2-1} + \frac{1}{\sqrt{p^2-1}}\right\}.$$

Integrating, we get

$$y\sqrt{p^2-1} = a\left\{p\sqrt{p^2-1} + \cosh^{-1} p\right\} + C$$

or, $$y = ap + \frac{\left(C + a\cosh^{-1} p\right)}{\sqrt{p^2-1}}.$$

Using this in the given equation, we get

$$x = \frac{p(C + a\cosh^{-1} p)}{\sqrt{p^2-1}}$$

The p eliminant of these two will give the general solution.

3. Solve: $y - 2px + ayp^2 = 0.$

Solution

We rewrite the equation as

$$x = \frac{y}{2p} + \frac{ayp}{2}.$$

Differentiating both sides with respect y, we get

$$\frac{1}{p} = \frac{dx}{dy} = \frac{1}{2p} - \frac{y}{2p^2}\frac{dp}{dy} + \frac{ap}{2} + \frac{ay}{2}\frac{dp}{dy}$$

or, $$\frac{dp}{dy}\left(\frac{ay}{2} - \frac{y}{2p^2}\right) = \frac{1}{2p} - \frac{ap}{2}$$

or, $$\frac{dp}{dy}\frac{y\left(ap^2-1\right)}{2p^2} = \frac{1-ap^2}{2p}$$

or, $$\frac{dp}{p} + \frac{dy}{y} = 0, \text{ if } 1 - ap^2 \neq 0.$$

Integrating, we get $\quad py = C.$

Eliminating p from the given equation with **the help of above** equation, we get

$$y - \frac{2xC}{y} + ay\frac{C^2}{y^2} = 0$$

or, $$y^2 = 2xC - aC^2.$$

If $ap^2 = 1$, then from the given equation we get

$$y = px \text{ or, } x = \sqrt{ay}.$$

This solution of the given equation neither contains any arbitrary constant nor can be derived from the general solution, given above, by assigning any particular value to the arbitrary constant. This is called singular solution of the equation.

4. Solve: $p^2 - 2xp + 1 = 0.$

Solution

We rewrite the equation as

$$x = \frac{1}{2}\left(p + \frac{1}{p}\right).$$

Differentiating with respect to y, we get

$$\frac{1}{p} = \frac{dx}{dy} = \frac{1}{2}\left(1 - \frac{1}{p^2}\right)\frac{dp}{dy}$$

or, $$dy = \frac{1}{2}\left(p - \frac{1}{p}\right)dp.$$

Integrating, we get

$$y = \frac{1}{4}p^2 - \frac{1}{2}\ln p + C.$$

Hence, p-eliminant of $x = \frac{1}{2}\left(p + \frac{1}{p}\right)$ and $y = \frac{1}{4}p^2 - \frac{1}{2}\ln p + C$ give the general solution.

5. Solve: $p^3 - 4xyp + 8y^2 = 0.$

Solution

We rewrite the equation as

$$x = \frac{p^2}{4y} + \frac{2y}{p}.$$

Differentiating with respect to y, we get

$$\frac{1}{p} = \frac{dx}{dy} = -\frac{1}{4}\frac{p^2}{y^2} + \left(\frac{p}{2y} - \frac{2y}{p^2}\right)\frac{dp}{dy} + \frac{2}{p}$$

or, $$\frac{1}{4}\frac{p^2}{y^2} - \frac{1}{p} = \left(\frac{p}{2y} - \frac{2y}{p^2}\right)\frac{dp}{dy}$$

or, $$\frac{dy}{2y} = \frac{dp}{p}, \text{ if, } p^3 \neq 4y^2.$$

Integrating, we get $p^2 = 4Cy.$

Using this in the given equation, we get

$$p(C - x) = -2y$$

or, $$4Cy(C - x)^2 = 4y^2$$

or, $$y = C(x - C)^2.$$

If $p^3 = 4y^2$, then from the given equation we get

$$3y = px \text{ or, } 27y = 4x^3.$$

This solution of the given equation neither contains any arbitrary constant nor can be derived from the general solution, given above, by assigning any particular value to the arbitrary constant. This is called singular solution of the equation.

6. Solve: $x = 4p + 4p^3$.

Solution

Differentiating with respect to y, we get

$$\frac{1}{p} = \frac{dx}{dy} = 4\left(1+3p^2\right)\frac{dp}{dy}$$

or, $$dy = 4\left(p+3p^3\right)dp.$$

Integrating, we get $$y = 2p^2 + 3p^4 + C.$$

Hence, p-eliminant of $x = 4p + 4p^3$ and $y = 2p^2 + 3p^4 + C$ will give the general solution

4.4 Equations Solvable for y

In this case, equation (1) can be written as

$$y = f(x, p).$$

Differentiating both sides with respect to x, we get

$$p = \frac{dy}{dx} = \frac{\partial f}{\partial x} + \frac{\partial f}{\partial p}\frac{dp}{dx} = F\left(x, p, \frac{dp}{dx}\right).$$

This is an equation in x, p. If $x = g(p,c)$ be its solution, then using it in $y = f(x, p)$, we get $y = G(p, c), x = g(p, c)$.

These two give the parametric form of the general solution of the given equation. If p is eliminated from these two, we get the required solution in the form of a function in x, y.

Sometimes it is possible to express the function $f(x, p)$ in the equation $y = f(x, p)$, as $f(x, p) = px + g(p)$,

A differential equation of the form $y = px + g(p)$, is known as *Clairaut's equation*.

Differentiating this with respect to x, we get

$$p = \frac{dy}{dx} = p + x\frac{dp}{dx} + g'(p)\frac{dp}{dx} \quad \text{or, } \{x + g'(p)\}\frac{dp}{dx} = 0,$$

This gives either $\frac{dp}{dx} = 0$ or, $x + g'(p) = 0$. From $\frac{dp}{dx} = 0$, we get $p = C$ and eliminating p we see that the general solution of the differential equation of Clairaut's form is

$$y = Cx + g(C).$$

When $x + g'(p) = 0$, then eliminating from the p, with the aid of it, from the given equation we get a relation between x, y without containing any constant and this relation can not be obtained from the general solution by giving a particular value to the arbitrary constant. Such a solution is known as *singular solution*. It can be easily shown that these solutions are linearly independent.

It is to be mentioned that a differential equation, which is not given in Clairaut's form, can be reduced to Clairaut's equation under some suitable substitution. So the above method can be applied to solve the reduced equation and after solving it and using inverse substitution, we get the solution of the original equation.

Further, if the function $f(x, p)$ in the equation $y = f(x, p)$, has the following form

$$f(x, p) = x\phi(p) + g(p),$$

then the differential equation takes the form

$$y = x\phi(p) + g(p),$$

which is known as *Lagrange's equation*.

If $\phi(p) \neq p$, then differentiating with respect to x, we get

$$p = \frac{dy}{dx} = \phi(p) + \{x\phi'(p) + g'(p)\}\frac{dp}{dx}$$

or, $$\frac{dx}{dp} = \frac{x\phi'(p) + g'(p)}{p - \phi(p)}$$

or, $$\frac{dx}{dp} - \frac{\phi'(p)}{p - \phi(p)}x = \frac{g'(p)}{p - \phi(p)},$$

which is a linear differential equation and can be solved easily. The p eliminant of the solution of this equation and the given equation give the general solution.

If $\phi(p) = p$, then the above equation reduces to Clairaut's form. Thus *Lagrange's equation* is an extended form of the Clairaut's equation.

If the equation does not contain x and is solvable for y, then it can be expressed as

$$y = f(p).$$

So the above method can be applied to solve it. On the other hand, if it can be expressed as

$$p = f(y),$$

i.e., $$\frac{dy}{dx} = f(y),$$

then the solution is given by

$$\int \frac{dy}{f(y)} = x + C.$$

ILLUSTRATIVE EXAMPLES

1. Solve: $y = p^2 x + p$.

Solution

Differentiating the given equation with respect to x, we get

$$p = \frac{dy}{dx} = p^2 + (2px + 1)\frac{dp}{dx}$$

or, $$p(1-p)\frac{dx}{dp} = (2px+1)$$

or, $$\frac{dx}{dp} + \frac{2x}{p-1} = \frac{1}{p(1-p)}.$$

This is a linear differential equation whose integrating factor is

$$e^{\int \frac{2dp}{p-1}} = e^{2\ln(p-1)} = (p-1)^2.$$

Multiplying the above equation by $(p-1)^2$, we get

$$\frac{d}{dp}\left[x(p-1)^2\right] = \frac{1}{p} - 1.$$

Integrating, we get $$x(p-1)^2 = \ln p - p + C$$

or, $$x = \frac{\ln p - p + C}{(p-1)^2}.$$

Hence p-eliminant of $x = \dfrac{\ln p - p + C}{(p-1)^2}$ and $y = p^2 x + p$ will give the solution.

2. Solve: $y = 2px + p^2$

Solution

Differentiating both sides with respect to x, we get

$$p = \frac{dy}{dx} = 2p + 2(x+p)\frac{dp}{dx} \quad \text{or,} \quad \frac{dx}{dp} + \frac{2x}{p} = -2.$$

This is a linear differential equation whose integrating factor is $e^{\int \frac{2}{p}dp} = p^2$. Multiplying the above equation by p^2, we get

$$\frac{d}{dp}\left(xp^2\right) = -2p^2.$$

Integrating, we get

$$xp^2 = -\frac{2}{3}p^3 + C \text{ or, } x = -\frac{2}{3}p + Cp^{-2}.$$

Using this in the given equation, we get

$$y = -\frac{1}{3}p^2 + \frac{2C}{p}.$$

Hence, the p-eliminant of $x = -\frac{2}{3}p + Cp^{-2}$ and $y = -\frac{1}{3}p^2 + \frac{2C}{p}$ will give the general solution.

3. Solve: $y = \ln\left(p^3 + p\right)$.

Solution

Differentiating with respect to x, we get

$$p = \frac{dy}{dx} = \frac{3p^2+1}{p^3+p}\frac{dp}{dx} \quad \text{or,} \quad dx = \frac{3p^2+1}{\left(p^2+1\right)p^2}dp$$

Integrating, we get

$$x = 2\tan^{-1} p - \frac{1}{p} + C.$$

Hence, the p-eliminant of $x = 2\tan^{-1} p - \frac{1}{p} + C$ and $y = \ln\left(p^3 + p\right)$ will give the general solution.

4. Solve: $y = px + \sqrt{a^2p^2 + b^2}$.

Solution

The given equation is in Clairaut's form. Differentiating both sides with respect to x, we get

$$p = \frac{dy}{dx} = p + \left(x + \frac{a^2p}{\sqrt{a^2p^2+b^2}}\right)\frac{dp}{dx}$$

$$\text{or,} \qquad \left(x + \frac{a^2p}{\sqrt{a^2p^2+b^2}}\right)\frac{dp}{dx} = 0.$$

From $\frac{dp}{dx} = 0$, we get $p = C$. Using this in the given equation we get general solution as

$$y = Cx + \sqrt{a^2C^2 + b^2}.$$

From $x + \frac{a^2p}{\sqrt{a^2p^2+b^2}} = 0$, we get $x = -\frac{a^2p}{\sqrt{a^2p^2+b^2}}$. Eliminating p from the given equation with the aid of it we get the singular solution as

$$\frac{x^2}{a^2} + \frac{y^2}{b^2} = \frac{a^2p^2}{a^2p^2+b^2} + \left(-\frac{a^2p^2}{\sqrt{a^2p^2+b^2}} + \sqrt{a^2p^2+b^2}\right)^2 \frac{1}{b^2} = 1.$$

5. Reduce the differential equation $xy(y-px)=x+py$ to Clairaut's form by the substitution $x^2=u,\ y^2=v$ and then solve it.

Solution

From $x^2=u,\ y^2=v$, we get $\frac{ydy}{xdx}=\frac{dv}{du}$. Writing $\frac{dy}{dx}=p$ and $\frac{dv}{du}=P$, we get

$$xv-uxP=x+xP \text{ or, } v=uP+P+1.$$

This is Clairaut's form. So the general solution is

$$v=uC+C+1 \text{ or, } y^2-1=C(x^2+1).$$

6. Reduce the differential equation $(y+xp)^2=x^2p$ to Clairaut's form by the substitution $xy=v$ and then solve it.

Solution

From $xy=v$, we get $xp+y=\frac{dv}{dx}=P(\text{say})$. Using this in the given equation, we get

$$P^2=x(P-y)=xP-v \text{ or, } v=xP-P^2.$$

This is in Clairaut's form, So the general solution is

$$v=xC-C^2 \text{ or, } xy=xC-C^2.$$

7. Reduce the different equation $y^2\ln y=xyp+p^2$ to Clairaut's form by the substitution $\ln y=v$ and then solve it.

Solution

From $\ln y=v$, we get $\frac{p}{y}=\frac{dv}{dx}=P(\text{say})$. Using this in the given equation we get

$$y^2v=xy^2P+y^2P^2 \text{ or, } v=xP+P^2.$$

This is in Clairaut's form. So the general solution is

$$v=xC+C^2 \text{ or, } \ln y=xC+C^2.$$

8. Reduce the differential equation

$(x^2+y^2)(1+p)^2-2(x+y)(1+p)(x+yp)+(x+yp)^2=0$ to Clairaut's form by the substitution $x+y=u,\ x^2+y^2=v$ and then solve it.

Solution

Form $x+y=u,\ x^2+y^2=v$ we get

$$2xdx+2ydy=dv \text{ and } dx+dy=du.$$

Dividing we get

$$P=\frac{dv}{du}=\frac{2(x+yp)}{1+p}.$$

Using this in the given equation we get

$$v - uP + P^2 = 0 \text{ or, } v = uP - P^2.$$

This is in Clairaut's form. So the general solution is

$$v = uC - C^2 \text{ or, } x^2 + y^2 = (x+y)C - C^2.$$

9. Reduce the different equation $x^2p^2 + y(2x+y)p + y^2 = 0$ to Clairaut's form by the substitution $y = u, xy = v$ and then solve it. Obtain also the singular solution, if any.

Solution

From $y = u, xy = v$, we get $P = \frac{dv}{du} = x + y\frac{dx}{dy} = x\frac{y}{p}$ or $p = \frac{y}{p-x}$. Using this in the given equation, we get

$$\frac{x^2y^2}{(P-x)^2} + y(2x+y)\frac{y}{P-x} + y^2 = 0$$

$$\text{or,} \quad x^2 + (2x+y)(P-x) + (P-x)^2 = 0$$

$$\text{or,} \quad xy = yP + P^2$$

$$\text{or,} \quad v = uP + P^2.$$

Differentiating with respect to u, we get

$$(u + 2P)\frac{dP}{du} = 0.$$

If $\frac{dP}{du} = 0$, then $P = c$. So the general solution is

$$v = uc + c^2 \text{ or, } xy = yc + c^2.$$

If $u + 2P = 0$, then using this in $v = uP + P^2$ we get the singular solution as

$$v = -\frac{u^2}{2} + \frac{u^2}{4} = -\frac{u^2}{4}$$

$$\text{or, } u^2 + 4v = 0 \quad \text{or, } y^2 + 4xy = 0$$

$$\text{or, } y + 4x = 0, \text{ since } y \neq 0.$$

10. Solve: $y = xp^2 + p^4$.

Solution

The given equation is in Lagrange's form, where $\phi(p) = p^2, g(p) = p^4$. Differentiating both sides with respect to x, we get

$$\frac{dx}{dp} - \frac{2}{1-p}x = \frac{4p^2}{1-p}.$$

This is a linear differential equation whose integrating factor is

$$e^{\int \frac{2}{p-1}dp} = e^{2\ln(p-1)} = (p-1)^2.$$

Integrating, we get

$$x(p-1)^2 = \frac{4}{3}p^3 - p^4 + \frac{C}{3} \quad \text{or,} \quad x = \frac{4p^3 - 3p^4 + C}{3(p-1)^2}.$$

So, $$y = \frac{4p^3 - 3p^4 + C}{3(p-1)^2}p^2 + p^4.$$

Hence, p, eliminant of $x = \dfrac{4p^3 - 3p^4 + C}{3(p-1)^2}$ and $y = \dfrac{4p^3 - 3p^4 + C}{3(p-1)^2}p^2 + p^4$ will give the general solution.

4.5 Equations Homogeneous in *x*, *y*

If the equation $f(x, y, p) = 0$ is homogeneous in x, y, then it can be expressed as

$$F\left(\frac{y}{x}, p\right) = 0,$$

which in turn can be expressed either as $p = \phi\left(\dfrac{y}{x}\right)$ or, as $\dfrac{y}{x} = \psi(p)$.

We have already discussed the method of solving the equation $p = \phi\left(\dfrac{y}{x}\right)$. Now, differentiating

$$y = x\psi(p)$$

with respect to x, we get

$$p = \frac{dy}{dx} = \psi(p) + x\psi'(p)\frac{dp}{dx} \quad \text{or,} \quad \frac{dx}{x} = \frac{\psi'(p)}{p - \psi(p)}dp,$$

which can be integrated easily. The p-eliminant of the solution of this equation and the given equation give the general solution.

ILLUSTRATIVE EXAMPLES

1. Solve: $y = (p^2 + p)x$.

Solution

The given equation is of the form $y = x\psi(p)$, where $\psi(p) = p^2 + p$. Differentiating with respect to x, we get

$$\frac{dx}{x} = \frac{\psi'(p)}{p - \psi(p)}dp = -\frac{2p+1}{p^2}dp.$$

Integrating, we get

$$\ln x = -2\ln p + \frac{1}{p} + \ln C \quad \text{or,} \quad x = \frac{C}{p^2}e^{\frac{1}{p}}.$$

Using this in the given equation we get $y = \dfrac{C(p+1)}{p}e^{\frac{1}{p}}$. The p-eliminant of these two will give the general solution.

EXERCISE

- Solve the following equations:

1. $p^2+2xp-3x^2=0,$

2. $p^2-2p\cosh x+1=0,$

3. $p(p+x)=y(x+y),$

4. $p(p+y)=x(x+y),$

5. $p(p-y)=x(x+y),$

6. $px-y=py\sqrt{x^2-y^2},$

7. $x^2p^2+xyp-6y^2=0,$

8. $x^2p^2-2xyp+2y^2-x^2=0,$

9. $p^3-(x^2+xy+y^2)p^2+xy(x^2+xy+y^2)p-x^3y^3=0,$

10. $y=2px+yp^2,$

11. $y=2px+y^2p^3,$

12. $y^2\ln y=xpy+p^2,$

13. $y=x-p^2,$

14. $x(1+p^2)=1,$

15. $x=\tan^{-1}p+\dfrac{p}{1+p^2},$

16. $x=py+ap^2,$

17. $p^2-2xp+1=0,$

18. $p^3y^2-2px+y=0,$

19. $y=3px+6p^2y^2,$

20. $x=ap^2+bp^3,$

21. $y=x+p^3,$

22. $xp^2-2yp+ax=0,$

23. $y=2px+\tan^{-1}(xp^2),$

24. $y=2px+f(xp^2),$

25. $y+px=x^4p^2,$

26. $y=3px+4p^2,$

27. $y=\dfrac{x}{p}+p,$

28. $y=p\tan p+\ln(\cos p),$

29. $y=\sin p-p\cos p,$

30. $y=p\sin p+\cos p,$

31. $y=p-\ln(p^2-1),$

32. $y=a\sqrt{p^2+1},$

33. $y=2p+3p^2,$

34. $y=ap^2+bp^3,$

35. $y=a+bp+dp^2,$

36. $p^2y^2\cos^2\alpha-2px\sin^2\alpha+y^2-x^2\sin^2\alpha=0,$

37. $y=px+\dfrac{a}{p},$

38. $y=px+a\sqrt{p^2+1},$

39. $y=px+p(1-p),$

40. $y=px+\sin^{-1}p,$

41. $y=px-p^2,$

42. $y=px+p^n,$

43. $py=p^2(x-b)+a,$

44. $(x-a)p^2+(x-y)p-y=0.$

- Reduce the following equations to Clairaut's form using suitable substitutions and solve: [Use $x^2 = u$, $y^2 = v$ in 45-49]

45. $x^2(y - px) = p^2 y$ **46.** $(px - y)(x - py) = 2p$,

47. $xyp^2 - (x^2 + y^2 - 1)p + xy = 0$, **48.** $xy(y - px) = x + yp$,

49. $y^2 = \left(\frac{px}{x}\right)x^2 + f\left(\frac{py}{x}\right)$. [Use $x = u$, $y^2 = v$ in 50 and 51]

50. $y = 2px - p^2 y$, **51.** $y = 2px + y^2 p^3$,

[Use $x = u$, $xy = v$ in 52]

52. $(px + y)^2 = py^2$, **53.** $(px + y)^2 = px^2$, [Use $x = u$, $xy = v$]

- Reduce the following equations to Clairaut's form under the given substitutions and solve:

54. $e^{by}(a - bp) = f\left(pe^{by-ax}\right), \left[e^{ax} = u, e^{by} = v\right]$,

55. $e^{3x}(p-1) + p^3 e^{2y} = 0\left[e^x = u, e^y = v\right]$,

56. $e^{4x}(p-1) + p^2 e^{2y} = 0\left[e^{2x} = u, e^{2y} = v\right]$,

57. $y^2(y - xp) = x^4 p^2\left[x = \frac{1}{u}, y = \frac{1}{v}\right]$,

58. $(x^2 - a^2)p^2 - 2xyp - x^2 = 0\left[x^2 = u\right]$,

59. $y = 2px + ayp^2\left[y^2 = v\right]$,

60. $y = 2px + y^2 p^3\left[y^2 = v\right]$,

61. $y^2 \ln y = pxy + p^2\left[\ln y = v\right]$,

62. $(px^2 + y^2)(px + y) = (p+1)^2\left[u = xy, v = x + y\right]$,

63. $x^2 p^2 + yp(2x + y) + y^2 = 0$, $\left[y = u, xy = v\right]$,

64. $xp^2 - 2yp + x + 2y = 0$ $\left[x^2 = u, y - x = v\right]$.

ANSWERS

1. $(2y + 3x^2 - C)(2y - x^2 - C) = 0$, **2.** $(y - e^x - C)(y + e^{-x} - C) = 0$,

3. $(y - Ce^x)(y + x - Ce^{-x} - 1) = 0$, **4.** $(x^2 + y^2 - C)(2xy + x^2 - C) = 0$,

5. $(x^2+2y-C)(y+x-Ce^x+1)=0$, **6.** $x = y\cosh(y+C)$,

7. $(yx^3-C)\left(\dfrac{y}{x^2}-C\right)=0$, **8.** $\sin^{-1}\dfrac{y}{x}=\pm\ln Cx$,

9. $(x^3-3y+C)\left(e^{\frac{x^2}{2}}+Cy\right)(xy+Cy+1)=0$,

10. $y^2=2xC+C^2$, **11.** $y^2=2xC+C^3$,

12. $\ln y = Cx+C^2$,

13. $y=C-\dfrac{p^2}{2}-p-\ln(p-1), x=C-2\{p+\ln(p-1)\}$,

14. $x(1+p^2)=1, y=C-\tan^{-1}p+\dfrac{p}{1+p^2}$,

15. $x=\tan^{-1}p+\dfrac{p}{1+p^2}, y=C-\dfrac{1}{1+p^2}$,

16. $x=\dfrac{p(C+a\sin^{-1}p)}{\sqrt{1-p^2}}, y=\dfrac{C+a\sin^{-1}p}{\sqrt{1-p^2}}-ap$,

17. $x=\dfrac{p}{2}+\dfrac{1}{2p}, y=\dfrac{p^2}{4}-\dfrac{1}{2}\ln p+C$,

18. $y^2=2Cx-C^3$, **19.** $y^3=3Cx+6C^2$,

20. $x=ap^2+bp^3, y=\dfrac{2a}{3}p^3+\dfrac{3b}{4}p^4+C$,

21. $x=\dfrac{3}{2}p^2+3p+3\ln(p-1)+C, y=p^3+\dfrac{3}{2}p^2+3p+3\ln(p-1)+C$,

22. $C^2x^2-2yC+a=0$, **23.** $y=2\sqrt{Cx}+\tan^{-1}C$,

24. $y=2C\sqrt{x}+f(C^2)$, **25.** $xy+C=C^2x$,

26. $x=-\dfrac{8}{5}p+Cp^{-3/2}, y=\dfrac{3C}{\sqrt{p}}-\dfrac{4}{5}p^2$,

27. $x=\dfrac{p(C+\cosh^{-1}p)}{\sqrt{p^2-1}}, y=p+\dfrac{(C+\cosh^{-1}p)}{\sqrt{p^2-1}}$,

28. $x=\tan p+C, y=p\tan p+\ln(\cos p)$,

29. $x=C-\cos p, y=\sin p-p\cos p$,

30. $x=C+\sin p, y=p\sin p+\cos p$,

31. $x = \ln \dfrac{p(p+1)}{p-1} + C,\ y = p - \ln(p^2 - 1),$

32. $x = C + a \sinh^{-1} p,\ y = a\sqrt{p^2+1},$

33. $x = C + 6p + 2\ln p,\ y = 2p + 3p^2,$

34. $y = ap^2 + bp^3,\ x = 2ap + \frac{3}{2}bp^2 + C,$

35. $y = a + bp + dp^2,\ x = b\ln p + 2dp + C$

36. $x^2 + y^2 - 2Cx\sec\alpha + C^2 = 0,$

37. $y = Cx + \frac{a}{C}$, singular solution $y^2 = 4ax,$

38. $y = Cx + a\sqrt{C^2+1}$, singular solution $x^2 + y^2 = a^2,$

39. $y = Cx + C(1-C)$, singular solution $(x+1)^2 = 4y,$

40. $y = Cx + \sin^{-1} C$, singular solution $y = \sqrt{x^2-1} + \sin^{-1}\dfrac{\sqrt{x^2-1}}{x},$

41. $y = Cx - C^2$, singular solution $x^2 = 4y,$

42. $y = Cx + C^n$, singular solution $n^n y^{n-1} + x^n (n-1)^{n-1} = 0,$

43. $Cy = C^2(x-b) + a$, singular solution $y^2 = 4a(x-b),$

44. $(x-a)C^2 + (x-y)C - y = 0$, singular solution $(x+y)^2 = 4ay,$

45. $y^2 = Cx^2 + C^2,$

46. $C^2x^2 + C(2 - x^2 - y^2) + y^2 = 0,$

47. $y^2 = Cx^2 + \dfrac{C}{C-1},$

48. $y^2 = Cx^2 + C + 1,$

49. $y^2 = Cx^2 + f(C),$

50. $y^2 = Cx - \frac{1}{4}C^2,$

51. $y^2 = Cx + \frac{1}{8}C^3,$

52. $xy = Cy - C^2,$

53. $xy = Cx - C^2,$

54. $e^{by} = e^{ax}C + \frac{1}{a}f\left(\frac{a}{b}C\right),$

55. $e^y = Ce^x + C^3,$

56. $e^{2y} = Ce^{2x} + C^2,$

57. $x = Cy + xyC^2,$

58. $2Cy = x^2 - a^2 - C^2,$

59. $y^2 = xC - \frac{a}{4}C^2,$

60. $y^2 = xC + \frac{1}{8}C^3,$

61. $\ln y = Cx + C^2,$

62. $x + y = Cxy + C^2,$

63. $xy = yC + C^2,$

64. $2C^2x^2 - 2C(y-x) + 1 = 0.$

Chapter 5

SINGULAR SOLUTIONS

5.1 Introduction

In the chapter 1, we have already introduced the idea of singular solution of a differential equation and in the chapter 4, we have given examples on it. In this chapter, we present a general discussion on singular solution of a differential equation after giving a few definitions related to singular solution.

The *discriminant* of an equation involving single parameter is a function of the coefficients in the rational integral form. For the function $f(x, y, \alpha) = 0$, α being the parameter, the discriminant is the α eliminant of the equations $f(x, y, \alpha) = 0$ and $\frac{\partial f}{\partial \alpha} = 0$.

If $\varphi(x, y, c) = 0$ be the general solution of the differential equation $f(x, y, p) = 0$, where $p = \frac{dy}{dx}$. Then p-discriminant relation is obtained by eliminating p from $f(x, y, p) = 0, \frac{\partial f}{\partial p} = 0$ and represents the locus, for each point of which $f(x, y, p) = 0$ has equal values of p.

Similarly, the c-discriminant relation is obtained by eliminating c from $\varphi(x, y, c) = 0, \frac{\partial \varphi}{\partial c} = 0$ and represents the locus, for each point of which $\varphi(x, y, c) = 0$ has equal values of c.

The equation $f(x, y, p) = 0$ and its solution $\varphi(x, y, c) = 0$ are of same degree in p and c respectively, so if there be a p-discriminant relation, then there must be a c discriminant relation.

5.2 Working Rule for Finding Singular Solution

Let $F(x, y) = 0$ be the p-discriminant relation, i.e., the p-eliminant of $f(x, y, p) = 0$ $\frac{\partial f}{\partial p} = 0$. If $F(x, y) = 0$ satisfies the differential equation $f(x, y, p) = 0$, then it is called its singular solution. If not so then the factors of it, which satisfy the equation, will form the singular solution of the equation. Clearly, the equation $f(x, y, p) = 0$ will have no singular solution, if it can be resolved into linear factors in p.

Again, let $g(x, y) = 0$ be the c-discriminant relation, i.e., the c-eliminant of $\varphi(x, y, c) = 0, \frac{\partial \varphi}{\partial c} = 0$. If $g(x, y) = 0$ satisfies the differential equation $f(x, y, p) = 0$, then it is called its singular solution. If not so, then the factors of it, which satisfy the equation, will form the singular solution of the equation.

If the equation $f(x, y, p) = 0$ be given in Clairaut's form

$$y = px + h(p),$$

then its general solution is $\quad y = cx + h(c).$

In this case both p-discriminant relation and c-discriminant relation will be same, i.e., both of these will give us the required solution, which is not true for other forms of the equation $f(x, y, p) = 0$.

5.3 Extraneous Loci

If $s(x, y) = 0$ is a singular solution of the differential equation $f(x, y, p) = 0$ whose general solution is $\varphi(x, y, c) = 0$, then $s(x, y)$ is a factor of both p-discriminant and c-discriminant. These, discriminants, however, may have other factors which give rise to other loci associated with the general solution. Since, these equations of the loci do not satisfy the equation $f(x, y, p) = 0$, generally, these are called *extraneous loci*.

Tac locus: Let P be a point on p-discriminant relation, then there are two equal values of p at the point P. If it is found that these equal values of p belong to two distinct, non-consequtive curves touching at P, then this point will satisfy the p-discriminant but not c-discriminant. Again, the point P does not lie on the envelope; hence this will not satisfy the differential equation. Locus of such point is called *tac locus*. If $T(x, y) = 0$ denotes the equation of tac locus, then $T(x, y)$ is a factor of p-discriminant. In general, $T(x, y)$ is not a factor of c-discriminant and $T(x, y) = 0$ does not satisfy the differential equation.

Consider a system of circles of fixed radius and having centres on x-axis. Here, the x-axis is the locus of the point of contacts of the non consequtive circles of the system and represents the tac locus.

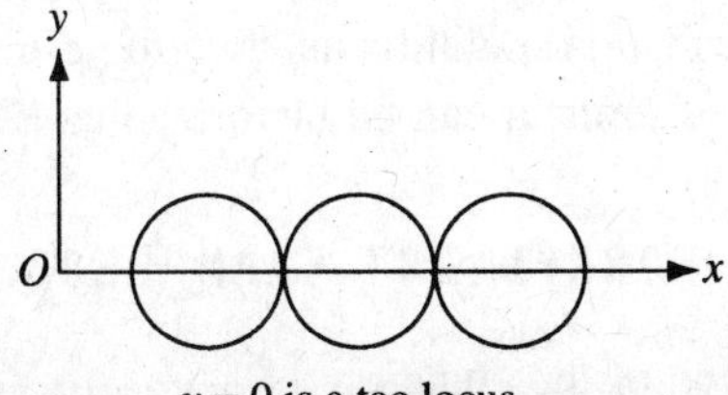

$y = 0$ is a tac locus.

Nodal locus: The c-discriminant relation represents the locus of the points for which two values of c are equal. If each of the curve of the family $\varphi(x, y, c) = 0$ has a node (a double point with distict tangents), then the locus of the nodes is called *nodal locus*. If $N(x, y) = 0$ denotes the equation of nodal locus, then $N(x, y)$ is a factor of

c-discriminant. In general, $N(x, y)$ is not a factor of p-discriminant and $N(x, y) = 0$ does not satisfy the differential equation.

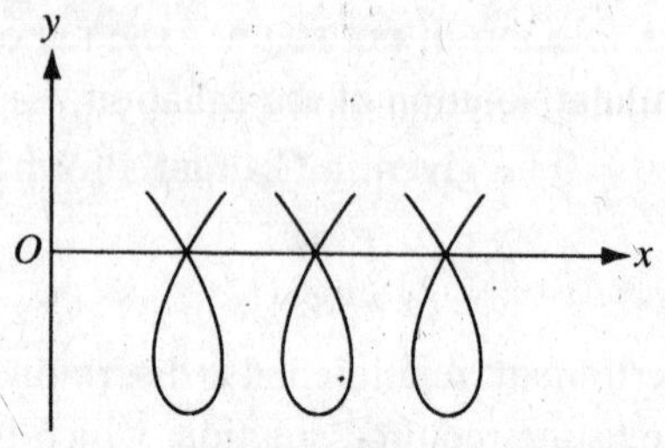

$y = 0$ is a nodal locus.

Cuspidal locus: If each member of the family of curves represented by the general solution $\varphi(x, y, c) = 0$ of the differential equation $f(x, y, p) = 0$ has a cusp (a double point with coincident tangent), then the locus of the cusps is called *cuspidal locus*. Since, a cusp is also the point of intersection of consecutive curves, both p-discriminant relation and c-discriminant relation will contain the cuspidal locus. If $C(x, y) = 0$ denotes the equation of cuspidal locus, then $C(x, y)$ is a factor of both c-discriminant and p-discriminant. In general, $C(x, y)$ does not satisfy the differential equation.

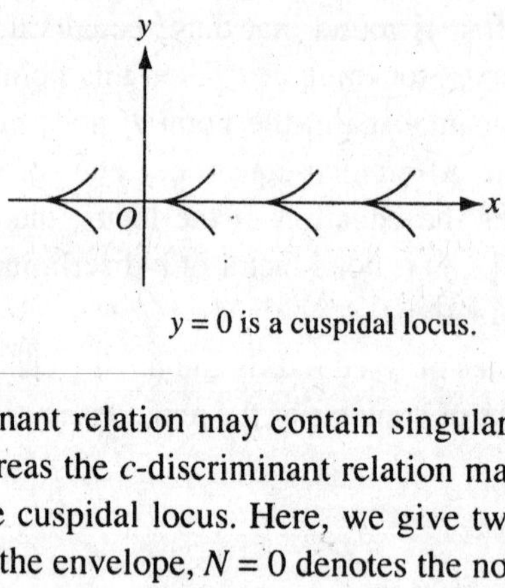

$y = 0$ is a cuspidal locus.

Thus the p-discriminant relation may contain singular solution, the tac locus and the cupsidal locus whereas the c-discriminant relation may contain singular solution, the nodal locus and the cuspidal locus. Here, we give two important results without proof. If $E = 0$ denotes the envelope, $N = 0$ denotes the nodal locus, $T = 0$ denotes the tac locus and $C = 0$ denotes the cuspidal locus, then the p-discriminant can be factorized as ET^2C and the c-discriminant can be factorized as EN^2C^3.

WORKED-OUT EXAMPLES

1. Find the singular solution of the equation $p^2 + y^2 - 1 = 0$.

Solution

From the equation we get $\dfrac{dy}{dx} = \sqrt{1 - y^2}$. Integrating, we get

$$\sin^{-1} y = x + c \quad \text{or,} \quad y = \sin(x + c).$$

This is the general solution.

Using the p-discriminant relation, we get

$$y^2 - 1 = 0 \quad \text{or,} \ y = \pm 1.$$

These solutions satisfy the given equation. Hence singular solutions of the given equation are $y = \pm 1$.

2. Find the singular solution of the equation $xp^2 - (x-a)^2 = 0$.

Solution

From $\quad xp^2 - (x-a)^2 = 0$, we get

$$p = \sqrt{x} - \frac{a}{\sqrt{x}}.$$

Integrating, we get $\quad y + c = \frac{2}{3}x^{\frac{3}{2}} - 2a\sqrt{x}$

$$\text{or,} \quad (y+c)^2 = \frac{4}{9}x(x-3a)^2$$

$$\text{or,} \quad c^2 + 2cy + y^2 - \frac{4}{9}x(x-3a)^2 = 0.$$

So, the c-discriminant is $\quad 4y^2 = 4y^2 - \frac{16}{9}x(x-3a)^2$

$$\text{or,} \quad x(x-3a)^2 = 0.$$

The p-discriminant is $\quad x(x-a)^2 = 0.$

So, singular solution is $x = 0$, tac locus is $x = a$ and nodal locus is $x = 3a$.

3. Find the singular solutions of the equation $(3y-1)^2 p^2 = 4y$.

Solution

From $(3y-1)^2 p^2 = 4y$, we get

$$p = \pm\frac{2\sqrt{y}}{3y-1}$$

$$\text{or,} \quad \frac{3y-1}{\sqrt{y}}dy = \pm 2dx$$

$$\text{or,} \quad \left(3\sqrt{y} - \frac{1}{\sqrt{y}}\right)dy = \pm 2dx.$$

Integrating, we get

$$2\left(y^{\frac{3}{2}} - \sqrt{y}\right) = 2(c \pm x)$$

$$\text{or,} \quad (x \pm c)^2 = y(y-1)^2$$

$$\text{or,} \quad c^2 \pm 2cx + x^2 - y(y-1)^2 = 0.$$

So, the c-discriminant is

$$4x^2 = 4x^2 - 4y(y-1)^2 \quad \text{or,} \quad y(y-1)^2 = 0.$$

The p-discriminant is $\qquad y(3y-1)^2 = 0.$

So, singular solution is $y = 0$, tac locus is $3y = 1$ and nodal locus is $y = 1$.

4. Find the singular solutions of the equation $x^3p^2 + x^2yp + a^3 = 0$.

Solution

From $x^3p^2 + x^2yp + a^3 = 0$, we get

$$y = -xp - \frac{a^3}{x^2p}.$$

Differentiating, we get

$$p = -p - x\frac{dp}{dx} + \frac{2a^3}{x^3p} + \frac{a^3}{x^2p^2}\frac{dp}{dx} \quad \text{or,} \quad \left(\frac{dp}{dx} + \frac{2p}{x}\right)\left(\frac{a^3}{x^2p^2} - x\right) = 0.$$

From $\qquad \dfrac{dp}{dx} + \dfrac{2p}{x} = 0,$

we get $\qquad \dfrac{dp}{p} + 2\dfrac{dx}{x} = 0.$

Integrating, we get $\qquad px^2 = c.$

Using this in the given equation, we get

$$c^2 + cxy + a^3x = 0.$$

The c-discriminant is

$$x^2y^2 - 4a^3x = 0 \quad \text{or,} \quad x\left(xy^2 - 4a^3\right) = 0.$$

The p-discriminant is

$$x^4y^2 - 4a^3x^3 = 0 \quad \text{or,} \quad x^2x\left(xy^2 - 4a^3\right) = 0.$$

Hence, $x\left(xy^2 - 4a^3\right) = 0$ is the singular solution and $x = 0$ is also tac locus.

5. Find the singular solutions of the equation $2p^2 - 2x^2p + 3xy = 0$.

Solution

From $2p^2 - 2x^2p + 3xy = 0$, we get $y = \dfrac{2}{3}px - \dfrac{2p^2}{3x}.$

Differentiating we get $p = \dfrac{2}{3}p + \dfrac{2}{3}x\dfrac{dp}{dx} + \dfrac{2p^2}{3x^2} - \dfrac{4p}{3x}\dfrac{dp}{dx}$

$$\text{or,} \quad \left(2\frac{dp}{dx} - \frac{p}{x}\right)\left(x - \frac{2p}{x}\right) = 0.$$

From $\qquad 2\dfrac{dp}{dx} - \dfrac{p}{x} = 0,$

we get $$p^2 = cx.$$

Using this in the given equation, we get

$$2cx + 3xy = 2x^2\sqrt{cx}$$

$$\text{or,} \quad (2c+3y)^2 = 4x^3c$$

$$\text{or,} \quad 4c^2 + 4c(3y - x^3) + 9y^2 = 0.$$

The c-discriminant is

$$16(3y - x^3)^2 - 144y^2 = 0 \text{ or, } x^3(x^3 - 6y) = 0.$$

The p-discriminant is

$$4x^4 - 24xy = 0 \text{ or, } x(x^3 - 6y) = 0.$$

Hence, $x^3 - 6y = 0$ is the singular solution and $x = 0$ is the cuspidal locus.

6. Find the singular solutions of the equation $p^2(1-x^2) = 1 - y^2$.

Solution

From $p^2(1-x^2) = 1 - y^2$, we have

$$p = \sqrt{\frac{1-y^2}{1-x^2}} \quad \text{or,} \quad \frac{dy}{\sqrt{1-y^2}} = \frac{dx}{\sqrt{1-x^2}}.$$

Integrating, we get $\quad \sin^{-1} y - \sin^{-1} x = c'$

$$\text{or,} \quad \cos^{-1} x - \cos^{-1} y = c'$$

$$\text{or,} \quad \cos^{-1}\left[xy + \sqrt{1-x^2}\sqrt{1-y^2}\right] = c'$$

$$\text{or,} \quad xy + \sqrt{1-x^2}\sqrt{1-y^2} = \cos c' = c$$

$$\text{or,} \quad (1-x^2)(1-y^2) = (c - xy)^2$$

$$\text{or,} \quad c^2 - 2xyc + x^2 + y^2 - 1 = 0.$$

So, the c-discriminant is

$$4x^2y^2 = 4(x^2 + y^2 - 1) \text{ or, } (1-x^2)(1-y^2) = 0.$$

Also, the p-discriminant is $\quad (1-x^2)(1-y^2) = 0.$

Hence, the singular solutions are $x = \pm 1, y = \pm 1$ and it has no extraneous loci.

7. Reduce the differential equation $(y + xp)^2 = x^2 p$ to Clairaut's form by the substitution $xy = v$ and then examine for singular solution.

Solution

From $xy = v$, we get $xp + y = \dfrac{dv}{dx} = P(\text{say})$. Using this in the given equation, we get

$$P^2 = x(P - y) = xP - v \text{ or, } v = xP - P^2.$$

This is in Clairaut's form. So the general solution is

$$v = xc - c^2 \text{ or, } xy = xc - c^2.$$

The c-discriminant is

$$x^2 - 4xy = 0 \text{ or, } x(x - 4y) = 0.$$

The p-discriminant is

$$x^2(x - 2y)^2 - 4x^2y^2 = 0 \text{ or, } x^3(x - 4y) = 0.$$

Hence, $x(x - 4y) = 0$ is the singular solution and $x = 0$ is also tac locus.

8. Reduce the differential equation $(xp - y)(x - py) = 2p$ to Clairaut's form by the substitution $x^2 = u$, $y^2 = v$ and then examine for singular solution.

Solution

From $x^2 = u$, $y^2 = v$, we get $\frac{y\,dy}{x\,dx} = \frac{dv}{du}$. Writing $\frac{dv}{du} = P$, we get

$$\left(\frac{x}{y}Px - y\right)\left(x - \frac{x}{y}Py\right) = 2\frac{x}{y}P$$

or, $\quad (uP - v)(1 - P) = 2P$

or, $\quad v = uP + \dfrac{2P}{P-1}$

which is Clairaut's form and its general solution is

$$v = uc + \frac{2c}{c-1}$$

or, $\quad y^2 = x^2c + \dfrac{2c}{c-1}$

or, $\quad x^2c^2 - c(x^2 + y^2 - 2) + y^2 = 0.$

The c-discriminant is $(x^2 + y^2 - 2)^2 - 4x^2y^2 = 0.$

Again, the given equation can be expressed as

$$xyp^2 - p(x^2 + y^2 - 2) + xy = 0.$$

The p-discriminant is $(x^2 + y^2 - 2)^2 - 4x^2y^2 = 0.$

Hence the singular solution is $(x^2 + y^2 - 2)^2 - 4x^2y^2 = 0.$

EXERCISE

- Examine for singular solutions (1-18):

1. $4p^2 = 9x$, **2.** $x^2 + y = p^2$,

3. $8p^3 = 27y$, **4.** $3y = 2px - \dfrac{2p^2}{x}$,

5. $xp^2 - 2yp + ax = 0$, **6.** $4xp^2 = (3x - a)^2$,

7. $y^2(1+p^2)=a^2$,

8. $y+px=x^4p^2$,

9. $y=px+e^p$,

10. $y=2px+p^2$,

11. $y=px+p^3$,

12. $y=2px-yp^2$,

13. $p^2+xp-y=0$,

14. $y=px+\frac{a}{p}$,

15. $y=px-\sqrt{p^2-1}$,

16. $y=px+p^3$,

17. $y=3px+6p^2y^2$,

18. $x^3p^2+x^2yp+1=0$.

- Reduce the following equation to Clairaut's form (19-24) by the substitutions indicated there and solve:

19. $p^4=4y(xp-2y)^2$, put $y=v^2$, $x=u$.
20. $p^3-4xyp+8y^2=0$, put $y=v^2$, $x=u$.
21. $x^2p^2+py(2x+y)+y^2=0$, put $y=u, xy=v$.
22. $xp^2-2yp+x+2y$, put $y-x=v$, $x^2=u$.
23. $y^2(y-xp)=x^4p^2$, put $x=\frac{1}{u}$, $y=\frac{1}{v}$.
24. $(px^2+y^2)(px+y)=(p+1)^2$, put $x+y=u$, $xy=v$.
25. Obtain the differential equation satisfied by the family of curves $c^2+2cy-x^2+1=0$, where c is the parameter. Find also the singular solution.
26. Obtain the differential equation satisfied by the family of curves $x^2+y^2+2cx+2c^2-1=0$, where c is the parameter. Find also the singular solution.

ANSWERS

1. No singular solution, cusp locus $x = 0$.
2. No singular solution. cusp locus $y + x^2 = 0$.
3. S.S.: $y = 0$; cusp locus: $y = 0$.
4. S.S.: $x^3 = 6y$; cusp locus: $x = 0$.
5. S.S.: $y^2 = ax^2$.
6. S.S.: $x = 0$, tac locus: $3x - a = 0$, nodal locus: $x - a = 0$.
7. S.S.: $y=\pm a$; tac locus: $y = 0$.
8. S.S.: $4x^2y+1=0$; tac locus: $x = 0$.
9. S.S.: $y=x(\ln x-1)$,
10. No singular solution, cusp locus: $x^2+y=0$.
11. S.S.: $4x^3+27y^2=0$.
12. S.S.: $y=\pm x$.
13. S.S.: $x^2+4y=0$.
14. S.S.: $y^2=4ax$.

15. S.S.: $y^2 = x^2 - 1.$

16. S.S.: $4x^3 + 27y^2 = 0.$

17. S.S.: $3x^2 + 8y^3 = 0.$

18. S.S.: $x\left(xy^2 - 4\right) = 0$; tac locus: $x = 0.$

19. G.S.: $y = c^2(x - c)^2$; S.S.: $x^4 = 16y$; tac locus: $y = 0.$

20. G.S.: $y = c(x - c)^2$; S.S.: $y\left(27y - 4x^3\right) = 0$; tac locus: $y = 0.$

21. G.S.: $xy = cy + c^2$; S.S.: $y(y + 4x) = 0$; tac locus: $y = 0.$

22. G.S.: $2c^2x^2 - 2c(y - x) + 1 = 0$; S.S.: $x^2 + 2xy - y^2 = 0.$

23. G.S.: $c^2xy + cy - x = 0$; S.S.: $y\left(y + 4x^2\right) = 0$; tac locus: $xy = 0.$

24. S.S.: $x^2y^2 + 4(x + y) = 0$; tac locus: $y = x.$

25. $\left(1 - x^2\right)p^2 + 2xyp + x^2 = 0$; S.S.: $x^2 + y^2 = 1$; tac locus: $x = 0.$

26. $2y^2p^2 + 2xyp + x^2 + y^2 - 1 = 0$; S.S.: $x^2 + 2y^2 = 2$; tac locus: $y = 0.$

Chapter 6

Linear Equations with Constant Coefficients

6.1 Introduction

Linear differential equations appear in the field of applied mathematics, mathematical physics frequently. In such equations, the coefficients may be functions of x or constants. In this chapter we will consider the linear equations of second and higher orders with constant coefficients only as we have already considered first order linear differential equations in chapter 2. The general form of an n-th order linear differential equation is

$$\frac{d^n y}{dx^n} + c_1 \frac{d^{n-1} y}{dx^{n-1}} + c_2 \frac{d^{n-2} y}{dx^{n-2}} + \cdots + c_{n-1} \frac{dy}{dx} + c_n y = X$$

or, $$\left[D^n + c_1 D^{n-1} + c_2 D^{n-2} + \cdots + c_{n-1} D + c_n\right] y = X, \quad (1)$$

where $c_k \,; 1 \le k \le n$ and X are constants or functions of x only and $D = \frac{d}{dx}$. Thus an n-th order linear differential equation can be written as $f(D)y = X$, where

$$f(D) = D^n + c_1 D^{n-1} + c_2 D^{n-2} + \cdots + c_{n-1} D + c_n.$$

Since D is a differential operator, $f(D)$ is also an operator and $\frac{1}{f(D)}$ is called its inverse operator.

If $X = 0$, then the above equation is called an n-th order homogeneous linear differential equation.

For the operator D, the following results are valid:

(i) $D^m + D^n = D^n + D^m$,

(ii) $D^m D^n = D^n D^m = D^{m+n}$,

(iii) $D(u + v) = Du + Dv$, where u, v are functions of x,

(iv) $(D - a)(D - b) = (D - b)(D - a)$, where a, b are constants,

(v) $D^{-k} f(x) = \iint \cdots \int f(x)(dx)^k$, where k is a positive integer,

(vi) $f_1(D) f_2(D) = f_2(D) f_1(D)$, if the operators $f_1(D), f_2(D)$ do not contain terms in x.

6.2 Linearly Dependent, Independent Functions and the Wronskian

The functions $y_k(x); 1 \le k \le n$ are said to be linearly dependent if there exists n scalars $c_1, c_2, \cdots, c_n$ (not all zero) such that

$$c_1 y_1(x) + c_2 y_2(x) + \cdots + c_n y_n(x) = 0. \tag{2}$$

Example: Since $e^x - \cosh x - \sinh x = 0$, the functions e^x, $\cosh x$, $\sinh x$ are linearly dependent.

On the other hand, if the identity (2) implies $c_k = 0 \forall k$, then the set of functions $\{y_1(x), y_2(x), \cdots, y_n(x)\}$ is said to be linear independent.

Example: The identity $ae^{-x} + be^x + ce^{2x} = 0$ implies $a = b = c = 0$. Hence, the functions e^{-x}, e^x, e^{2x} are linearly independent.

The Wronskian of n functions $y_1(x), y_2(x), \cdots, y_n(x)$ is denoted as $W(y_1, y_2, \cdots, y_n)$ and is defined as

$$W(y_1, y_2, \cdots, y_n) = \begin{vmatrix} y_1 & y_2 & \cdots & y_n \\ y_1' & y_2' & \cdots & y_n' \\ \cdots & \cdots & \cdots & \cdots \\ \cdots & \cdots & \cdots & \cdots \\ y_1^{n-1} & y_2^{n-1} & \cdots & y_n^{n-1} \end{vmatrix}$$

Here, we give two important theorems (without proof).

Theorem I

If $p_1(x), p_2(x), \cdots, p_n(x)$ be real or complex valued continuous functions defined on an interval I and $\varphi_1(x), \varphi_2(x), \cdots, \varphi_n(x)$ be n solutions of the equation

$$y^n(x) + p_1(x)\, y^{n-1}(x) + p_2(x)\, y^{n-2}(x) + \cdots + p_n(x)\, y(x) = 0$$

in I, then the solutions will be linearly independent, if and only if

$$W(\varphi_1(x), \varphi_2(x), \cdots, \varphi_n(x)) = 0 \text{ for all } x \in \text{I}.$$

Consider the functions e^x, $\cosh x$, $\sinh x$. Wronskian of these function is

$$\begin{vmatrix} e^x & \cosh x & \sinh x \\ e^x & \sinh x & \cosh x \\ e^x & \cosh x & \sinh x \end{vmatrix}$$

which vanishes identically. Hence, the functions e^x, cosh x, sinh x are linearly dependent. Again, for the functions e^{-x}, e^x, e^{2x}, the Wronskian is

$$\begin{vmatrix} e^{-x} & e^x & e^{2x} \\ -e^{-x} & e^x & 2e^{2x} \\ e^{-x} & e^x & 4e^{2x} \end{vmatrix} = \begin{vmatrix} e^{-x} & e^x & e^{2x} \\ -e^{-x} & e^x & 2e^{2x} \\ 0 & 0 & 3e^{2x} \end{vmatrix} = 6e^{2x} \neq 0.$$

Hence, the functions e^{-x}, e^x, e^{2x} are linearly independent.

Theorem II

An n-th degree homogeneous linear equation

$$p_0(x)y^n(x) + p_1(x)y^{n-1}(x) + p_2(x)y^{n-2}(x) + \cdots + p_n(x)y(x) = 0$$

always has n linearly independent solutions $\varphi_1(x), \varphi_2(x), \cdots, \varphi_n(x)$ and its general solution is given by

$$y = c_1\varphi_1(x) + c_2\varphi_2(x) + \cdots + c_n\varphi_n(x),$$

where $c_k's, 1 \le k \le n$ are arbitrary constants.

6.3 Method of Solution

We first consider the homogeneous form of the equation (1). Let $y = e^{mx}$, which is the solution of first order homogeneous linear equation, be a trial solution of the equation

$$\left[D^n + c_1D^{n-1} + c_2D^{n-2} + \cdots + c_{n-1}D + c_n\right]y = 0. \tag{3}$$

Then we get

$$m^n + c_1m^{n-1} + c_2m^{n-2} + \cdots c_{n-1}m + c_n = 0, \text{ as } y \neq 0.$$

This is called *auxiliary equation* (A.E.) associated with the differential equation given by (1). If the roots of the auxiliary equation are $m_k, 1 \le k \le n$, then the following cases arise:

(i) If $m_k, 1 \le k \le n$ are distinct, then the equation (3) can be written as

$$(D - m_1)(D - m_2)\cdots(D - m_n)y = 0.$$

This shows that solutions of the equations $(D - m_k)y = 0, 1 \le k \le n$ are also the solution of the equation (3).

Hence, $e^{m_1x}, e^{m_2x}, \cdots, e^{m_nx}$ are all linearly independent solutions of the given equation. Thus the general solution of the equation (3) is

$$y = a_1e^{m_1x} + a_2e^{m_2x} + \cdots + a_ne^{m_nx}, \tag{4}$$

where $a_k, 1 \le k \le n$ are arbitrary constants. The general solution of the homogeneous form of the equation (1) is called *complementary function* (C.F.) of it.

(ii) If $m_k, 1 \le k \le n$ are not all distinct, say, $m_{n-p+1} = m_{n-p+2} = \cdots = m_n = m$, then the equation (3) can be written as

$$(D-m)^p (D-m_1)(D-m_2)\cdots(D-m_{n-p})y = 0.$$

Now solutions of the equation $(D-m)^p y = 0$ are also solutions of the equation (3). Let $y = e^{mx}v$, where v is a function of x, is a solution of the equation

$$(D-m)^p y = 0.$$

Using $y = e^{mx}v$ in the above equation, we see that

$$(D-m)^{p-1}(D-m)e^{mx}v = 0$$

or, $$(D-m)^{p-1}\left[me^{mx}v + e^{mx}Dv - me^{mx}v\right] = 0$$

or, $$(D-m)^{p-1}\left[e^{mx}Dv\right] = 0$$

or, $$(D-m)^{p-2}(D-m)\left[e^{mx}Dv\right] = 0$$

or, $$(D-m)^{p-2}\left[me^{mx}Dv + e^{mx}D^2v - me^{mx}Dv\right] = 0$$

or, $$(D-m)^{p-2}\left[e^{mx}D^2v\right] = 0$$

or, $$e^{mx}D^p v = 0$$

or, $$D^p v = 0 \text{ as } e^{mx} \ne 0.$$

This gives that v is a function of x of $(p-1)$th degree. Thus the general solution of the equation (3), in this case, is

$$y = \left(b_1 + b_2x + b_3x^2 + \cdots + b_px^{p-1}\right)e^{mx} + a_1e^{m_1x} + a_2e^{m_2x} + \cdots + a_{n-p}e^{m_{n-p}x}$$

where $b_k, 1 \le k \le p$ and $a_k, 1 \le k \le n-p$ are arbitrary constants.

(iii) If two roots of the auxiliary equation be complex, then taking

$$m_1 = \alpha + i\beta \text{ and } m_2 = \alpha - i\beta$$

we see that the general solution of the equation (1) becomes

$$\begin{aligned} y &= a_1e^{(\alpha+i\beta)x} + a_2e^{(\alpha-i\beta)x} + a_3e^{m_3x} + \cdots + a_ne^{m_nx} \\ &= e^{\alpha x}\left[a_1e^{i\beta x} + a_2e^{-i\beta x}\right] + a_3e^{m_3x} + \cdots + a_ne^{m_nx} \\ &= e^{\alpha x}\left[a_1(\cos\beta x + i\sin\beta x) + a_2(\cos\beta x - i\sin\beta x)\right] + a_3e^{m_3x} + \cdots + a_ne^{m_nx} \\ &= e^{\alpha x}\left[b_1\cos\beta x + b_2\sin\beta x\right] + a_3e^{m_3x} + \cdots + a_ne^{m_nx}, \end{aligned}$$

where $b_1 = a_1 + a_2;\ b_2 = i(a_1 - a_2)$.

(iv) If the auxiliary equation has repeated pair of imaginary roots $\alpha \pm i\beta$, then the general solution of the equation (3) can be obtained using (ii) and (iii) as

$$y = [(c_1 + xc_2)\cos\beta x + (c_3 + xc_4)\sin\beta x]e^{ax} + e_5 e^{5x} + \cdots + c_n e^{m_n x}.$$

ILLUSTRATIVE EXAMPLES

1. Solve: $[D^2 - (a+b)D + ab]y = 0.$

Solution

Let $y = e^{mx}$ be a trial solution, then the auxiliary equation is

$$m^2 - (a+b)m + ab = 0.$$

It's roots are $m = a, b$. Hence the general solution is

$$y = Ae^{ax} + Be^{bx}, \text{ where } A, B \text{ are arbitrary constants.}$$

2. Solve: $[D^2 - 6D + 9]y = 0.$

Solution

Let $y = e^{mx}$ be a trial solution, then the auxiliary equation is

$$m^2 - 6m + 9 = 0.$$

Its roots are $m = 3,3$. Hence the general solution is

$$y = (A + Bx)e^{3x}, \text{ where } A, B \text{ are arbitrary constants.}$$

3. Solve: $[D^2 - 2mD + (m^2 + n^2)]y = 0.$

Solution

Let $y = e^{kx}$ be a trial solution, then the auxiliary equation is

$$k^2 - 2km + (m^2 + n^2) = 0.$$

Its roots are $k = m \pm in$. Hence the general solution is

$$y = e^{mx}(A\cos nx + B\sin nx), \text{ where } A, B \text{ are arbitrary constants.}$$

4. Solve: $[D^2 + kD + \mu]y = 0$ if $k^2 < 4\mu$ and $n^2 = \mu - \frac{1}{4}k^2.$

Solution

Let $y = e^{mx}$ be a trial solution, then the auxiliary equation is

$$m^2 + km + \mu = 0.$$

Solving, we get $\quad m = \dfrac{-k \pm \sqrt{k^2 - 4\mu}}{2} = \dfrac{-k \pm 2in}{2}.$

So, the general solution is

$$y = e^{-\frac{kx}{2}}[A\cos nx + B\sin nx], \text{ where } A, B \text{ are two arbitrary constants.}$$

5. Solve: $(D^2 + p^2)v = 0$, given that $v(l) = 0$, $v(x_0) = v_0$.

Solution

Let $v = e^{mx}$ be a trial solution, then the auxiliary equation is

$$m^2 + p^2 = 0.$$

Solving, we get $m = \pm ip$. So, the general solution is

$$v = A\cos px + B\sin px, \text{ where } A, B \text{ are two arbitrary constants.}$$

Using the given conditions $v(l) = 0$, $v(x_0) = v_0$, we get

$$0 = A\cos pl + B\sin pl$$

and
$$v_0 = A\cos px_0 + B\sin px_0.$$

Solving we get

$$v_0 = A\cos px_0 - A\cot pl \sin px_0$$

or,
$$A = \frac{v_0 \sin pl}{\sin p(l - x_0)}.$$

So,
$$B = -\frac{v_0 \cos pl}{\sin p(l - x_0)}.$$

Using these values of A, B in the general solution, we get

$$v = \frac{v_0 \sin pl}{\sin pl(l - x_0)} \cos px - \frac{v_0 \cos pl}{\sin p(l - x_0)} \sin px = \frac{v_0 \sin p(l - x)}{\sin p(l - x_0)}.$$

6. Solve: $[D^3 - 2D^2 - 5D + 6]y = 0$.

Solution

Let $y = e^{mx}$ be a trial solution, then the auxiliary equation is

$$m^3 - 2m^2 - 5m + 6 = 0.$$

Solving we get $m = -2, 1, 3$. So the general solution is

$$y = Ae^{-2x} + Be^x + Ce^{3x}, \text{ where } A, B, C \text{ are three arbitrary constants.}$$

7. Solve: $[D^4 - 4D^3 + 3D^2 + 4D - 4]y = 0$.

Solution

Let $y = e^{mx}$ be a trial solution, then the auxiliary equation is

$$m^4 - 4m^3 + 3m^2 + 4m - 4 = 0$$

or,
$$m^4 - 4m^3 + 4m^2 - m^2 + 4m - 4 = 0$$

or,
$$(m^2 - 1)(m - 2)^2 = 0.$$

So, roots of the auxiliary equation are $m = \pm 1, 2, 2$. So the general solution is

$$y = (A + Bx)e^{2x} + Ce^x + Ee^{-x}, \text{ where } A, B, C, E \text{ are four arbitrary constants.}$$

8. Solve: $(D^2-1)(D^2-9)(D^2+2D-3)y=0.$

Solution

Let $y=e^{mx}$ be a trial solution, then the auxiliary equation is

$$(m^2-1)(m^2-9)(m^2+2m-3)=0$$

or, $$(m-1)^2(m+3)^2(m+1)(m-3)=0.$$

So, roots of the auxiliary equation are $m-3,-3,-1,1,1,3$. So the general solution is

$y=(A+Bx)e^{-3x}+Ce^{-x}+(E+Fx)e^{x}+Ge^{3x}$, where A, B, C, E, F, G are arbitrary constants.

9. Solve: $(D^3+1)y=0.$

Solution

Let $y=e^{mx}$ be a trial solution, then the auxiliary equation is $m^3+1=0$ and its solutions are

$$m=-1,\tfrac{1}{2}(1\pm i\sqrt{3}).$$

So, the general solution is

$y=c_1e^{-x}+e^{\frac{x}{2}}\left[c_2\cos\frac{\sqrt{3}x}{2}+c_3\sin\frac{\sqrt{3}x}{2}\right]$, where c_1, c_2, c_3 are arbitrary constants.

EXERCISE

- Solve the following equations:

1. $[2D^2-3D+1]y=0,$

2. $[D^2+4D+4]y=0,$

3. $[D^2+8D+25]y=0,$

4. $[D^3-6D^2+11D-6]y=0,$

5. $[D^3+2D^2-5D-6]y=0,$

6. $[D^3-3D^2+4]y=0,$

7. $[D^3-5D^2+8D-4]y=0,$

8. $[D^4+2D^3+3D^2+2D+1]y=0,$

9. $(D^4+6D^2+8)y=0,$

10. $[D^4-1]y=0,$

11. $(D^4+1)y=0,$

12. $(D^4+4D^3-5D^2-36D-36)y=0,$

13. $(D^4-7D^3+18D^2-20D+8)y=0,$

14. $D^2y-3Dy+2y=0$, given that $y=Dy=0$ when $x=0$,

15. Solve $l\dfrac{d^2\theta}{dt^2}+g\theta=0$, given that $\theta=\alpha, \dfrac{d\theta}{dt}=0$ when $t=0$.

16. Solve $\dfrac{d^2x}{dt^2}+n^2x=0$, given that $x=\alpha, \dfrac{dx}{dt}=u$ when $t=0$. Also find the maximum value of x.

17. Solve $\left(D^4 - n^4\right)y = 0$ completely, if $Dy = y = 0$ when $x = 0$ and $x = l$, and prove that $y = A(\cos nx - \cosh nx) + B(\sin nx - \sinh nx)$ and $\cosh nl \cos nl = 1$.

[Hints: From $y = A(\cos nx - \cosh nx) + B(\sin nx - \sinh x)$, we get

$$Dy = -An(\sin nx + \sin nx) + Bn(\cos nx - \cosh nx).$$

using $Dy = 0 = y$ when $x = l$, we get

$$A(\cos nl - \cosh nl) + B(\sin nl - \sinh nl) = 0,$$

$$-A(\sin nl + \sinh nl) + B(\cos nl - \cosh nl) = 0.$$

Eliminating, A, B we get $\cosh nl \cos nl = 1$].

ANSWERS

1. $y = c_1 e^x + c_2 e^{\frac{x}{2}}$, **2.** $y = (c_1 + xc_2)e^{-2x}$,

3. $y = e^{-4x}(c_1 \cos 3x + c_2 \sin 3x)$, **4.** $y = c_1 e^x + c_2 e^{2x} + c_3 e^{3x}$,

5. $y = c_1 e^{-x} + c_2 e^{2x} + c_3 e^{-3x}$, **6.** $y = (c_1 + xc_2)e^{2x} + c_3 e^{-x}$,

7. $y = c_1 e^x + (c_2 + xc_3)e^{2x}$,

8. $y = e^{-\frac{x}{2}}\left[(c_1 + xc_2)\cos\frac{\sqrt{3}x}{2} + (c_3 + xc_4)\sin\frac{\sqrt{3}x}{2}\right]$,

9. $y = c_1 \cos 2x + c_2 \sin 2x + c_3 \cos\sqrt{2}x + c_4 \sin\sqrt{2}x$,

10. $y = c_1 \cos x + c_2 \sin x + c_3 \cosh x + c_4 \sinh x$,

11. $y = e^{\frac{x}{\sqrt{2}}}\left[c_1 \cos\frac{x}{\sqrt{2}} + c_2 \sin\frac{x}{\sqrt{2}}\right] + e^{-\frac{x}{\sqrt{2}}}\left[c_3 \cos\frac{x}{\sqrt{2}} + c_4 \sin\frac{x}{\sqrt{2}}\right]$.

12. $y = c_1 e^{-3x} + c_2 e^{3x} + (c_3 + xc_4)e^{-2x}$,

13. $y = c_1 e^x + (c_2 + xc_3 + x^2 c_4)e^{2x}$,

14. $y = 0$, **15.** $\theta = \alpha \cos\left(t\sqrt{\frac{g}{l}}\right)$,

16. $x = a\cos nt + \frac{u}{n}\sin nt,\ x_{\text{max}} = \sqrt{a^2 + \frac{u^2}{n^2}}$.

6.4 Particular Integral

Consider an n-th order linear differential equation

$$f(D)y = X.$$

If $g(x)$ be a general solution of the associated homogeneous equation, then we have $f(D)\,[g(x)] = 0$ and if $h(x)$ be a solution of the above equation, then $f(D)\,[h(x)] = X$. Now

$$f(D)[g(x) + h(x)] = f(D)[g(x)] + f(D)[h(x)] = 0 + X = X.$$

This shows that $g(x) + h(x)$ is also a solution of the equation $f(D)y = X$, i.e., the general solution of a non-homogeneous linear differential equation consists of two parts:

(i) The first part, which is the general solution of associated homogeneous equation, i.e., $g(x)$ is called its complementary function.

(ii) The second part, which is due to the presence of the function X, is called *particular integral* (P.I.) and is given by $\frac{1}{f(D)}X$.

Again, if $\phi(x), \psi(x)$ be two solutions of the equation $f(D)y = X$, then we have

$$f(D)[\phi(x)] = X \text{ and } f(D)[\psi(x)] = X.$$

Subtracting, we get

$$f(D)[\phi(x) - \psi(x)] = 0,$$

i.e., $\phi(x) - \psi(x)$ is a solution of the associated homogeneous equation.

6.5 Rules for Determining Particular Integral

For certain forms of X, formulae for evaluation of P.I. are as follows:

Rule I: If X is of the form e^{ax}, P.I. $= \frac{1}{F(D)}X = \frac{1}{F(a)}X$, provided $F(a) \neq 0$. If $F(a) = 0$, then apply rule IV.

Illustration: If we take $F(D) = D^n + c_1D^{n-1} + c_2D^{n-2} + \cdots + c_{n-1}D + c_n$, then

$$\begin{aligned} F(D)e^{ax} &= [D^n + c_1D^{n-1} + c_2D^{n-2} + \cdots + c_{n-1}D + c_n]e^{ax} \\ &= [a^n + c_1a^{n-1} + c_2a^{n-2} + \cdots + c_{n-1}a + c_n]e^{ax} = F(a)e^{ax}. \end{aligned}$$

Now $e^{ax} = \frac{1}{F(D)}[F(D)e^{ax}] = \frac{1}{F(D)}[F(a)e^{ax}] = F(a)\frac{1}{F(D)}e^{ax}$

or, $\frac{1}{F(D)}e^{ax} = \frac{1}{F(a)}e^{ax}$.

Rule II. If X is of the form $\sin ax$ or $\cos ax$, P.I. $= \frac{1}{F(D^2)}X = \frac{1}{F(-a^2)}X$, provided $F(-a^2) \neq 0$.

Illustration: Since $D^{2n}\sin ax = (-a^2)\sin ax$ for integral values of n, we see that

$$F(D^2)\sin ax = F(-a^2)\sin ax.$$

So, $\sin ax = \frac{1}{F(D^2)}[F(D^2)\sin ax] = \frac{1}{F(D^2)}[F(-a^2)\sin ax] = F(-a^2)\frac{1}{F(D^2)}[\sin ax]$.

or, $\frac{1}{F(D^2)}[\sin ax] = \frac{1}{F(-a^2)}\sin ax$, provided $F(-a^2) \neq 0$.

Rule III: If X is of the form x^m,

P.I. $= \frac{1}{F(D)}X = [a_0 + a_1D + a_2D^2 + \cdots + a_mD^m]X, a_0 \neq 0$ where $\frac{1}{F(D)}$ is expanded in ascending powers of D and neglecting all terms after D^m.

Illustration: It is easy to note that

$$(D-a)x^m = mx^{m-1} - ax^m,$$

$$(D-a)\frac{m}{a}x^{m-1} = \frac{m(m-1)}{a}x^{m-2} - mx^{m-1},$$

$$(D-a)\frac{m(m-1)}{a^2}x^{m-2} = \frac{m(m-1)(m-2)}{a^2}x^{m-3} - \frac{m(m-1)}{a}x^{m-2},$$

....

$$(D-a)\frac{m(m-1)\cdots 3\cdot 2}{a^{m-1}}x = \frac{m!}{a^{m-1}} - \frac{m(m-1)\cdots 3\cdot 2}{a^{m-2}}x,$$

$$(D-a)\frac{m!}{a^m} = -\frac{m!}{a^{m-1}}.$$

Now adding all these, we get

$$(D-a)\left[x^m + \frac{m}{a}x^{m-1} + \frac{m(m-1)}{a^2}x^{m-2} + \cdots + \frac{m(m-1)\cdots 3\cdot 2}{a^{m-1}}x + \frac{m!}{a^m}\right] = -ax^m \quad (1)$$

Again, $\frac{1}{D-a}x^m = -\frac{1}{a}\left(1-\frac{D}{a}\right)^{-1}x^m$

$$= -\frac{1}{a}\left[1+\frac{1}{a}D+\frac{1}{a^2}D^2+\cdots+\frac{1}{a^m}D^m+\cdots\right]x^m$$

$$= -\frac{1}{a}\left[x^m + \frac{m}{a}x^{m-1} + \frac{m(m-1)}{a^2}x^{m-2} + \cdots + \frac{m(m-1)\cdots 3\cdot 2}{a^{m-1}}x + \frac{m!}{a^m}\right]$$

or, $(D-a)\left[x^m + \frac{m}{a}x^{m-1} + \frac{m(m-1)}{a^2}x^{m-2} + \cdots + \frac{m(m-1)\cdots 3\cdot 2}{a^{m-1}}x + \frac{m!}{a^m}\right] = -ax^m,$

which is same as (1).

Rule IV. If X is of the form $e^{ax}V(x)$, P.I.$=\frac{1}{F(D)}e^{ax}V(x) = e^{ax}\frac{1}{F(D+a)}V(x)$.

Illustration: We have

$D(e^{ax}U) = e^{ax}DU + Uae^{ax} = e^{ax}(D+a)U$, where U is a function of x.

$$D^2(e^{ax}U) = D\left[e^{ax}(D+a)U\right]$$

$$= ae^{ax}(D+a)U + e^{ax}D(D+a)U = e^{ax}(D+a)^2U.$$

Let $D^m(e^{ax}U) = e^{ax}(D+a)^mU$, where m is a positive integer, then

$$D^{m+1}(e^{ax}U) = D\left[D^m(e^{ax}U)\right] = D\left[e^{ax}(D+a)^mU\right]$$

$$= ae^{ax}(D+a)^mU + e^{ax}D(D+a)^mU = e^{ax}(D+a)^{m+1}U.$$

Thus by method of induction,

$D^n(e^{ax}U) = e^{ax}(D+a)^nU$ for all positive integer n.

Hence, we have $F(D)(e^{ax}U) = e^{ax}F(D+a)U$.

Taking $U = \dfrac{1}{F(D+a)}V$, i.e., $F(D+a)U = V$, we see that

$$F(D)\left(e^{ax}\frac{1}{F(D+a)}V\right) = e^{ax}V,$$

or, $\dfrac{1}{F(D)}(e^{ax}V) = e^{ax}\dfrac{1}{F(D+a)}V.$

Note: If $F(D) = (D-a)^p\phi(D)$ and $\varphi(a) \neq 0$

$$\frac{1}{F(D)}e^{ax} = \frac{1}{(D-a)^p\varphi(D)}e^{ax}$$

$$= \frac{1}{(D-a)^p\varphi(a)}e^{ax} = \frac{1}{\varphi(a)}\frac{1}{(D-a)^p}e^{ax}$$

$$= \frac{1}{\varphi(a)}e^{ax}\frac{1}{(D+a-a)^p}1 = \frac{e^{ax}}{\varphi(a)}\frac{x^p}{p!}.$$

Rule V. If X is of the form $\sin ax$ or $\cos ax$ and $F(-a^2) = 0$, then write

$$Z = \frac{1}{F(D^2)}\cos ax, \; Y = \frac{1}{F(D^2)}\sin ax.$$

So, $$Z + iY = \frac{1}{F(D)^2}e^{iax} = e^{iax}\frac{1}{F(D+ia)^2}1 = P(x) + iQ(x).$$

Equating both sides we get the P.I.

Rule VI. If X is of the form $xV(x)$, P.I. $= \dfrac{1}{F(D)}xV(x) = \left[x - \dfrac{F'(D)}{F(D)}\right]\dfrac{1}{F(D)}V(x).$

Illustration: We have

$$D(xU) = xDU + U, \text{ where } U \text{ is a function of } x.$$

$$= xDU + \left[\frac{d}{dD}D\right]U.$$

$$D^2(xU) = D[xDU + U] = DU + xD^2U + DU$$

$$= xD^2U + 2DU = xD^2U + \left[\frac{d}{dD}D^2\right]U.$$

$$D^3(xU) = D\left[xD^2U + 2DU\right] = D^2U + xD^3U + 2D^2U$$

$$= xD^3U + 3D^2U = xD^3U + \left[\frac{d}{dD}D^3\right]U.$$

Proceeding in this way, we get $D^n(xU) = xD^nU + \left[\dfrac{d}{dD}D^n\right]U.$

Hence, $F(D)(xU) = xF(D)U + \left[\dfrac{d}{dD}F(D)\right]U = xF(D)U + F'(D)U.$

Taking $U = \frac{1}{F(D)}V$, i.e., $F(D)U = V$, we get

$$F(D)\left[x\frac{1}{F(D)}V\right] = xV + \frac{F'(D)}{F(D)}V$$

or, $$\frac{1}{F(D)}xV = x\frac{1}{F(D)}V - \frac{1}{F(D)}\frac{F'(D)}{F(D)}V = \left[x - \frac{F'(D)}{F(D)}\right]\frac{1}{F(D)}V.$$

Rule VII. If X is of the form $x^k V(x)$, where k is a positive integer, then

P.I. $= \frac{1}{F(D)}x^k V(x) = \left[x - \frac{F'(D)}{F(D)}\right]^k \frac{1}{F(D)}V(x)$, where k is a positive integer.

Illustration: Using $V = xU$ in the rule VI, we get

$$\frac{1}{F(D)}x^2 U = \frac{1}{F(D)}xV = \left[x - \frac{F'(D)}{F(D)}\right]\frac{1}{F(D)}V$$

$$= \left[x - \frac{F'(D)}{F(D)}\right]\left[x - \frac{F'(D)}{F(D)}\right]\frac{1}{F(D)}U$$

$$= \left[x - \frac{F'(D)}{F(D)}\right]^2 \frac{1}{F(D)}U.$$

Again, taking $U = xW$ and using the rule VI, we get

$$\frac{1}{F(D)}x^3 W = \frac{1}{F(D)}x^2 U = \left[x - \frac{F'(D)}{F(D)}\right]^2 \frac{1}{F(D)}U$$

$$= \left[x - \frac{F'(D)}{F(D)}\right]^2 \frac{1}{F(D)}xW$$

$$= \left[x - \frac{F'(D)}{F(D)}\right]^2 \left[x - \frac{F'(D)}{F(D)}\right]\frac{1}{F(D)}W$$

$$= \left[x - \frac{F'(D)}{F(D)}\right]^3 \frac{1}{F(D)}W.$$

Proceeding in this way we get

$$\frac{1}{F(D)}x^k V(x) = \left[x - \frac{F'(D)}{F(D)}\right]^k \frac{1}{F(D)}V(x),$$ where k is a positive integer.

WORKED-OUT EXAMPLES

1. Solve: $[D^2 - D]y = e^{2x}$.

Solution

Let $y = e^{mx}$ be a trial solution of the equation $[D^2 - D]y = 0$. So the auxiliary equation is

$$m^2 - m = 0.$$

Solving, we get $m = 0, 1$. Hence the C.F. is $A + Be^x$. Now

$$\text{P.I.} = \frac{1}{D^2 - D} e^{2x} = \frac{1}{2^2 - 2} e^{2x} = \frac{1}{2} e^{2x}.$$

Hence the general solution is

$$y = A + Be^x + \frac{1}{2} e^{2x}, \text{ where } A, B \text{ are arbitrary constants.}$$

2. Solve: $[D^2 + D - 6]\, y = x$.

Solution

Let $y = e^{mx}$ be a trial solution of the equation $[D^2 + D - 6]\, y = 0$. So the auxiliary equation is

$$m^2 + m - 6 = 0.$$

Solving, we get $m = -3, 2$. Hence the C.F. is $Ae^{-3x} + Be^{2x}$. Now

$$\text{P.I.} = \frac{1}{D^2 + D - 6} x = -\frac{1}{6}\left[1 - \frac{D^2 + D}{6}\right]^{-1} x = -\frac{1}{6}\left[1 + \frac{D}{6}\right] x = -\frac{1}{6} x - \frac{1}{36}.$$

So, the general solution of the given equation is

$$y = Ae^{-3x} + Be^{2x} - \frac{1}{6} x - \frac{1}{36}, \text{ where } A, B \text{ are arbitrary constants.}$$

3. Solve: $[D^2 - a^2]\, y = \sin ax$.

Solution

In this case, the auxiliary equation is $m^2 - a^2 = 0$. Its solutions are $m = \pm a$. So the C.F. is $y = Ae^{ax} + Be^{-ax}$.

Now, the P.I. $= \dfrac{1}{D^2 - a^2} \sin ax = \dfrac{1}{-a^2 - a^2} \sin ax = -\dfrac{1}{2a^2} \sin ax.$

So, the general solution of the given equation is

$$y = Ae^{ax} + Be^{-ax} - \frac{1}{2a^2} \sin ax, \text{ where } A, B \text{ are arbitrary constants.}$$

4. Solve: $[D^2 - 6D + 9]\, y = \sin x$.

Solution

Let $y = e^{mx}$ be a trial solution of the equation $[D^2 - 6D + 9]\, y = 0$. So the auxiliary equation is

$$m^2 - 6m + 9 = 0.$$

Solving we get $m = 3, 3$. Hence the C.F. is $(A + xB)\, e^{3x}$. Now,

$$\begin{aligned}
\text{P.I.} &= \frac{1}{D^2 - 6D + 9} \sin x \\
&= \frac{1}{-1^2 - 6D + 9} \sin x = \frac{1}{8 - 6D} \sin x \\
&= (8 + 6D) \frac{1}{64 - 36D^2} \sin x = \frac{8 + 6D}{64 + 36} \sin x \\
&= \frac{1}{50}(4 \sin x + 3 \cos x).
\end{aligned}$$

So the general solution is

$$y=(A+xB)e^{3x}+\frac{1}{50}(4\sin x+3\cos x),\text{ where } A,\ B \text{ are arbitrary constants.}$$

5. Solve: $(D-3)^3y = e^{3x}$.

Solution

Let $y = e^{mx}$ be a trial solution of the equation $(D-3)^3y = 0$. So the auxiliary equation is

$$(m-3)^3 = 0.$$

Solving we get $m = 3, 3, 3$. Hence the C.F. is $(A + xB + x^2C)\, e^{3x}$. Now,

$$\text{P.I.} =\frac{1}{(D-3)^3}e^{3x}$$

$$=e^{3x}\frac{1}{(D+3-3)^3}1=e^{3x}\frac{1}{D^3}1=\frac{x^3}{6}e^{3x}.$$

So, the general solution is

$$y=(A+xB+x^2C)e^{3x}+\frac{x^3}{6}e^{3x},\text{ where } A,\ B,\ C \text{ are arbitrary constants.}$$

6. Solve: $[D^2 - 1]\, y = \cosh x$.

Solution

Let $y = e^{mx}$ be a trial solution of the equation $[D^2 - 1]\, y = 0$. So the auxiliary equation is

$$m^2 - 1 = 0.$$

Solving we get $m = \pm 1$. Hence the C.F. is $Ae^x + Be^{-x}$. Now,

$$\text{P.I.} =\frac{1}{D^2-1}\cosh x=\frac{1}{2}\frac{1}{D-1}\frac{1}{D+1}e^x+\frac{1}{2}\frac{1}{D+1}\frac{1}{D-1}e^{-x}$$

$$=\frac{1}{4}\frac{1}{D-1}e^x-\frac{1}{4}\frac{1}{D+1}e^{-x}=\frac{1}{4}e^x\frac{1}{D+1-1}1-\frac{1}{4}e^{-x}\frac{1}{D-1+1}1$$

$$=\frac{x}{4}\left(e^x-e^{-x}\right)=\frac{x}{2}\sinh x.$$

So the general solution is $y=Ae^x+Be^{-x}+\frac{x}{2}\sinh x$, where A, B are arbitrary constants.

7. Solve: $[D^2 - 5D + 6]\, y = e^{3x}$.

Solution

Here the A.E. is $m^2 - 5m + 6 = 0$, whose solution are $m = 3, 2$. So, the C.F. is

$$Ae^{2x}+Be^{3x}.$$

Now, $$\text{P.I.}=\frac{1}{D-3}\frac{1}{D-2}e^{3x}=\frac{1}{D-3}\frac{1}{3-2}e^{3x}$$

$$=e^{3x}\frac{1}{D+3-3}1=xe^{3x}.$$

Hence, the general solution is

$$y=Ae^{2x}+Be^{3x}+xe^{3x},\text{ where } A,\ B \text{ are arbitrary constants.}$$

8. Solve: $[D^2-1]\,y = xe^{2x}$.

Solution

Here the C.F. is $Ae^x + Be^{-x}$. Now,

$$\text{P.I.} = \frac{1}{D^2-1}xe^{2x} = e^{2x}\frac{1}{(D+2)^2-1}x = e^{2x}\frac{1}{3}\left[1+\frac{D^2+4D}{3}\right]^{-1}x$$

$$= e^{2x}\frac{1}{3}\left[1-\frac{4D}{3}\right]x = \frac{1}{9}[3x-4]e^{2x}.$$

So, the general solution is

$$y = Ae^x + Be^{-x} + \frac{1}{9}[3x-4]e^{2x}, \text{ where } A, B \text{ are arbitrary constants.}$$

9. Solve: $[D^2 + a^2]\,y = \sin ax$.

Solution

Here the A.E. is $m^2 + a^2 = 0$, so the C.F. is $A\cos ax + B\sin ax$. Now, let

$$X = \frac{1}{D^2+a^2}\cos ax \text{ and } Y = \frac{1}{D^2+a^2}\sin ax.$$

Then $$X+iY = \frac{1}{D^2+a^2}e^{iax} = e^{iax}\frac{1}{(D+ia)^2+a^2}1$$

$$= e^{iax}\frac{1}{D^2+2iaD}1 = e^{iax}\frac{1}{2iaD}\left[1+\frac{D}{2ia}\right]^{-1}1$$

$$= e^{iax}\frac{1}{2iaD}\left[1-\frac{D}{2ia}\right]1 = e^{iax}\frac{1}{2iaD}1$$

$$= \frac{x}{2ia}(\cos ax + i\sin ax) = \frac{x}{2a}(\sin ax - i\cos ax).$$

Comparing both sides, we get

$$\text{P.I.} = \frac{1}{D^2+a^2}\sin ax = -\frac{x}{2a}\cos ax.$$

Hence, the general solution is

$$y = A\cos ax + B\sin ax - \frac{x}{2a}\cos ax, \text{ where } A, B \text{ are arbitrary constants.}$$

10. Solve: $[D^2-1]\,y = xe^x\sin x$.

Solution

Here, the C.F. is $Ae^x + Be^{-x}$. Now,

$$\text{P.I.} = \frac{1}{D^2-1}xe^x\sin x$$

$$= e^x\frac{1}{(D+1)^2-1}x\sin x = e^x\frac{1}{D^2+2D}x\sin x$$

$$= e^x\left[x-\frac{2(D+1)}{D^2+2D}\right]\frac{1}{-1+2D}\sin x = e^x\left[x-\frac{2(D+1)}{D^2+2D}\right]\frac{2D+1}{4D^2-1}\sin x$$

$$= -\frac{e^x}{5}\left[x - \frac{2(D+1)}{D^2+2D}\right](2D+1)\sin x = -\frac{e^x}{5}\left[x - \frac{2(D+1)}{D^2+2D}\right](2\cos x + \sin x)$$

$$= -\frac{e^x}{5}\left[x(2\cos x + \sin x) + \frac{2(2D^2+3D+1)}{5}(2\cos x + \sin x)\right]$$

$$= -\frac{e^x}{25}[5x(2\cos x + \sin x) + 2(\cos x - 7\sin x)]$$

$$= -\frac{e^x}{25}[(10x+2)\cos x + (5x-14)\sin x]$$

Hence, the general solution is

$$y = Ae^x + Be^{-x} - \frac{e^x}{25}[(10x+2)\cos x + (5x-14)\sin x],$$ where A, B are abritrary constants.

11. Solve: $D^2y = x^2 \sin x$.

Solution

Here the C.F. is $A + Bx$. Now,

$$\text{P.I.} = \left[x - \frac{2}{D}\right]^2 \frac{1}{D^2}\sin x$$

$$= -\left[x - \frac{2}{D}\right]\left[x - \frac{2}{D}\right]\sin x = -\left[x - \frac{2}{D}\right](x\sin x + 2\cos x)$$

$$= -x(x\sin x + 2\cos x) + 2(-x\cos x + 3\sin x) = (6 - x^2)\sin x - 4x\cos x.$$

Hence, the general solution is

$$y = A + Bx + (6 - x^2)\sin x - 4x\cos x.$$

12. Solve: $(D^3 + 1)\,y = \cos x$.

Solution

❑ First consider the equation $(D^3 + 1)\,y = 0$. The auxiliary equation of this equation is $m^3 + 1 = 0$ and its solutions are $m = -1, \frac{1}{2}(1 \pm i\sqrt{3})$. So the C.F. is

$$y = c_1e^{-x} + e^{\frac{x}{2}}\left[c_2\cos\frac{\sqrt{3}x}{2} + c_3\sin\frac{\sqrt{3}x}{2}\right],$$ where c_k's = 1, 2, 3, are arbitrary constants.

Now the P.I. is

$$\frac{1}{D^3+1}\cos x = \frac{1}{-D+1}\cos x = \frac{1+D}{1-D^2}\cos x = \frac{1}{2}(1+D)\cos x = \frac{1}{2}(\cos x - \sin x).$$

So the general solution is

$$y = c_1e^{-x} + e^{\frac{x}{2}}\left[c_2\cos\left(\frac{\sqrt{3}x}{2}\right) + c_3\sin\left(\frac{\sqrt{3}x}{2}\right)\right] + \frac{1}{2}(\cos x - \sin x)$$

13. Solve: $(D^4 + 6D^2 + 8)\, y = x^6$.

Solution

- Hence the auxiliary equation of the equation $(D^4 + 6D^2 + 8)\, y = 0$ is $m^4 + 6m^2 + 8 = 0$ whose roots are $m = \pm i\sqrt{2}, \pm 2i$. So the C.F. is

$$c_1 \cos 2x + c_2 \sin 2x + c_3 \cos \sqrt{2}x + c_4 \sin \sqrt{2}x.$$ where $c_k's,\ k = 1, 2, 3, 4$, are arbitrary constants.

Now, P.I.

$$\frac{1}{D^4 + 6D^2 + 8} x^6 = \frac{1}{8}\left[1 + \frac{D^4 + 6D^2}{8}\right]^{-1} x^6 = \frac{1}{8}\left[1 - \frac{D^4 + 6D^2}{8} + \frac{D^8 - 15D^6 + 36D^2}{64} - \dots\right] x^6$$

$$= \frac{1}{32}\left[4x^6 - 90x^4 + 630x^2 - 675\right].$$

So the general solution is

$$y = c_1 \cos 2x + c_2 \sin 2x + c_3 \cos \sqrt{2}x + c_4 \sin \sqrt{2}x + \frac{1}{32}\left[4x^6 - 90x^4 + 63x^2 - 675\right].$$

14. Solve: $(D^3 - 5D^2 + 8D - 4)\, y = e^{2x} + e^x + 3e^{-x}$.

Solution

The auxiliary equation is

$$m^3 - 5m^2 + 8m - 4 = 0$$

and its roots are $m = 1, 2, 2$. So the C.F. is

$$c_1 e^x + (c_2 + xc_3) e^{2x}.$$

where $c_k' s, k = 1, 2, 3$, are arbitrary constants.

Now the P.I. $= \dfrac{1}{(D-1)(D-2)^2}\left[e^{2x} + e^x + 3e^{-x}\right] = \dfrac{1}{(D-2)^2} e^{2x} + \dfrac{1}{D-1} e^x - \dfrac{e^{-x}}{6}$

$$= e^{2x} \frac{1}{D^2} 1 + e^x \frac{1}{D} 1 - \frac{e^{-x}}{6}$$

$$= \frac{1}{2} x^2 e^{2x} + xe^x - \frac{e^{-x}}{6}$$

So the general solution is

$$y = c_1 e^x + (c_2 + xc_3) e^{2x} + \frac{1}{2} x^2 e^{2x} + xe^x - \frac{e^{-x}}{6}$$

15. Solve: $(D^4 + 1)\, y = \cos 2x \sin x$.

Solution

The auxiliary equation is $m^4 + 1 = 0$ and its solution are $m = \pm \dfrac{1 \pm i}{\sqrt{2}}$. So the C.F. is

$$e^{\frac{x}{\sqrt{2}}}\left[c_1 \cos \frac{x}{\sqrt{2}} + c_2 \sin \frac{x}{\sqrt{2}}\right] + e^{-\frac{x}{\sqrt{2}}}\left[c_3 \cos \frac{x}{\sqrt{2}} + c_4 \sin \frac{x}{\sqrt{2}}\right]$$ where $c_k's,\ k = 1, 2, 3, 4$, are arbitrary constants.

Now the

$$\text{P.I.} = \frac{1}{D^4+1}\cos 2x \sin x = \frac{1}{2}\frac{1}{D^4+1}(\sin 3x - \sin x) = \frac{1}{164}(\sin 3x - 41\sin x)$$

So the general solution is

$$y = e^{\frac{x}{\sqrt{2}}}\left[c_1 \cos\frac{x}{\sqrt{2}} + c_2 \sin\frac{x}{\sqrt{2}}\right] + e^{-\frac{x}{\sqrt{2}}}\left[c_3 \cos\frac{x}{\sqrt{2}} + c_4 \sin\frac{x}{\sqrt{2}}\right] + \frac{1}{164}(\sin 3x - 41\sin x)$$

16. Solve: $(D^3 - 2D^2 + 9D - 18)y = 6\cos 3x + 48\sin 3x.$

Solution

The auxiliary equation is

$$(m^2 + 9)(m - 2) = 0$$

and its solutions are $m = 2, \pm 3i$. So the C.F. is

$$y = c_1 e^{2x} + c_2 \cos 3x + c_3 \sin 3x.$$

where $c_k' s, k = 1, 2, 3,$ are arbitrary constants.

$$\text{Now the P.I.} = \frac{1}{(D^2+9)(D-2)}(6\cos 3x + 48\sin 3x)$$

$$= \frac{D+2}{(D^2+9)(D^2-4)}(6\cos 3x + 48\sin 3x)$$

$$= -\frac{D+2}{13(D^2+9)}(6\cos 3x + 48\sin 3x)$$

$$= -\frac{6}{(D^2+9)}(2\cos 3x + \sin 3x)$$

$$= -6\left[\frac{x}{3}\sin 3x - \frac{x}{6}\cos 3x\right] = x[\cos 3x - 2\sin 3x]$$

Hence the general solution is

$$y = c_1 e^{2x} + c_2 \cos 3x + c_3 \sin 3x + x[\cos 3x - 2\sin 3x]$$

17. Solve: $(D^3 + 3D - 4)y = xe^{-2x}.$

Solution

The auxiliary equation is

$$m^3 + 3m^2 - 4 = 0 \quad (m-1)(m+2)^2 = 0$$

and its solutions are $m = 1, -2, -2$. So the C.F. is

$y = c_1 e^x + (c_2 + xc_3)e^{-2x}$, where $c_k' s, k = 1, 2, 3,$ are arbitrary constants.

$$\text{Now, P.I.} = \frac{1}{(D-1)(D+2)^2}xe^{-2x} = e^{-2x}\frac{1}{(D-3)D^2}x = -e^{-2x}\frac{1}{3D^2}\left[1 - \frac{D}{3}\right]^{-1}x$$

$$= -e^{-2x}\frac{1}{3D^2}\left(x + \frac{1}{3}\right) = -\frac{e^{-2x}}{18}(x^3 + x^2 + xc_4 + c_5).$$

So the general solution is

$$y = c_1e^x + (c_2 + xc_3)e^{-2x} - \frac{e^{-2x}}{18}(x^3 + x^2 + xc_4 + c_5)$$

$$= c_1e^x + (d_2 + xd_3)e^{-2x} - \frac{e^{-2x}}{18}(x^3 + x^2).$$

18. Solve: $(D^4 - 2D^3 + D^2)\,y = x^3$.

Solution

Here the auxiliary equation is

$$m^2(m^2 - 2m + 1) = 0,$$

and its solutions are $m = 0, 0, 1, 1$. So the C.F. is

$$y = c_1 + xc_2 + (c_3 + xc_4)e^x.$$

where $c_k's$, $k = 1, 2, 3, 4$ are arbitrary constants.

Now,

$$\text{P.I.} = \frac{1}{D^2(1-D)^2}x^3 = \frac{1}{D^2}(1-D)^{-2}x^3 = \frac{1}{D^2}\left[1 + 2D + 3D^2 + 4D^3 + \cdots\right]x^3$$

$$= \frac{1}{D^2}(x^3 + 6x^2 + 18x + 24) = \frac{x^5}{20} + \frac{x^4}{2} + 3x^3 + 12x^2 + xc_4 + c_5.$$

So the general solution is

$$y = c_1 + xc_2 + (c_3 + xc_4)e^x + \frac{x^4}{20} + \frac{x^4}{2} + 3x^3 + 12x^2 + xc_4 + c_5$$

$$= d_1 + xd_2 + (c_3 + xc_4)e^x + \frac{x^5}{20} + \frac{x^4}{2} + 3x^3 + 12x^2.$$

19. Solve: $(D^3 + 3D^2 + 2D)\,y = x^2$.

Solution

Here the auxiliary equation is $m\,(m^2 + 3m + 2) = 0$.
Solving we get $m = -2, -1, 0$. So the complementary function is

$$y = A + Be^{-x} + Ce^{-2x},$$

where A, B, C are arbitrary constants. Now the particular integral is

$$\frac{1}{D^3 + 3D^2 + 2D}x^2$$

$$= \frac{1}{2D}\left(1 + \frac{D^2 + 3D}{2}\right)^{-1}x^2 = \frac{1}{2D}\left(1 - \frac{D^2 + 3D}{2} + \frac{9D^2}{4} + \cdots\right)x^2$$

$$= \frac{1}{2D}\left(1 - \frac{3D}{2} + \frac{7D^2}{4} + \cdots\right)x^2 = \frac{1}{2D}\left(x^2 - 3x + \frac{7}{2}\right)$$

$$= \frac{1}{2}\left(\frac{x^3}{3} - \frac{3}{2}x^2 + \frac{7}{2}x\right).$$

So the general solution is

$$y = A + Be^{-x} + Ce^{-2x} + \frac{1}{12}(2x^3 - 9x^2 + 21x).$$

20. Solve: $(D^3-7D-6)y=e^{2x}(1+x)$.

Solution

Here the auxiliary equation is $m^3-7m-6=0$. Solving we get $m=-2,-1,3$. So the complementary function is $y=Ae^{3x}+Be^{-x}+Ce^{-2x}$, where A, B, C are arbitrary constants.
Now the particular integral is

$$\frac{1}{D^3-7D-6}e^{2x}(1+x)$$

$$=\frac{1}{(D+2)(D+1)(D-3)}e^{2x}(1+x)=-\frac{1}{12}e^{2x}+e^{2x}\frac{1}{(D+4)(D+3)(D-1)}x$$

$$=-\frac{1}{12}e^{2x}-\frac{1}{12}e^{2x}\left(1+\frac{5}{12}D+\cdots\right)x=-\frac{1}{12}e^{2x}-\frac{1}{12}e^{2x}\left(x+\frac{5}{12}\right)$$

$$=-\frac{17}{144}e^{2x}-\frac{1}{12}xe^{2x}.$$

So the general solution is

$$y=Ae^{3x}+Be^{-x}+Ce^{-2x}-\frac{1}{12}e^{2x}\left(x+\frac{17}{12}\right).$$

21. Solve: $(D^3-1)y=\cos^2\frac{x}{2}$.

Solution

Here the auxiliary equation is $m^3-1=0$. Solving we get $m=1,\frac{1\pm i\sqrt{3}}{2}$.
So the complementary function is

$$y=Ae^x+e^{\frac{x}{2}}\left[B\cos\left(\frac{\sqrt{3}}{2}x\right)+C\sin\left(\frac{\sqrt{3}}{2}x\right)\right].$$

where A, B, C are arbitrary constants.
Now the particular integral is

$$=\frac{1}{D^3-1}\cos^2\frac{x}{2}$$

$$=-\frac{1}{2}(1-D^3)^{-1}1-\frac{1}{2}(1-D^3)^{-1}\cos x.$$

$$=-\frac{1}{2}-\frac{1}{2(1+D)}\cos x$$

$$=-\frac{1}{2}-\frac{1-D}{2(1-D^2)}\cos x=-\frac{1}{2}-\frac{1}{4}(\cos x+\sin x).$$

So the general solution is

$$y=Ae^x+e^{\frac{x}{2}}\left[B\cos\left(\frac{\sqrt{3}}{2}x\right)+C\sin\left(\frac{\sqrt{3}}{2}x\right)\right]-\frac{1}{2}-\frac{1}{4}(\cos x+\sin x).$$

22. Solve: $(D-2)^2 (D-3)^2 y = x^2$.

Solution

Here the auxiliary equation is

$$(m-2)^2 (m-3)^2 = 0.$$

Solving we get $m = 2, 2, 3, 3$. So the complementary function is

$$y = (A+Bx)e^{2x} + (C+Dx)e^{3x}.$$

where A, B, C, D are arbitrary constants.

Now the particular integral is

$$= \frac{1}{(D-2)^2(D-3)^2} x^2 = \frac{1}{36}\left[1 + \frac{D^2-5D}{6}\right]^{-2} x^2$$

$$= \frac{1}{36}\left(1 - 2\frac{D^2-5D}{6} + 3\frac{25D^2}{36} + \cdots\right)x^2 = \frac{1}{216}(6x^2+20x+21).$$

So the general solution is

$$y = (A+Bx)e^{2x} + (C+Dx)e^{3x} + \frac{1}{216}(6x^2+20x+21).$$

23. Solve: $(D^5 - D)\,y = e^x + \sin x - x$.

Solution

Here the auxiliary equation is $m(m^4-1) = 0$. Solving we get $m = 0, \pm 1, \pm i$. So the complementary function is

$$y = A + Be^x + Ce^{-x} + D\cos x + E\sin x,$$

where A, B, C, D, E are arbitrary constants. Now the particular integral is

$$\frac{1}{D^5 - D}(e^x + \sin x - x)$$

$$= \frac{1}{(D-1)D(D+1)(D^2+1)}e^x + \frac{D}{(D^2+1)D^2(D^2-1)}\sin x + \frac{1}{D(1-D^4)}x$$

$$= \frac{1}{4(D-1)}e^x + \frac{1}{2(D^2+1)}\cos x + \frac{1}{D}(1+D^4+\cdots)x$$

$$= \frac{e^x}{4}\frac{1}{D}1 + \frac{x\sin x}{4} + \frac{1}{D}x = \frac{x(e^x+\sin x+2x)}{4}.$$

So the general solution is

$$y = A + Be^x + Ce^{-x} + D\cos x + E\sin x + \frac{x(e^x+\sin x+2x)}{4}.$$

24. Solve: $(D^2+1)^2 y = x^2 \cos x$.

Solution

Here the auxiliary equation is $(m^2+1)^2 = 0$.

Solving we get $m = \pm i, \pm i$. So the complementary function is

$$y = (A+Bx)\cos x + (C+Dx)\sin x,$$

where A, B, C, D are arbitrary constants.

Now the particular integral is

$$\frac{1}{(D^2+1)^2}x^2\cos x$$

$$= \text{Real part of } \frac{1}{(D^2+1)^2}x^2e^{ix} = \text{Real part of } e^{ix}\frac{1}{((D+i)^2+1)^2}x^2$$

$$= \text{Real part of } e^{ix}\frac{1}{(D^2+2iD)^2}x^2 = -\frac{1}{4}\text{Real part of } e^{ix}\frac{1}{D^2}\left(1+\frac{D}{2i}\right)^{-2}x^2$$

$$= -\frac{1}{4}\text{Real part of } e^{ix}\frac{1}{D^2}\left(1+iD-\frac{3}{4}D^2+\cdots\right)x^2$$

$$= -\frac{1}{4}\text{Real part of } e^{ix}\frac{1}{D^2}\left(x^2+2ix-\frac{3}{2}\right)$$

$$= -\frac{1}{4}\text{Real part of } e^{ix}\left(\frac{x^4}{12}+\frac{ix^3}{3}-\frac{3}{4}x^2\right)$$

$$= -\frac{1}{4}\left(\frac{x^4}{12}-\frac{3}{4}x^2\right)\cos x+\frac{x^3}{12}\sin x.$$

So the general solution is

$$y=(A+Bx)\cos x+(C+Dx)\sin x-\frac{1}{4}\left(\frac{x^4}{12}-\frac{3}{4}x^2\right)\cos x+\frac{x^3}{12}\sin x.$$

25. Solve: $(D^2 + a^2)\, y = \tan ax$.

Solution

Here the C.F. is $A\cos ax + B\sin ax$. Now,

$$\text{P.I.} = \frac{1}{D^2+a^2}\tan ax = \frac{1}{2ai}\left(\frac{1}{D-ai}-\frac{1}{D+ai}\right)\tan ax.$$

Now, let $z=\frac{1}{D-ai}\tan ax$.

Then $Dz - aiz = \tan ax$.

This is a linear differential equation whose I.F. $e^{-ai\int dx} = e^{-iax}$. Multiplying the equation by e^{-iax}, we get

$$z = e^{iax}\int e^{-iax}\tan ax\,dx$$

$$= e^{iax}\int(\cos ax - i\sin ax)\tan ax\,dx = e^{iax}\int(\sin ax + i\cos ax - i\sec ax)\,dx$$

$$= ie^{iax}\int(e^{-iax}-\sec ax)\,dx = -\frac{1}{a}-\frac{i}{a}e^{iax}\ln\tan\left(\frac{\pi}{4}+\frac{ax}{2}\right).$$

Similarly, $$\frac{1}{D+ai}\tan ax = -\frac{1}{a}+\frac{i}{a}e^{-iax}\ln\tan\left(\frac{\pi}{4}+\frac{ax}{2}\right).$$

So, $$\frac{1}{D^2+a^2}\tan ax = -\frac{1}{2a^2}(e^{iax}+e^{-iax})\ln\tan\left(\frac{\pi}{4}+\frac{ax}{2}\right)$$

$$= -\frac{1}{a^2}\cos ax\ln\tan\left(\frac{\pi}{4}+\frac{ax}{2}\right).$$

So the general solution is

$$y = A\cos ax + B\sin ax - \frac{1}{a^2}\cos ax \ln\tan\left(\frac{\pi}{4} + \frac{ax}{2}\right).$$

EXERCISE

- Solve the following differential equations

1. $(D^2 - 5D + 6)\,y = x^2$,
2. $(D^2 + 4)\,y = 2x + 3$,
3. $(D^2 - 3D + 2)\,y = e^{3x}$,
4. $(D^2 - 3D + 2)\,y = e^{2x}$,
5. $(D - 2)^2 y = 2e^{2x}$,
6. $(D^2 - 3D + 2)\,y = \sin 3x$,
7. $(D^2 - D - 2)\,y = \sin 2x$,
8. $(D^2 - 2D + 5)\,y = \sin x$,
9. $(D^2 + 1)\,y = \sin x \sin 2x$,
10. $(D^2 + 4)\,y = \sin 2x$,
11. $(D^2 + 1)\,y = xe^{2x}$,
12. $(D^2 - 1)\,y = e^x(1 + x^2)$,
13. $(D^2 + 2D + 2)\,y = xe^{-x}$,
14. $(D + 2)^2 y = 4\sinh 2x$,
15. $(D^2 - 1)\,y = x^2\cos x$,
16. $(D^2 - 4)\,y = \cos^2 x$,
17. $(D + 1)^2 y = x\cos x$,
18. $(D - 1)^2 y = xe^x \sin x$,
19. $(D^2 + 3D + 2)\,y = e^{e^x}$,
20. $(D^2 - 1)\,y = \cosh x \cos x$,

[*Hints*: P.I. $= \frac{1}{2}\frac{1}{(D+1)(D-1)}e^x\cos x + \frac{1}{2}\frac{1}{(D+1)(D-1)}e^{-x}\cos x$

$= \frac{1}{2}e^x\frac{1}{D^2 + 2D}\cos x + \frac{1}{2}e^{-x}\frac{1}{D^2 - 2D}\cos x$

$= \frac{1}{2}e^x\frac{2D+1}{4D^2 - 1}\cos x - \frac{1}{2}e^{-x}\frac{2D-1}{4D^2 - 1}\cos x$

$= -\frac{1}{10}e^x(\cos x - 2\sin x) - \frac{1}{10}e^{-x}(2\sin x + \cos x)$]

21. $(D^2 + 1)\,y = \sin x \sin 2x$,
22. $(D^2 + 1)\,y = \cos x \sin 3x$,
23. $(D^2 - 2D)\,y = e^{2x} - e^x$,
24. $(D^2 + a^2)\,y = \sec ax$,
25. $(D^2 + a^2)\,y = \operatorname{cosec} ax$,
26. $(D^2 + a^2)\,y = \cot ax$,
27. $(D^2 - a^2)\,y = \cosh ax$,
28. $(D^2 - 3D + 2)\,y = \sin e^{-x}$,
29. $(D^2 - 9D + 18)\,y = e^{e^{-3x}}$,
30. $(D^2 - 4)\,y = x\sin x$,
31. $(D^2 - 1)\,y = x^2 \sin x$,
32. $(D^2 + 4)\,y = x\sin^2 x$,

33. $D(D^2+3D+2)y = x^2$,

34. $D(D^2-1)y = e^x + e^{-x}$,

$$\left[\text{P.I.} = \frac{1}{(D-1)D(D+1)}e^x + \frac{1}{(D+1)D(D-1)}e^{-x} = \frac{1}{2}\frac{1}{D-1}e^x + \frac{1}{2}\frac{1}{D+1}e^{-x}\right.$$
$$\left. = \frac{1}{2}e^x\frac{1}{D}1 + \frac{1}{2}e^{-x}\frac{1}{D}1 = x\cosh x\right]$$

35. $(D^3+1)y = \cos 2x$,

36. $D(D^2+a^2)y = \sin ax$,

37. $(D^3+2D^2+D)y = e^{2x}$,

38. $(D^3+D^2-D-1)y = \cosh x$,

$$\left[\textit{Hints}\text{: P.I.} = \frac{1}{2}\frac{1}{(D-1)(D+1)^2}e^x + \frac{1}{2}\frac{1}{(D+1)^2(D-1)}e^{-x}\right.$$
$$\left. = \frac{1}{8}\frac{1}{D-1}e^x - \frac{1}{4}\frac{1}{(D+1)^2}e^{-x} = \frac{1}{8}e^x\frac{1}{D}1 - \frac{1}{4}e^{-x}\frac{1}{D^2}1\right]$$

39. $(D^3+D^2-D-1)y = \cos 2x$,

40. $(D^3-5D^2+7D-3)y = e^{2x}\cosh x$,

$$\left[\textit{Hints}\text{: P.I.} = \frac{1}{2}\frac{1}{(D-3)(D-1)^2}e^{3x} + \frac{1}{2}\frac{1}{(D-1)^2(D-3)}e^x\right.$$
$$\left. = \frac{1}{8}\frac{1}{D-3}e^{3x} - \frac{1}{4}\frac{1}{(D-1)^2}e^x = \frac{1}{8}e^{3x}\frac{1}{D}1 - \frac{1}{4}e^x\frac{1}{D^2}1\right]$$

41. $(D^3-3D^2+3D-1)y = xe^x + e^x$,

$$\left[\textit{Hints}\text{: P.I.} = \frac{1}{(D-1)^3}xe^x + \frac{1}{(D-1)^3}e^x = e^x\frac{1}{D^3}x + e^x\frac{1}{D^3}1\right]$$

42. $(D^3-5D^2+8D-4)y = e^{2x} + e^x + 3e^{-x}$,

$$\left[\textit{Hints}\text{: P.I.} = \frac{1}{(D-1)(D-2)^2}e^x + \frac{1}{(D-2)^2(D-1)}e^{2x} - \frac{1}{6}e^{-x}\right.$$
$$= \frac{1}{D-1}e^x + \frac{1}{(D-2)^2}e^{2x} - \frac{1}{6}e^{-x}$$
$$\left. = e^x\frac{1}{D}1 + e^{2x}\frac{1}{D^2}1 - \frac{1}{6}e^{-x} = xe^x + \frac{x^2}{2}e^{2x} - \frac{1}{6}e^{-x}\right]$$

43. $(D^3-2D^2-5D+6)y = (e^{2x}+3)^2 + e^{3x}\cosh x$.

44. $(D^3-3D^2+4)y = xe^{2x}$

45. $(D^3+1)y = e^{\frac{x}{2}}\sin\frac{\sqrt{3}x}{2}$,

46. $D^2(D-1)^2 y = x,$

[*Hints*: $\text{P.I.} = \dfrac{1}{D^2(1-D)^2}x = \dfrac{1}{D^2}(1+2D+3D^2+\cdots)x = \dfrac{1}{D^2}(x+2) = \dfrac{x^3}{6}+x^2$]

47. $(D^4-2D^3+D^2)y = x^2$

48. $(D^4-4D^3+8D^2-8D+4)y = e^x$

49. $(D^4+D^3-3D^2-5D-2)y = 3xe^{-x}$

50. $(D^4-8D)y = x^2+e^{3x}$

51. $(D^4-4D^3+6D^2-4D+1)y = e^{2x}\cos x$.

52. $(D^4+2D^3+3D^2+2D+1)y = xe^x$

53. $(D^4+10D^2+9)y = \cos(2x+3),$

54. $(D^3-3D^2-6D+8)y = xe^{-3x},$

55. $(D^4-1)y = e^x\cos x,$

56. $(D^4+2D^2+1)y = x^2\cos x,$

57. $(D-1)^2(D^2+1)^2 y = e^x + x.$

ANSWERS

1. $y = c_1e^{2x} + c_2e^{3x} + \frac{1}{6}\left(x^2+\frac{5x}{3}+\frac{19}{18}\right)$

2. $y = c_1\cos 2x + c_2\sin 2x + \frac{1}{4}(2x+3),$

3. $y = c_1e^{2x} + c_2e^x + \frac{1}{2}e^{3x},$

4. $y = c_1e^{2x} + c_2e^x + xe^{2x},$

5. $y = (c_1+xc_2)e^{2x} + x^2e^{2x},$

6. $y = c_1e^{2x} + c_2e^x + \frac{1}{130}(9\cos 3x - 7\sin 3x),$

7. $y = c_1e^{2x} + c_2e^{-x} + \frac{1}{20}(\cos 2x - 3\sin 2x),$

8. $y = (c_1\cos 2x + c_2\sin 2x)e^x + \frac{1}{10}(2\sin x + \cos x),$

9. $y = c_1\cos x + c_2\sin x + \frac{1}{4}x\sin x + \frac{1}{16}\cos 3x,$

10. $y = c_1\cos 2x + c_2\sin 2x - \frac{1}{4}x\cos 2x,$

11. $y = c_1\cos x + c_2\sin x + \frac{1}{25}(5x-4)e^{2x},$

12. $y = c_1e^{-x} + c_2e^x + \frac{1}{12}xe^x(2x^2-3x+9),$

13. $y = (c_1\cos x + c_2\sin x + x)e^{-x},$

14. $y = (c_1+xc_2)e^{-2x} + \frac{1}{8}e^{2x} - x^2e^{-2x}$

15. $y = c_1 e^{-x} + c_2 e^{2x} + \frac{1}{2}(1 - x^2)\cos x + x\sin x,$

16. $y = c_1 e^{-2x} + c_2 e^{2x} - \frac{1}{16}(2 + \cos 2x),$

17. $y = (c_1 + xc_2)e^{-x} + \frac{1}{2}[(x-1)\sin x + \cos x],$

18. $y = (c_1 + xc_2)e^{x} - (2\cos x + x\sin x)e^{x},$

19. $y = c_1 e^{-2x} + c_2 e^{-x} + e^{-2x} e^{e^{x}},$

20. $y = c_1 e^{x} + c_2 e^{-x} + \frac{2}{5}\sin x \sinh x - \frac{1}{5}\cos x \cosh x,$

21. $y = c_1 \cos x + c_2 \sin x + \frac{x}{4}\sin x + \frac{1}{16}\cos 3x,$

22. $y = c_1 \cos x + c_2 \sin x - \frac{1}{30}\sin 4x - \frac{1}{6}\sin 2x,$

23. $y = c_1 + \left(c_2 + \frac{x}{2}\right)^2 e^{2x} + e^{x},$

24. $y = c_1 \cos ax + c_2 \sin ax + \frac{x}{a}\sin ax + \frac{1}{a^2}\cos ax \ln \cos ax,$

25. $y = c_1 \cos ax + c_2 \sin ax - \frac{x}{a}\cos ax + \frac{1}{a^2}\sin ax \ln \sin ax,$

26. $y = c_1 \cos ax + c_2 \sin ax + \frac{1}{a^2}\sin ax \ln \tan\left(\frac{ax}{2}\right),$

27. $y = c_1 e^{ax} + c_2 e^{-ax} + \frac{x}{2a}\sinh ax,$

28. $y = c_1 e^{x} + c_2 e^{-x} - e^{2x}\sin e^{-x},$

29. $y = c_1 e^{2x} + c_2 e^{6x} + \frac{1}{9}e^{6x} e^{e^{-3x}},$

30. $y = c_1 \cos 2x + c_2 \sin 2x + \frac{x}{3}\sin x - \frac{2}{9}\cos x,$

31. $y = c_1 e^{x} + c_2 e^{-x} - x\cos x - \frac{1}{2}(x^2 - 1)\sin x,$

32. $y = c_1 \cos 2x + c_2 \sin 2x + \frac{x}{8}\left(1 - \frac{x}{2}\sin 2x - \frac{1}{4}\cos 2x\right),$

33. $y = c_0 + c_1 e^{-x} + c_2 e^{-2x} + \frac{1}{6}x^3 - \frac{3}{4}x^2 + \frac{7}{4}x,$

34. $y = c_0 + c_1 e^{x} + c_2 e^{-x} + x\cosh x,$

35. $y = c_1 e^{-x} + e^{\frac{x}{2}}\left[c_2 \cos\frac{\sqrt{3}x}{2} + c_3 \sin\frac{\sqrt{3}x}{2}\right] + \frac{1}{65}(\cos 2x - 8\sin 2x),$

36. $y = c_0 + c_1 \sin ax + c_2 \cos ax - \frac{x}{2a^2} \sin ax,$

37. $y = c_1 + (c_2 + xc_3)e^{-x} + \frac{e^{2x}}{18},$

38. $y = c_1 e^x + (c_2 + xc_3)e^{-x} + \frac{xe^x - x^2 e^{-x}}{8},$

39. $y = c_1 e^x + (c_2 + xc_3)e^{-x} - \frac{1}{25}(2\sin 2x + \cos 2x),$

40. $y = (c_1 + xc_2)e^x + c_3 e^{3x} + \frac{x}{8}e^{3x} - \frac{x^2}{8}e^x,$

41. $y = (c_1 + xc_2 + x^2 c_3)e^x + \frac{1}{24}(x+4)x^3 e^x,$

42. $y = c_1 e^x + (c_2 + xc_3)e^{2x} + \frac{1}{2}x^2 e^{2x} + xe^x - \frac{e^{-x}}{6},$

43. $y = c_1 e^x + c_2 e^{3x} + c_3 e^{-2x} - \frac{13}{8}e^{2x} + \frac{3}{2} + \frac{1}{12}e^{4x},$

44. $y = (c_1 + xc_2)e^{2x} + c_3 e^{-x} + \frac{(x-1)x^2 e^{2x}}{18}$

45. $y = c_1 e^x + e^{-\frac{x}{2}}\left[c_2 \cos\frac{\sqrt{3}x}{2} + c_3 \sin\frac{\sqrt{3}x}{2}\right] - \frac{x}{6}e^{\frac{x}{2}}\left[\sqrt{3}\cos\frac{\sqrt{3}x}{2} + \sin\frac{\sqrt{3}x}{2}\right]$

46. $y = c_1 + xc_2 + (c_3 + xc_4)e^x + x^2 + \frac{1}{6}x^3,$

47. $y = c_1 + xc_2 + (c_3 + xc_4)e^x + \frac{x^4}{12} + \frac{2}{3}x^3 + 3x^2,$

48. $y = e^x[(c_1 + xc_2)\cos x + (c_3 + xc_4)\sin x + 1]$

49. $y = c_1 e^{2x} + (c_2 + xc_3 + x^2 c_4)e^{-x} - \frac{e^{-x}}{72}(3x^4 + 4x^3)$.

50. $y = c_1 + c_2 e^{2x} + e^{-x}[c_3 \cos\sqrt{3}x + c_4 \sin\sqrt{3}x] - \frac{x^4}{24} + \frac{e^{3x}}{57},$

51. $y = (c_1 + xc_2 + x^2 c_3 + x^3 c_4)e^x - \frac{1}{4}e^{2x}\cos x$

52. $y = e^{-\frac{x}{2}}\left[(c_1 + xc_2)\cos\frac{\sqrt{3}x}{2} + (c_3 + xc_4)\sin\frac{\sqrt{3}x}{2}\right] + \frac{(x-2)e^x}{9}$

53. $y = c_1 \cos x + c_2 \sin x + c_3 \cos 3x + c_4 \sin 3x - \frac{1}{15}\cos(2x+3)$

54. $y = c_1 e^x + c_2 e^{-2x} + c_3 e^{4x} - \frac{(28x+39)e^{-3x}}{784},$

55. $y = c_1 e^x + c_2 e^{-x} + c_3 \cos x + c_4 \sin x - \frac{e^x}{5}\cos x,$

56. $y=(c_1+xc_2)\cos x+(c_3+xc_4)\sin x-\frac{1}{4}\left(\frac{x^4-9x^2}{12}\cos x-\frac{x^3}{3}\sin x\right)$,

57. $y=(c_1+xc_2)\cos x+(c_3+xc_4)\sin x+(c_5+xc_6)e^x+\frac{x^2e^x}{8}+x+2.$

6.6 Method of Variation of Parameter

In Chapter 2, we have already described the method of variation of parameter to solve a first order linear differential equation. In fact, method of variation of parameter is a powerful method to find complete primitive of a linear differential equation of any order, if its complementary function is known. Here we consider a second order linear differential equation only, but the method can be applied to linear differential equation of any order. Consider the following second order linear differential equation

$$y''+P(x)y'+Q(x)y=R(x), \tag{i}$$

where $P(x)$, $Q(x)$ need not be constants. To find a particular integral this method can be applied in all the cases regardless the nature of the coefficients $P(x)$, $Q(x)$ and the right-hand side term $R(x)$, if the complementary function is known.

Assume the general solution of $y''+P(x)y'+Q(x)y=0$ as

$$y=c_1y_1(x)+c_2y_2(x),$$

where c_1, c_2 are constants and $y_1(x)$, $y_2(x)$ are two linearly independent solution of the above equation.

In this method, we replace c_1, c_2 by $v_1(x)$, $v_2(x)$. Our problem is to find $v_1(x)$, $v_2(x)$ so that

$$y_p=v_1y_1(x)+v_2y_2(x)$$

satisfies (i). This will be a particular solution of (i). To determine $v_1(x)$, $v_2(x)$ we need two equations. Differentiating (i), we get

$$y_p'=v_1'y_1(x)+v_2'y_2(x)+v_1y_1'(x)+v_2y_2'(x). \tag{ii}$$

We shall choose $v_1(x)$, $v_2(x)$ in such a way that

$$v_1'y_1(x)+v_2'y_2(x)=0. \tag{iii}$$

Again differentiating the above results, we get

$$y_p''=v_1'y_1'(x)+v_2'y_2'(x)+v_1y_1''(x)+v_2y_2''(x). \tag{iv}$$

Putting the values of y_p'', y_p', y_p in (i), we get

$$v_1'y_1'(x)+v_2'y_2'(x)=R(x). \tag{v}$$

Solving (iii) and (v), we get

$$v_1'(x)=-\frac{y_2(x)R(x)}{W(x)} \text{ and } v_2'(x)=\frac{y_1(x)R(x)}{W(x)},$$

where $W(x)=y_2'y_1-y_1'y_2\neq 0$ as y_1, y_2 are independent. Integrating, we get

$$v_1(x)=-\int\frac{y_2(x)R(x)}{W(x)}dx \text{ and } v_2=\int\frac{y_1(x)R(x)}{W(x)}dx.$$

Hence, we obtain a particular solution of (i) as

$$y_p = v_1(x)y_1(x) + v_2(x)y_2(x).$$

So, the general solution is

$$y = [c_1 + v_1(x)]y_1(x) + [c_2 + v_2(x)]y_2(x).$$

For the third order linear differential equation

$$y''' + P(x)\,y'' + Q(x)y' + R(x)\,y = S(x). \tag{vi}$$

Let the complementary function be

$$y = Ay_1(x) + By_2(x) + Cy_3(x),$$

where A, B, C are arbitrary constants and $y_1(x)$, $y_2(x)$, $y_3(x)$ are three linearly independent solutions of the equation (vi).

So we assume the particular integral of the equation as

$$y = v_1(x)\,y_1(x) + v_2(x)\,y_2(x) + v_3(x)\,y_3(x). \tag{vii}$$

We choose $v_1(x)$, $v_2(x)$, $v_3(x)$ in such a way that

$$v_1'(x)\,y_1(x) + v_2'(x)\,y_2(x) + v_3'(x)\,y_3(x) = 0, \tag{viii}$$

i.e.,
$$y' = v_1(x)\,y_1'(x) + v_2(x)\,y_2'(x) + v_3(x)\,y_3'(x). \tag{ix}$$

Differentiating (ix) and setting

$$v_1'(x)\,y_1'(x) + v_2'(x)\,y_2'(x) + v_3'(x)\,y_3'(x) = 0, \tag{x}$$

we get
$$y'' = v_1(x)\,y_1''(x) + v_2(x)\,y_2''(x) + v_3(x)\,y_3''(x). \tag{xi}$$

Again differentiating (xi) with respect to x we get

$$\begin{aligned} y''' &= v_1(x)\,y_1'''(x) + v_2(x)\,y_2'''(x) + v_3(x)\,y_3'''(x) \\ &\quad + v_1'(x)\,y_1''(x) + v_2'(x)\,y_2''(x) + v_3'(x)\,y_3''(x). \end{aligned} \tag{xii}$$

Using the results (vii), (ix), (xii) in the equation (vi), we obtain

$$v_1'(x)\,y_1''(x) + v_2'(x)\,y_2''(x) + v_3'(x)\,y_3''(x) = S(x) \tag{xiii}$$

Solving the equations (viii), (xiii), we get the values of $v_1'(x), v_2'(x), v_3'(x)$ and integrating these, we get $v_1(x), v_2(x), v_3(x)$. So the general solution of the equation (vi) is

$$y = \{A + v_1(x)\}\,y_1(x) + \{B + v_2(x)\}\,y_2(x) + \{C + v_3(x)\}\,y_3(x).$$

Thus in the method of variation of parameter, we first find n linearly independent solutions of the associated homogeneous equation, i.e., the complementary function of the given linear equation of nth order, then replace the arbitrary constants occurring in the complementary function by unknown functions to write the particular integral. Now differentiating the particular integral successively $(n - 1)$ times and equating the terms associated to the derivatives of the unknown functions to zero each time and finally setting the terms associated to the derivatives of the unknown functions to right-hand side member of the equation, we get n equation, in n functions $v_1'(x), v_2'(x), \cdots, v_n'(x)$, solving which we get the values of the functions and hence the particular integral of the given nth order linear differential equation.

WORKED-OUT EXAMPLES

1. Using the method of variation of parameter, solve the equation $y'' + y = \operatorname{cosec} x$.

Solution

Let $y = e^{mx} \neq 0$, be a trial solution of the equation

$$y'' + y = 0; \tag{1}$$

then the auxiliary equation is $m^2 + 1 = 0$.

Solving, we get $m = \pm i$. So, C.F. is $c_1 \sin x + c_2 \cos x$. Since $\sin x$ and $\cos x$ are two linearly independent solutions of (1), we can assume the particular solution of the given equation as

$$y_p = v_1(x)\sin x + v_2(x)\cos x.$$

Then $$y_p' = v_1'(x)\sin x + v_2'(x)\cos x + v_1(x)\cos x - v_2(x)\sin x.$$

Next we choose $v_1(x), v_2(x)$ in such a way that

$$v_1'(x)\sin x + v_2'(x)\cos x = 0 \tag{2}$$

so that $$y_p' = v_1(x)\cos x - v_2(x)\sin x$$

and again differentiating we get

$$y_p'' = v_1'(x)\cos x - v_2'(x)\sin x - v_1(x)\sin x - v_2(x)\cos x$$

$$= v_1'(x)\cos x - v_2'(x)\sin x - y_p,$$

i.e., $$y_p'' + y_p = v_1'(x)\cos x - v_2'(x)\sin x = \operatorname{cosec} x. \tag{3}$$

Solving (2) and (3), we get

$$v_1'(x) = \cot x, \; v_2'(x) = -1.$$

This on integration gives

$$v_1(x) = \log|\sin x|, \; v_2(x) = -x.$$

Hence $$y_p = \sin x \log|\sin x| - x\cos x.$$

So, the complete primitive is

$$y = \{c_1 + \log|\sin x|\}\sin x + (c_2 - x)\cos x.$$

2. Using the method of variation of parameter solve the equation $y'' + a^2y = \tan ax$.

Solution

Let $y = e^{mx} \neq 0$, be a trial solution of the equation

$$y'' + a^2 y = 0; \tag{1}$$

then the auxiliary equation is

$$m^2 + a^2 = 0.$$

Solving, we get $m = \pm ai$. So, C.F. is $c_1 \sin ax + c_2 \cos ax$. Since $\sin ax$ and $\cos ax$ are two linearly independent solutions of (1), we can assume the particular solution of the given equation as

$$y_p = v_1(x)\cos ax + v_2(x) \sin ax.$$

Then $\qquad y_p' = v_1'(x)\cos ax + v_2'(x)\sin ax - av_1(x)\sin ax + av_2(x)\cos ax.$

Next we choose $v_1(x), v_2(x)$ in such a way that

$$v_1'(x)\cos ax + v_2'(x)\sin ax = 0 \tag{2}$$

so that $\qquad y_p' = -av_1(x)\sin ax + av_2(x)\cos ax,$

and again differentiating, we get

$$y_p'' = -av_1'(x)\sin ax + av_2'(x)\cos ax - a^2[v_1(x)\cos ax + v_2(x)\sin ax],$$

i.e., $\qquad y_p'' + a^2 y_p = -av_1'(x)\sin ax + av_2'(x)\cos ax = \tan ax. \qquad (3)$

Solving (2) and (3), we get

$$v_1'(x) = -\frac{1}{a}\frac{\sin^2 ax}{\cos ax};\ v_2'(x) = \frac{1}{a}\sin ax,$$

which on integration give

$$v_1(x) = -\frac{1}{a^2}\ln|\sec ax + \tan ax| + \frac{1}{a^2}\sin ax,\ v_2(x) = -\frac{1}{a^2}\cos ax.$$

Hence,

$$y_p = \left[-\frac{1}{a^2}\ln|\sec ax + \tan ax| + \frac{1}{a^2}\sin ax\right]\cos ax - \frac{1}{a^2}\cos ax \sin ax$$

$$= -\frac{1}{a^2}\cos ax \ln|\sec ax + \tan ax|.$$

So the complete primitive is

$$y = c_1 \cos ax + c_2 \sin ax - \frac{1}{a^2}\cos ax \ln|\sec ax + \tan ax|.$$

3. Using the method of variation of parameter, solve the equation $y'' - 6y' + 9y = \dfrac{e^{3x}}{x^2}$.

Solution

Let $y = e^{mx} \neq 0$ be a trial solution of the equation

$$y'' - 6y' + 9y = 0\,; \tag{1}$$

then the auxiliary equation is

$$m^2 - 6m + 9 = 0.$$

Solving, we get m = 3, 3, So, C.F. is $(xc_1 + c_2)e^{3x}$. Since xe^{3x}, e^{3x} are two linearly independent solutions of (1), we can assume the particular solution of the given equation as

$$y = v_1(x)e^{3x} + xv_2(x)e^{3x}.$$

Then $\quad y_p' = v_1'(x)e^{3x} + xv_2'(x)e^{3x} + [3v_1(x) + v_2(x)]e^{3x} + 3xv_2(x)e^{3x}.$

Next we choose $v_1(x), v_2(x)$ in such a way that

$$v_1'(x)e^{3x} + xv_2'(x)e^{3x} = 0 \tag{2}$$

so that $\quad y_p' = [3v_1(x) + v_2(x)]e^{3x} + 3xv_2(x)e^{3x}$

and again differentiating, we get

$$y_p'' = [3v_1'(x) + v_2'(x)]e^{3x} + [9v_1(x) + 6v_2(x)]e^{3x} + 9xv_2(x)e^{3x} + 3xv_2'(x)e^{3x}.$$

Now using the values of y_p, y_p' and y_p'' in the given equation, we get

$$[3v_1'(x) + v_2'(x)]e^{3x} + 3xv_2'(x)e^{3x} = \frac{e^{3x}}{x^2}. \tag{3}$$

Solving (2) and (3), we get

$$v_1'(x) = -\frac{1}{x}; \quad v_2'(x) = \frac{1}{x^2},$$

which on integration give

$$v_1(x) = -\ln x; \quad v_2(x) = -\frac{1}{x}.$$

Hence, $\quad y_p = -[1 + \ln x]e^{3x}.$

So, the complete primitive is

$$y = (xc_1 + c_3)e^{3x} - e^{3x}\ln x.$$

4. Using the method of variation of parameter solve the equation $y'' - y = \dfrac{2}{1+e^x}$.

Solution

Let $y = e^{mx} \neq 0$ be a trial solution of the equation

$$y'' - y = 0; \tag{1}$$

then the auxiliary equation is

$$m^2 - 1 = 0.$$

Solving, we get $m = \pm 1$. So, C.F. is $c_1e^x + c_2e^{-x}$. Since e^x, e^{-x} are two linearly independent solutions of (1), we can assume the particular solution of the given equation as

$$y_p = v_1(x)e^x + v_2(x)e^{-x}.$$

Then $\quad y_p' = v_1'(x)e^x + v_2'(x)e^{-x} + v_1(x)e^x - v_2(x)e^{-x}.$

Next we choose $v_1(x), v_2(x)$ in such a way that

$$v_1'(x)e^x + v_2'(x)e^{-x} = 0 \tag{2}$$

so that $\quad y_p' = v_1(x)e^x - v_2(x)e^{-x}$

and again differentiating, we get

$$y_p'' = v_1'(x)e^x - v_2'(x)e^{-x} + v_1(x)e^x + v_2(x)e^{-x}.$$

Now using the values of y_p and y_p'' in the given equation, we get

$$v_1'(x)e^x - v_2'(x)e^{-x} = \frac{2}{1+e^x}. \tag{3}$$

Solving (2) and (3), we get

$$v_1'(x) = \frac{e^{-x}}{1+e^x},\ v_2'(x) = -\frac{e^x}{1+e^x},$$

which on integration give

$$v_1(x) = -e^{-x} + \ln(1+e^{-x}),\ v_2(x) = -\ln(1+e^x).$$

Hence,

$$y_p = -1 + e^x \ln(1+e^{-x}) - e^{-x}\ln(1+e^x).$$

So the complete primitive is

$$y = c_1 e^x + c_2 e^{-x} - 1 + e^x \ln(1+e^{-x}) - e^{-x}\ln(1+e^x).$$

5. Solve: $y_2 - 2y_1 = e^x \cos x$ by the method of variation of parameter.

Solution

Let $y = e^{mx} \neq 0$ be a trial solution of the equation

$$y_2 - 2y_1 = 0; \tag{1}$$

then the auxiliary equation is

$$m^2 - 2m = 0.$$

Solving, we get m = 0, 2. So, C.F. is $c_2 e^{2x} + c_1$. Since $e^{2x}, 1$ are two linearly independent solutions of (1), we can assume the particular solution of the given equation as

$$y_p = v_2(x)e^{2x} + v_1(x).$$

Then $$y_p' = v_1'(x) + v_2'(x)e^{2x} + 2v_2(x)e^{2x}.$$

Now we choose $v_1(x), v_2(x)$ in such a way that

$$v_1'(x) + v_2'(x)e^{2x} = 0$$

so that $$y_p' = 2v_2(x)e^{2x}.$$

Again differentiating, we get

$$y_p'' = 4v_2(x)e^{2x} + 2v_2'(x)e^{2x}.$$

Using the values of y_p', y_p'' in the given equation, we get

$$2v_2'(x)e^{2x} = e^x \cos x$$

or, $$v_2'(x) = \frac{1}{2}e^{-x}\cos x.$$

Using this in $v_1'(x)+v_2'(x)e^{2x}=0,$

we get $v_1'(x)=-\frac{1}{2}e^x\cos x.$

Now integrating, we get $v_1(x)=-\frac{1}{2}\int e^x\cos x\,dx=-\frac{1}{4}e^x(\cos x+\sin x)$ and

$$v_2(x)=\frac{1}{4}e^{-x}(-\cos x+\sin x).$$

Hence, $y_p=-\frac{1}{4}e^x(\cos x+\sin x)+\frac{1}{4}e^x(-\cos x+\sin x)=-\frac{1}{2}e^x\cos x.$

So, the complete primitive is

$$y=c_1+c_2e^{2x}-\frac{1}{2}e^x\cos x.$$

6. Solve: $y_2+2y_1+y=e^{-x}\ln x$ by the method of variation of parameter.

Solution

Let $y=e^{mx}\neq 0$ be a trial solution of the equation

$$y_2+2y_1+y=0; \tag{1}$$

then the auxiliary equation is

$$(m+1)^2=0.$$

Solving, we get $m=-1,-1$. So, C.F. is $(c_2x+c_1)e^{-x}$. Since xe^{-x}, e^{-x} are two linearly independent solutions of (1), we can assume the particular solution of the given equation as

$$y_p=xv_2(x)e^{-x}+v_1(x)e^{-x}.$$

Then $y_p'=v_1'(x)e^{-x}+xv_2'(x)e^{-x}-v_1(x)e^{-x}+v_2(x)e^{-x}-xv_2(x)e^{-x}.$

Now we choose $v_1(x), v_2(x)$ in such a way that

$$v_1'(x)e^{-x}+xv_2'(x)e^{-x}=0$$

so that $y_p'=-v_1(x)e^{-x}+v_2(x)e^{-x}-xv_2(x)e^{-x}.$

Again differentiating, we get

$$y_p''=v_1(x)e^{-x}-v_1'(x)e^{-x}-2v_2(x)e^{-x}+v_2'(x)e^{-x}-xv_2'(x)e^{-x}+xv_2(x)e^{-x}.$$

Using the values of y_p', y_p'' in the given equation, we get

$$-v_1'(x)e^{-x}+(1-x)v_2'(x)e^{-x}=e^{-x}\ln x.$$

Solving for $v_1'(x), v_2'(x)$, we get

$$v_2'(x)=\ln x \text{ and } v_1'(x)=-x\ln x$$

Integrating, we get

$$v_2(x)=x(\ln x-1) \text{ and } v_1(x)=-\frac{x^2}{2}\ln x+\frac{x^2}{4}.$$

Hence,

$$y_p = -\frac{x^2e^{-x}}{4}(2\ln x - 1) + x^2e^{-x}(\ln x - 1) = \frac{x^2e^{-x}}{2}\ln x - \frac{3}{4}x^2e^{-x}.$$

So, the complete primitive is

$$y = c_1e^{-x} + c_2xe^{-x} + \frac{x^2e^{-x}}{2}\ln x - \frac{3}{4}x^2e^{-x}.$$

7. Solve $x^2y_2 + xy_1 - y = x^2\log x$ by the method of variation of parameter, given that $y = x, y = \frac{1}{x}$ are two linearly independent solutions of the corresponding homogeneous equation.

Solution

We rewrite the equation as

$$y_2 + \frac{1}{x}y_1 - \frac{1}{x^2}y = \log x.$$

Now we assume the particular integral as

$$y_p = v_1(x)x + \frac{v_2(x)}{x}, \qquad (1)$$

where $v_1(x), v_2(x)$ are unknown functions. Differentiating this with respect to x, we get

$$y_p' = v_1(x) - \frac{v_2(x)}{x^2} \qquad (2)$$

provided

$$v_1'(x)x + \frac{v_2'(x)}{x} = 0. \qquad (3)$$

Again differentiating (2) with respect to x, we get

$$y_p'' = v_1'(x) - \frac{v_2'(x)}{x^2} + \frac{2v_2(x)}{x^3}. \qquad (4)$$

Using (1), (2), (4) in the given equation, we get

$$v_1'(x) - \frac{v_2'(x)}{x^2} = \log x. \qquad (5)$$

Solving (3) and (5), we get

$$v_2'(x) = -\frac{1}{2}x^2\log x, \quad v_1'(x) = \frac{1}{2}\log x.$$

Integrating, we get

$$v_1(x) = \frac{x}{2}(\log x - 1), \quad v_2(x) = -\frac{x^3}{18}(3\log x - 1).$$

Hence, the P.I. is

$$y = \frac{x^2}{2}(\log x - 1) - \frac{x^2}{18}(3\log x - 1) = \frac{1}{3}x^2\log x - \frac{4}{9}x^2.$$

So, the complete solution is

$$y = ax + \frac{b}{x} + \frac{1}{3}x^2\log x - \frac{4}{9}x^2.$$

8. Solve: $(x^2+x)y_2+(2-x^2)y_1-(2+x)y=x(x+1)^2$ by method of variation of parameter, given that $y=e^x, y=\frac{1}{x}$ are two linearly independent solutions of the corresponding homogeneous equation.

Solution

We rewrite the equation as

$$y_2+\frac{2-x^2}{x^2+x}y_1-\frac{2+x}{x^2+x}y=x+1.$$

Now assume the particular integral as

$$y_p=v_1(x)e^x+\frac{v_2(x)}{x}, \tag{1}$$

where $v_1(x), v_2(x)$ are unknown functions. Differentiating this with respect to x, we get

$$y_p'=e^x v_1(x)-\frac{v_2(x)}{x^2} \tag{2}$$

provided
$$e^x v_1'(x)+\frac{v_2'(x)}{x}=0. \tag{3}$$

Again differentiating (2) with respect to x, we get

$$y_p''=e^x v_1(x)+e^x v_1'(x)-\frac{v_2'(x)}{x^2}+\frac{2v_2(x)}{x^3}. \tag{4}$$

Using (1), (2), (4) in the given equation, we get

$$e^x v_1'(x)-\frac{v_2'(x)}{x^2}=x+1. \tag{5}$$

Solving (3) and (5), we get

$$v_2'(x)=-x^2,\ v_1'(x)=xe^{-x}.$$

Integrating, we get

$$v_1(x)=-xe^{-x}-e^{-x},\ v_2(x)=-\frac{x^3}{3}.$$

Hence, the P.I. is

$$y_p=-x-1-\frac{x^2}{3}.$$

So, the complete solution is

$$y=ae^x+\frac{b}{x}-x-1-\frac{x^2}{3}.$$

EXERCISE

- Solve the following differential equations using the method of variation of parameter.

1. $y'' + y = \cos x$,
2. $y'' + y = \sin x$,
3. $y'' + 4y = \tan 2x$,
4. $y'' + y = e^x$,
5. $y'' + a^2y = \sec ax$,
6. $y'' + a^2y = \cot ax$,
7. $y'' - 2y' = e^x \cos x$,
8. $y''' + y' = \operatorname{cosec} x$,
9. $y'' + 2y' + y = \dfrac{1}{x^2e^x}$,
10. $y'' + 4y = \operatorname{cosec} 2x$,
11. $y'' - 6y' + 9y = \dfrac{e^{3x}}{x^2}$,
12. $y'' - 4y' + 4y = e^{2x}$,
13. $y'' - y' - 2y = x^2$
14. $y'' + 4y' + 5y = e^{-2x} \sec x$,
15. $y'' - 2y' + y = e^x \sin x$,
16. $y'' + 2y' + y = e^{-x} \cos x$,
17. $y'' - y = \dfrac{2}{1+e^x}$,
18. $y'' - 3y' + 2y = \dfrac{e^x}{1+e^x}$,
19. $y'' - 4y' + 3y = \dfrac{e^x}{1+e^x}$,
20. $y'' - 2y' + y = xe^x \log x$,
21. $y'' - 2y' + 2y = e^x \tan x$,
22. $y'' + y = \dfrac{1}{1+\sin x}$,
23. $y'' + 3y' + 2y = 2e^x$.
24. $y'' - y = xe^x$,
25. $y'' - 2y' = e^x \sin x$,
26. $y'' - 2y' + y = xe^x \sin x$,
27. Solve $(2x + 1)(x + 1)\, y_2 + 2xy_1 - 2y = (2x + 1)^2$ by the method of variation of parameter, given that $y = x,\ y = \dfrac{1}{x+1}$ are two linearly independent solutions of the corresponding homogeneous equation.
28. Solve $x^2y_2 - x(x + 2)\, y_1 + (x + 2)\, y = x^3$ by the method of variation of parameter, given that $y = x$, $y = xe^x$ are two linearly independent solutions of the corresponding homogeneous equation.
29. Solve $x^2y_2 - xy_1 - 3y = x^2\log x$ by the method of variation of parameter, given that $y = x^3,\ y = \dfrac{1}{x}$ are two linearly independent solutions of the corresponding homogeneous equation.
30. Solve $x^2y_2 + xy_1 - y = x^2e^x$ by the method of variation of parameter, given that $y = x,\ y = \dfrac{1}{x}$ are two linearly independent solutions of the corresponding homogeneous equation.
31. Solve $x^2y_2 - xy_1 = x^3e^x$ by the method of variation of parameter, given that $y = x^2$, $y = 1$ are two linearly independent solutions of the corresponding homogeneous equation.

32. Solve $x^2y_2 - 2x(x+1)y_1 + 2(1+x)y = x^3$ by method of variation of parameter, given that $y = x$, $y = xe^{2x}$ are two linearly independent solutions of the corresponding homogeneous equation.

33. Solve $x^3y_3 + 3x^2y_2 = 1$ by method of variation of parameter, given that $y = 1$, $y = x$, $y = \frac{1}{x}$ are three linearly independent solutions of the corresponding homogeneous equation.

ANSWERS

1. $y = c_1 \cos x + c_2 \sin x + \frac{1}{2} x \sin x$, **2.** $y = c_1 \cos x + c_2 \sin x - \frac{1}{2} x \cos x$,

3. $y = c_1 \cos 2x + c_2 \sin 2x - \frac{1}{4} \cos 2x \ln|\sec 2x + \tan 2x|$

4. $y = c_1 \cos x + c_2 \sin x + \frac{1}{2} e^x$,

5. $y = c_1 \cos ax + c_2 \sin ax - \frac{1}{a^2} \cos ax \ln|\sec ax| + \frac{x}{a} \sin ax$,

6. $y = c_1 \cos ax + c_2 \sin ax + \frac{1}{a^2} \sin ax \ln \tan\left(\frac{ax}{2}\right)$,

7. $y = c_1 + c_2 e^{2x} - \frac{1}{2} e^x \cos x$,

8. $y = c_1 \cos x + c_2 \sin x + c_3 - \cos x \ln|\sin x| - x \sin x + \ln|\text{cosec}\, x - \cot x|$

9. $y = e^{-x}[c_1 x + c_2 - \ln x]$

10. $y = c_1 \cos 2x + c_2 \sin 2x + \frac{1}{4} \sin 2x \ln|\sin 2x| - \frac{x}{2} \cos 2x$,

11. $y = e^{3x}[c_1 x + c_2 - \ln x]$ **12.** $y = e^{2x}\left[c_1 x + c_2 + \frac{x^2}{2}\right]$

13. $y = c_1 e^{-x} + c_2 e^{2x} - \frac{x^2}{2} + \frac{x}{2} - \frac{3}{4}$,

14. $y = e^{-2x}(c_1 \cos x + c_2 \sin x) + xe^{-2x} \sin x + e^{-2x} \cos x \ln(\cos x)$,

15. $y = (c_1 + c_2 x - \sin x)e^x$, **16.** $y = (c_1 + c_2 x - \cos x)e^{-x}$,

17. $y = ae^x + be^{-x} - 1 - xe^x + (e^x - e^{-x})\ln(1 + e^x)$,

18. $y = ae^x + be^{2x} - xe^x - xe^{2x} + e^x \log(e^x + 1) + e^{2x} \log(e^x + 1)$,

19. $y = ae^x + be^{3x} + \frac{1}{2} e^{2x} + \frac{1}{2}(1 - e^{2x}) e^x \log(1 + e^{-x})$,

20. $y = (a + bx)e^x + \frac{1}{6} x^3 e^x \log x - \frac{5}{36} x^3 e^x$,

21. $y = e^x[a\cos x + b\sin x - \cos x \log(\sin x + \tan x)]$,

22. $y = a\cos x + b\sin x - 1 + \sin x - x\cos x + \sin x \log(1 + \sin x)$,

23. $y = \frac{a}{x} + \frac{b}{x^2} + \frac{1}{3}e^x$,

24. $y = Ae^x + Be^{-x} + \frac{x(x-1)e^x}{4}$,

25. $y = a + be^{2x} - \frac{1}{2}e^x \sin x$,

26. $y = ae^x + bxe^x - xe^x \sin x - 2e^x \cos x$,

27. $y = ax + \frac{b}{x+1} + \frac{x^2}{6}\frac{4x+3}{x+1}$,

28. $y = ax + bxe^x - x(x+1)$,

29. $y = ax^3 + \frac{b}{x} - \frac{1}{3}x^2\left(\log x + \frac{2}{3}\right)$,

30. $y = ax + \frac{b}{x} + e^x\left(1 - \frac{1}{x}\right)$,

31. $y = a + bx^2 + (x-1)e^x$,

32. $y = ax + bxe^{2x} - \frac{x}{4} - \frac{x^2}{2}$,

33. $y = a + bx + \frac{c}{x} - \log x$.

6.7 Method of Undetermined Coefficients

The method of undetermined coefficients is also applied to find a particular solution of a differential equation. Though this method is simpler and economical than the method of variation of parameter, this method cannot be applied in a few cases which are discussed later on. We consider the problem of solving an equation of the form

$$F(D)y = [D^n + P_1 D^{n-1} + P_2 D^{n-2} + \cdots + P_{n-1} D + P_n]y = Q, \tag{i}$$

where $D = \frac{d}{dx}$.

In this method we assume the particular integral as

$$y_p = A_1 q_1(x) + A_2 q_2(x) + \cdots + A_s q_s(x), \tag{ii}$$

where $q_1(x), q_2(x), \cdots, q_s(x)$ are the terms of Q and those arising from these terms by differentiation, for example, if the equation is $F(D)y = \sin x$, we take for (ii),

$$y_p = A\cos x + B\sin x,$$

if the equation is $F(D)y = x^2$, we take for (ii),

$$y_p = Ax^2 + Bx + C,$$

if the equation is $F(D)y = e^{2x} + e^{4x}$, we take for (ii),

$$y_p = Ae^{2x} + Be^{4x},$$

since no new term is obtained by differentiating e^{2x}, e^{4x}, and A_i's are undetermined constants to be so chosen that (ii) satisfies (i). If Q is $\sec x$, $\operatorname{cosec} x$, $\tan x$ or $\cot x$, the method fails, since the number of terms obtained by differentiation of Q is infinite.

If a term of Q, say L, is also a term of the complementary function corresponding to an s-fold root m, then in (ii) we add a term $x^s L$ and the terms arising from it by differentiation. For example, for the equation $(D-1)^2(D+2)y = e^x + \sin x$, we take the particular integral as

$$y_p = Ax^2e^x + Bxe^x + Ce^x + D\sin x + E\cos x.$$

If a term of Q is $x^r L$ and L is also a term of the complementary function corresponding to an s-fold root m, then in (ii) we add a term $x^{r+s}L$ and the terms arising from it by differentiation. For example, for the equation $(D-1)^2(D+2)y = xe^x + x^3$, we take the particular integral as

$$y_p = Ax^3e^x + Bx^2e^x + Cxe^x + De^x + Ex^3 + Fx^2 + Gx + H.$$

When Q is an algebraic function then the particular integral of (6.7.1) will be of the form of (ii), if all P_i are non-zero. If $P_n = 0$, then we take the particular integral as

$$y_p = x[A_1q_1(x) + A_2q_2(x) + \cdots + A_sq_s(x)].$$

If $P_n = 0$ and $P_{n-1} = 0$, then we take the particular integral as

$$y_p = x^2[A_1q_1(x) + A_2q_2(x) + \cdots + A_sq_s(x)] \text{ and so on.}$$

WORKING RULE

We know that the general solution of the differential equation

$$y_2 + P(x)y_1 + Q(x)y = R(x)$$

has the form $\qquad y = \text{C.F.} + \text{P.I.}$

The method of undetermined coefficients leads the way of finding the P.I. in the following ways.

R1. If $R(x) = e^{\lambda x}$ and λ is a multiple root of the auxiliary equation of order k, then we take the P.I. as

$$y_p = Ax^ke^{\lambda x},$$

where A is the undetermined constant. Obviously, $k = 0$, if λ is not a root of the auxiliary equation.

R2. If $R(x) = a_0 + a_1x + a_2x^2 + \ldots\ a_kx^k$, i.e., a polynomial in x and $P(x)$, $Q(x)$ do not vanish, we take the P.I. as

$$y_p = A_0 + A_1x + A_2x^2 + \cdots + A_kx^k.$$

When $Q(x) = 0$, we take

$$y_p = x(A_0 + A_1x + A_2x^2 + \cdots + A_kx^k)$$

and if $P(x) = 0 = Q(x)$, then we take

$$y_p = x^2(A_0 + A_1x + A_2x^2 + \cdots + A_kx^k).$$

R3. If $R(x) = \sin ax$ or, $\cos ax$ then we take

$$y_p = A\sin ax + B\cos ax,$$

if $\sin ax$ or $\cos ax$ is absent in the complementary function.

If $\sin ax$ or $\cos ax$ is present in the complementary function then we take

$$y_p = x(A\sin ax + B\cos ax).$$

R4. If $R(x) = e^{ax}\sin bx$ or $e^{ax}(a_0 + a_1x + a_2x^2 + \ldots + a_kx^k)$ or $\sin bx\,(a_0 + a_1x + a_2x^2 + \ldots + a_kx^k)$, then the P.I. will be modified according to the above mentioned rules.

For example, if $R(x) = e^{\lambda x}\sin ax$ and λ is not a root of the auxiliary equation, then we take the P.I. as

$$R(x) = e^{\lambda x}(A\sin ax + B\cos ax).$$

If $R(x) = e^{\lambda x}(a_0 + a_1 + \ldots + a_kx^k)$ and λ is a multiple root of the auxiliary equation of order k, then we take

$$y_p = x^k e^{\lambda x}(A_0 + A_1x + \ldots + A_kx^k).$$

If $R(x) = \sin bx\,(a_0 + a_1x + \ldots + a_kx^k)$ and $\sin bx$ is present in the complementary function, then we take

$$y_p = x(A\cos bx + B\sin bx)(A_0 + A_1x + \ldots + A_kx^k).$$

WORKED-OUT EXAMPLE

1. Solve the following equation by the method of undetermined coefficients:

$$y_2 + 2y_1 + 2y = e^{2x}.$$

Solution

Let $y = e^{mx}$ be a trial solution, then the auxiliary equation is

$$m^2 + 2m + 2 = 0.$$

Solving, we get $m = -1 \pm i$. So the C.F. is

$$e^{-x}(A\cos x + B\sin x).$$

Since 2 is not a root of the auxiliary equation, we take the P.I. as

$$y_p = Ce^{2x}.$$

Using this in the given equation, we get $C = \frac{1}{10}$. Hence the general solution of the given equation is

$$y = e^{-x}(A\cos x + B\sin x) + \frac{e^{2x}}{10}.$$

2. Solve the following equation by the method of undetermined coefficients:

$$y_2 + 4y_1 + 4y = 16e^{-2x}$$

Solution

Let $y = e^{mx}$ be a trial solution, then the auxiliary equation is

$$m^2 + 4m + 4 = 0.$$

Solving, we get $m = -2, -2$. So the C.F. is

$$(c_1 + c_2x)e^{-2x}.$$

Since –2 is a double root of the auxiliary equation, we take the P.I. as

$$y_p = Cx^2e^{-2x}.$$

Differentiating this, we get

$$y'_p = 2Cx(1-x)e^{-2x} \text{ and } y''_p = -4Cx(1-x)e^{-2x} + 2C(1-2x)e^{-2x}.$$

Using the above results in the given equation, we get $C = 8$. Hence the general solution of the given equation is

$$y = (c_1 + c_2x)e^{-2x} + 8x^2e^{-2x}.$$

3. Solve the following equation by the method of undetermined coefficients:

$$y_2 + 4y = \sin x.$$

Solution

Let $y = e^{mx}$ be a trial solution, then the auxiliary equation is

$$m^2 + 4 = 0.$$

Solving, we get $m = \pm 2i$. So the C.F. is

$$c_1 \cos 2x + c_2 \sin 2x.$$

Since, $\sin x$ is not a part of the complementary function, we take the P.I. as

$$y_p = A\cos x + B\sin x.$$

Using it in the given equation, we get

$$3A\cos x + 3B\sin x = \sin x.$$

This gives $B = \frac{1}{3}, A = 0$. Hence the general solution of the given equation is

$$y = c_1 \cos 2x + c_2 \sin 2x + \frac{1}{3}\sin x.$$

4. Solve the following equation by the method of undetermined coefficients:

$$y_2 + y = 2\cos x.$$

Solution

Here the C.F. is $C\cos x + D\sin x$. Now we see that $\cos x$ is a part of Q and also occur in C.F. corresponding to a root of multiplicity 1. So, we assume the particular integral as

$$y_p = x(A\cos x + B\sin x).$$

Putting these values in the given equation and comparing both sides, we get $A = 0$, $B = 1$.

Hence, the general solution is

$$y = C\cos x + D\sin x + x\sin x.$$

5. Solve the following equation by the method of undetermined coefficients:

$$y_2 - 4y_1 + y = x^2 - 2x + 2.$$

Solution

Let $y = e^{mx}$ be a trial solution, then the auxiliary equation is

$$m^2 - 4m + 1 = 0.$$

Solving, we get $m = 2 \pm \sqrt{3}$. So the C.F. is

$$e^{2x}[De^{\sqrt{3}x} + Ee^{-\sqrt{3}x}].$$

We take the P.I. as

$$y_p = Ax^2 + Bx + C,$$

which gives $(y_p)_1 = 2Ax + B$ and $(y_p)_2 = 2A$.

Putting these values in the given equation and comparing both sides, we get $A = 1$, $B = 6$, $C = 24$. So the general solution is

$$y = e^{2x}\left[De^{\sqrt{3}x} + Ee^{-\sqrt{3}x}\right] + x^2 + 6x + 24.$$

6. Solve the following equation by the method of undetermined constants:

$$y_2 - 2y_1 = e^x \cos x.$$

Solution

Here the C.F. is $C + De^{2x}$. We take the particular integral as

$$y_p = Ae^x \cos x + Be^x \sin x.$$

This gives

$$(y_p)_1 = [A + B]e^x \cos x + [B - A]e^x \sin x \text{ and } (y_p)_2 = 2[B\cos x - A\sin x]e^x.$$

Putting these values in the given equation and comparing both sides, we get $A = -\frac{1}{2}$ and $B = 0$. So the general solution is

$$y = C + De^{2x} - \frac{1}{2}e^x \cos x.$$

7. Solve the following equation by the method of undetermined coefficients:

$$y_2 - y_1 - 2y = x^2 e^x.$$

Solution

Here the C.F. is $De^{-x} + Ee^{2x}$. We assume the particular integral as

$$y_p = [Ax^2 + Bx + C]e^x,$$

which gives

$$(y_p)_1 = [Ax^2 + (2A + B)x + B + C]e^x \text{ and } (y_p)_2 = [Ax^2 + (4A + B)x + 2A + 2B + C]e^x$$

Putting these values in the given equation and comparing both sides we get

$$A = -\frac{1}{2},\ B = -\frac{1}{2} \text{ and } C = -\frac{3}{4}.$$

Hence, the general solution is

$$y = De^{-x} + Ee^{2x} - \frac{1}{4}[2x^2 + 2x + 3]e^x.$$

8. Solve the following equation by the method of undetermined coefficients:

$$y_2 - 4y_1 + 4y = x^3 e^{2x}.$$

Solution

Here the C.F. is $[E + xF]e^{2x}$. Now we see that e^{2x} is a part of Q and also occur in C.F. corresponding to a root of multiplicity 2. So, we assume the particular integral as

$$y_p = [Ax^5 + Bx^4 + Cx^3 + Dx^2]e^{2x},$$

leaving the terms xe^{2x}, e^{2x} as those appear in the C.F. Putting these values in the given equation and comparing both sides, we get $A = \frac{1}{20}, B = C = D = 0.$ Hence the general solution is

$$y = [E + xF]e^{2x} + \frac{1}{20}e^{2x}x^5.$$

9. Solve by the method of undetermined coefficients:

$$y_3 + 3y_2 + 2y_1 = x^2 + 4x + 8.$$

Solution

Here the auxiliary equation is

$$m^3 + 3m^2 + 2m = 0,$$

whose solutions are $m = 0, -1, -2$. So the C.F. is

$$C_1 + C_2e^{-x} + C_3e^{-2x}.$$

Here $P_3 = 0$, so we take the particular integral

$$y_p = x(A + Bx + Cx^2).$$

Now $\quad y_p' = A + 2Bx + 3Cx^2;\ y_p'' = 2B + 6Cx;\ y''' = 6C.$

Using these values in $y_p''' + 3y_p'' + 2y_p' = x^2 + 4x + 8$, we get

$$6C + 3(2B + 6Cx) + 2(A + 2Bx + 3Cx^2) = x^2 + 4x + 8$$

Comparing both sides, we get

$$6C = 1;\ 4B + 18C = 4 \text{ and } 2A + 6B + 6C = 8.$$

Solving, we get

$$A = \frac{11}{4};\ B = \frac{1}{4};\ C = \frac{1}{6}.$$

So the general solution is

$$y = C_1 + C_2e^{-x} + C_3e^{-2x} + x\left(\frac{11}{4} + \frac{x}{4} + \frac{x^2}{6}\right).$$

10. Solve by the method of undetermined coefficients: $y_2 = x$.

Solution

Here the auxiliary equation is $m^2 = 0$,
whose solutions are $m = 0, 0$. So the C.F. is $C_1 + C_2x$.
We note here $P_2 = 0$ and $P_3 = 0$, so we take the particular integral

$$y_p = x^2(A + Bx).$$

Differentiating with respect to x, we get

$$y_p' = 2Ax + 3Bx^2 \text{ and } y_p'' = 2A + 6Bx.$$

So, $$y_p'' = 2A + 6Bx = x \text{ gives } A = 0, B = \frac{1}{6}.$$

Hence, the general solution is

$$y = C_1 + C_2 x + \frac{x^3}{6}.$$

11. Solve by the method of undetermined coefficients: $y_2 - 7y_1 + 6y = (x-2)e^x$.

Solution

Here the C.F. is $Ee^x + Fe^{6x}$. Now we see that e^x is a part of Q and also occurs in C.F. corresponding to a root of multiplicity 1. So, we assume the particular integral as

$$y_p = (A + Bx)xe^x.$$

Differentiating with respect to x, we get

$$y_p' = (A + Ax + 2Bx + Bx^2)e^x \text{ and } y_p'' = (2A + 2B + Ax + 4Bx + Bx^2)e^x.$$

Putting these values in $y_p'' - 7y_p' + 6y_p = (x-2)e^x$ and comparing both sides, we get $-10B = 1$ and $5A - 2B = 2$.

Solving we get $A = \frac{9}{25}, B = -\frac{1}{10}$. So

$$y_p = \left(\frac{9}{25} - \frac{1}{10}x\right)xe^x.$$

Hence, the general solution is

$$y = Ee^x + Fe^{6x} + \left(\frac{9}{25} - \frac{1}{10}x\right)xe^x.$$

12. Solve by the method of undetermined coefficients: $y_2 - 4y_1 + 4y = (x^2+1)xe^{2x}$.

Solution

Here the C.F. is $(F + xG)e^{2x}$. Now we see that e^{2x} is a part of Q and also occurs in C.F. corresponding to a root of multiplicity 2. So, we assume the particular integral as

$$y_p = (Ax^3 + Bx^2 + Cx + E)x^2e^{2x}.$$

Differentiating with respect to x, we get

$$y_p' = [2Ax^5 + (5A+2B)x^4 + (4B+2C)x^3 + (3C+2E)x^2 + 2Ex]e^{2x} \text{ and}$$

$$y_p'' = [4Ax^5 + 4(5A+B)x^4 + 4(5A+4B+C)x^3 + 4(3B+3C+E)x^2 + (6C+8E)x + 2E]e^{2x}.$$

Putting these values in $y_p'' - 4y_p' + 4y_p = (x^2+1)xe^{2x}$ and comparing both sides, we get

$$20A = 1, 12B - 5E = 0, 6C = 1, E = 0.$$

Solving we get $A = \frac{1}{20}, B = 0, C = \frac{1}{6}, E = 0.$

So, $$y_p = \left(\frac{1}{20}x^3 + \frac{1}{6}x\right)x^2e^{2x}.$$

Hence, the general solution is

$$y = (F + xG)e^{2x} + \left(\frac{1}{20}x^3 + \frac{1}{6}x\right)x^2e^{2x}.$$

13. Solve by the method of undetermined coefficients: $y_2 - 3y_1 = x + e^x \sin x.$

Solution

Here the C.F. is $A + Be^{3x}$. Since $P_3 = 0$ and Q has an algebraic part, we take the particular integral as

$$y_p = x(E + xF) + e^x(G \sin x + H \cos x).$$

Differentiating with respect to x, we get

$$y_p' = E + 2Fx + e^x(G - H)\sin x + e^x(G + H)\cos x,$$

and $$y_p'' = 2F + 2Ge^x \cos x - 2He^x \sin x.$$

Putting these values in $y_p'' - 3y_p' = x + e^x \sin x$ and comparing both sides, we get

$$2F - 3E = 0, -6F = 1, H - 3G = 1, G + 3H = 0.$$

Solving, we get

$$E = -\frac{1}{9}, F = -\frac{1}{6}, G = -\frac{3}{10}, H = \frac{1}{10}.$$

So, $$y_p = x\left(-\frac{1}{9} - \frac{x}{6}\right) - \frac{e^x}{10}(3\sin x - \cos x).$$

14. Solve by the method of undetermined coefficients:

$$y_2 + y_1 - 6y = 10e^{2x} - 18e^{3x} - 6x - 11.$$

Solution

Here the C.F. is $Ae^{2x} + Be^{-3x}$. Since e^{2x} is a part of Q and also occur in C.F. corresponding to a root of multiplicity 1 we assume the particular integral as

$$y_p = Cxe^{2x} + Ee^{3x} + Fx + G.$$

Differentiating it with respect to x, we get

$$y_p' = (2x+1)Ce^{2x} + 3Ee^{3x} + F, \text{ and } y_p'' = 4(x+1)Ce^{2x} + 9Ee^{3x}.$$

Using these values in $y_p'' + y_p' - 6y_p = 10e^{2x} - 18e^{3x} - 6x - 11$, we get

$$5Ce^{2x} + 6Ee^{3x} - 6Fx + F - 6G = 10e^{2x} - 18e^{3x} - 6x - 11.$$

Comparing, we get

$$C = 2, E = -3, F = 1, G = 2.$$

So, $$y_p = 2xe^{2x} - 3e^{3x} + x + 2.$$

Hence, the general solution is

$$y = Ae^{2x} + Be^{-3x} + 2xe^{2x} - 3e^{3x} + x + 2.$$

EXERCISE

- Solve the following equations by the method of undetermined coefficients:

1. $y_2 + 2y_1 + 2y = 8e^{2x}$,

2. $y_2 + 2y = e^x + 2$,

3. $y_2 - y_1 - 6y = 20e^{-2x}$,

4. $y_2 + y = 2\sin x$,

5. $y_2 - 4y_1 = x^2$,

6. $y_2 - 3y_1 = 2x^2 + 1$,

7. $y_2 - 9y = x + e^{2x} - \sin 2x$,

8. $y_2 - y = e^x \sin 2x$

9. $y_2 + 4y = \sin 2x$

10. $y_2 - 3y_1 + 2y = \sin 2x$,

11. $y_2 + 6y_1 + 9y = e^{-3x}$,

12. $y_2 - 2y_1 + 3y = x^3 + \sin x$,

13. $y_3 + 2y_2 - y_1 - 2y = x^2 + e^x$,

14. $y_3 - y_1 = x$,

15. $y_2 - 2y_1 + y = xe^x$,

16. $y_2 = x^2$,

17. $y_2 - 2y_1 + 5y = 25x^2 + 12$,

18. $y_2 + 4y = x^2 \sin 2x$,

19. $y_2 - 2y_1 = e^x \sin x$,

20. $y_2 + 3y_1 + 2y = e^x \sin x$,

21. $y_2 - 4y_1 + 3y = 12x^2 + x - 6$, given $y = 6; \dfrac{dy}{dx} = 11$, when $x = 0$.

22. $y_2 + 4y_1 = 12x^2 + e^x$, given that $y = 1; \dfrac{dy}{dx} = 1$, when $x = 0$.

23. $y_2 + y = 10e^{2x}$, given that $y = 0; \dfrac{dy}{dx} = 0$, when $x = 0$.

24. $y_2 + y = \cos 2x$, given that $y = 0; \dfrac{dy}{dx} = 1$, when $x = 0$.

25. $y_2 + 4y_1 + 5y = 5$, given that $y = 0; \dfrac{dy}{dx} = 0$, when $x = 0$.

ANSWER

1. $e^{-x}[A\cos x + B\sin x] + \frac{4}{5}e^{2x}$,

2. $A\cos\sqrt{2}x + B\sin\sqrt{2}x + 1 + \frac{1}{3}e^x$,

3. $Ae^{3x} + Be^{-2x} - 4xe^{-2x}$,

4. $A\cos x + B\sin x - x\cos x$,

5. $A+Be^{4x}-\frac{x^3}{12}-\frac{x^2}{16}-\frac{x}{32}$, **6.** $A+Be^{3x}-\frac{x}{27}(6x^2+6x+13)$,

7. $Ae^{3x}+Be^{-3x}-\frac{x}{9}-\frac{e^{2x}}{5}+\frac{1}{13}\sin 2x$,

8. $Ae^{x}+Be^{-x}-\frac{1}{8}[\sin 2x+\cos 2x]e^{x}$,

9. $A\cos 2x+B\sin 2x-\frac{x}{4}\cos 2x$,

10. $Ae^{x}+Be^{2x}-\frac{1}{20}\sin 2x+\frac{3}{20}\cos 2x$,

11. $(A+xB)e^{-3x}+\frac{1}{2}x^2e^{-3x}$,

12. $e^{x}[A\cos\sqrt{2}x+B\sin\sqrt{2}x]+\frac{2x}{9}+\frac{2x^2}{3}+\frac{x^3}{3}-\frac{8}{27}+\frac{1}{4}(\sin x+\cos x)$,

13. $Ae^{x}+Be^{-x}+Ce^{-2x}-\frac{x^2}{2}+\frac{x}{2}-\frac{5}{4}+\frac{xe^{x}}{6}$,

14. $A+Be^{x}+Ce^{-x}-\frac{1}{2}x^2$,

15. $e^{x}[A+Bx]+\frac{1}{6}e^{x}x^3$,

16. $A+Bx+\frac{x^4}{12}$,

17. $e^{x}(A\cos 2x+B\sin 2x)+5x^2+4x+2$,

18. $A\cos 2x+B\sin 2x-\frac{1}{12}x^3\cos 2x+\frac{1}{16}x^2\sin 2x+\frac{1}{32}x\cos 2x$,

19. $A+Be^{2x}-\frac{1}{2}e^{x}\sin x$,

20. $Ae^{-x}+Be^{-2x}+\frac{e^{x}}{10}(\sin x-\cos x)$,

21. $-6e^{x}+2e^{2x}+4x^2+11x+10$,

22. $\frac{29}{32}+\frac{3x}{8}-\frac{3x^2}{4}+x^3+\frac{e^{x}}{5}-\frac{17}{160}e^{-4x}$,

23. $2e^{2x}-2\cos x-4\sin x$,

24. $\sin x+\frac{1}{3}(\cos x-\cos 2x)$,

25. $1-e^{-2x}\cos x-2e^{-2x}\sin x$.

Chapter 7

Euler's Homogeneous Linear Equation

7.1 Introduction

If P_i's are functions of x and of the form $\frac{1}{x^i}$ then the equation (i) of the article 6.7 is known as *Cauchy-Euler's* homogeneous linear equation. The interesting feature of this equation is that each term in the left hand side of the equation is of the form $x^i \frac{dy^i}{dx^i}$. To convert it to a linear equation with constant coefficients, we put

$$x = e^z, \text{ i.e., } z = \ln x.$$

Then $Dy = \frac{dy}{dx} = \frac{1}{x}\delta y$, where $D = \frac{d}{dx}$ and $\delta = \frac{d}{dz}$.

i.e., $xDy = \delta y$.

Again, $D^2 y = D\left(\frac{1}{x}\delta y\right) = -\frac{1}{x^2}\delta y + \frac{1}{x^2}\delta^2 y$

i.e., $x^2 D^2 y = \delta(\delta - 1)y$.

Similarly, $D^3 y = D\left(-\frac{1}{x^2}\delta y + \frac{1}{x^2}\delta^2 y\right) = -\frac{2}{x^3}(\delta^2 y - \delta y) + \frac{1}{x^3}(\delta^3 y - \delta^2 y)$.

i.e., $x^3 D^3 y = \delta(\delta - 1)(\delta - 2)y$

Let us assume

$$x^m D^m y = \delta(\delta - 1)(\delta - 2)\cdots(\delta - m + 1)y.$$

We put $y = (\delta - m)u$, then

$$D^m y = D^m[xDu - mu] = xD^{m+1}u + mD^m u - mD^m u = xD^{m+1}u$$

$$x^m(xD^{m+1}u) = x^m D^m y = \delta(\delta - 1)(\delta - 2)\ldots(\delta - m + 1)(\delta - m)u$$

$$x^{m+1}D^{m+1}u = \delta(\delta - 1)(\delta - 2)\cdots(\delta - m + 1)(\delta - m)u.$$

So, if the result is true for m, then it is also true for $m + 1$, hence it is true for $n = 1, 2, 3, \cdots$.

So, using the result $\quad x^n D^n y = \delta(\delta - 1)(\delta - 2)\cdots(\delta - n + 1)y$

in an equation with variable coefficients, the equation can be converted to the equation with constant coefficients and hence the methods discussed in the previous chapter can be applied.

Note:

$$D(xDu) = Du + xD^2u;$$

$$D^2(xDu) = D(Du + xD^2u) = 2D^2u + xD^3u.$$

Let $\quad D^m(xDu) = mD^mu + xD^{m+1}u,$ then

$$D^{m+1}(xDu) = D(mD^mu + xD^{m+1}u) = (m+1)D^{m+1}u + xD^{m+2}u.$$

Hence, by the method of induction,

$$D^n(xDu) = nD^nu + xD^{n+1}u.$$

Otherwise, by Leibniz's rule

$$D^n(xDu) = xD^n(Du) + {}^nC_1 D^{n-1}(Du).1$$

$$= xD^{n+1}u + nD^nu.$$

7.2 Equation Reducible to Homogeneous Linear Form

An equation of the form

$$(a+bx)^n \frac{d^ny}{dx^n} + P_1(a+bx)^{n-1}\frac{d^{n-1}y}{dx^{n-1}} + \ldots + P_{n-1}(a+bx)\frac{dy}{dx} + P_ny = X,$$

where $P_1, P_2, \cdots, P_n$ are constants and X is a function of x only, can be reduced to a linear homogeneous equation as follows:

Putting $z = a + bx$, we get

$$\frac{dy}{dx} = \frac{dy}{dz}\frac{dz}{dx} = b\frac{dy}{dz},$$

$$\frac{d^2y}{dx^2} = \frac{d}{dx}\left(b\frac{dy}{dz}\right) = \frac{d}{dz}\left(b\frac{dy}{dz}\right)\frac{dz}{dx} = b^2\frac{d^2y}{dz^2},$$

....

....

$$\frac{d^ny}{dx^n} = b^n\frac{d^ny}{dz^n}.$$

Using these results in the above equation, we get

$$z^n\frac{d^ny}{dz^n} + \frac{P_1}{b}z^{n-1}\frac{d^{n-1}y}{dz^{n-1}} + \ldots + \frac{P_{n-1}}{b^{n-1}}z\frac{dy}{dz} + \frac{P_n}{b^n}y = Z,$$

where Z is a functions of z only.

WORKED-OUT EXAMPLES

1. Solve: $(x^2D^2 + xD + 1)y = 0.$

Solution

Putting $x = e^z$ in the given equation, we get

$$[\delta(\delta-1)+\delta+1]y = 0 \Rightarrow [\delta^2+1]y = 0.$$

So the auxiliary equation is $m^2 + 1 = 0$. Solving, we get $m = \pm i$. So the general solution is

$$y = A\cos z + B\sin z = A\cos(\ln x) + B\sin(\ln x).$$

2. Solve: $(x^2D^2 + 1)y = x^2.$

Solution

Putting $x = e^z$ in the given equation, we get

$$[\delta(\delta-1)+1]y = e^{2z} \qquad [\delta^2-\delta+1]y = e^{2z}.$$

So the auxiliary equation is $m^2 - m + 1 = 0$. Solving, we get

$$m = \frac{1 \pm i\sqrt{3}}{2}.$$

So, the C.F. is $e^{\frac{z}{2}}\left[A\cos\frac{\sqrt{3}z}{2} + B\sin\frac{\sqrt{3}z}{2}\right]$. Now P.I. $= \frac{1}{\delta^2-\delta+1}e^{2z} = \frac{1}{3}e^{2z}.$

Hence, the general solution is

$$y = e^{\frac{z}{2}}\left[A\cos\frac{\sqrt{3}z}{2} + B\sin\frac{\sqrt{3}z}{2}\right] + \frac{1}{3}e^{2z}$$

$$= \sqrt{x}\left[A\cos\left(\frac{\sqrt{3}\ln x}{2}\right) + B\sin\left(\frac{\sqrt{3}\ln x}{2}\right)\right] + \frac{1}{3}x^2.$$

3. Solve: $(x^2D^2 - xD + 1)y = \sin(\ln x^2).$

Solution

Putting $x = e^z$ in the given equation, we get

$$[\delta(\delta-1)-\delta+1]y = \sin 2z \qquad (\delta-1)^2 y = \sin 2z.$$

So the auxiliary equation is $(m-1)^2 = 0$. Solving we get $m = 1, 1$. So, the C.F. is

$$(A + zB)e^z.$$

Now, $\qquad$ P.I. $= \frac{1}{(\delta-1)^2}\sin 2z$

$$= -\frac{1}{2\delta+3}\sin 2z = -\frac{2\delta-3}{4\delta^2-9}\sin 2z$$

$$= \frac{1}{25}(2\delta-3)\sin 2z = \frac{1}{25}(4\cos 2z - 3\sin 2z).$$

Hence, the general solution is

$$y=(A+zB)e^{z}+\frac{1}{25}(4\cos 2z-3\sin 2z)$$

$$=(A+B\ln x)x+\frac{4}{25}\cos(2\ln x)-\frac{3}{25}\sin(2\ln x).$$

4. Solve: $(x^3D^3 - 3x^2D^2 + 7xD - 8)y = 0$.

Solution

Putting $x = e^z$ in the given equation, we get

$$\delta(\delta-1)(\delta-2)y-3\delta(\delta-1)y+7\delta y-8y=0 \Rightarrow [\delta^3-6\delta^2+12\delta-8]y=0.$$

So the auxiliary equation is

$$m^3-6m^2+12m-8=0$$

and its solutions are $m = 2, 2, 2$. So the general solution is

$$y=(c_1+zc_2+z^2c_3)e^{2z}=[c_1+c_2\ln x+c_3(\ln x)^2]x^2.$$

where $c_i, i=1, 2, 3,$ arbitrary constant.

5. Solve: $(x^3D^3 - x^2D^2 + 2xD - 2)y = x^3$.

Solution

Putting $x = e^z$ in the given equation, we get

$$\delta(\delta-1)(\delta-2)y-\delta(\delta-1)y+2\delta y-2y=e^{3z} \Rightarrow [\delta^3-4\delta^2+5\delta-2]y=e^{3z}.$$

So the auxiliary equation is

$$m^3-4m^2+5m-2=0$$

and its solutions are $m = 1, 1, 2$. So the C.F. is $(c_1+zc_2)e^{z}+c_3e^{2z}$.

Now the P.I. $=\dfrac{1}{(\delta-2)(\delta-1)^2}e^{3z}=\dfrac{1}{4}e^{3z}.$

So the general solution is

$$y=(c_1+zc_2)e^{z}+c_3e^{2z}+\frac{1}{4}e^{3z}=(c_1+c_2\ln x)x+c_3x^2+\frac{1}{4}x^3,$$

where $c_i, i=1, 2, 3,$ arbitrary constants.

6. Solve $(x^3D^3+2x^2D^2+2)y=10\left(x+\frac{1}{x}\right)$.

Solution

Putting $x = e^z$ in the given equation, we get

$$\delta(\delta-1)(\delta-2)y+2\delta(\delta-1)y+2y=10(e^{z}+e^{-z})$$

$$[\delta^3-\delta^2+2]y=10(e^{z}+e^{-z}).$$

So the auxiliary equation is

$$m^3-m^2+2=0$$

and its solutions are $m = -1, 1 \pm i$. So the C.F. is

$$c_1 e^{-z} + e^z (c_2 \cos z + c_3 \sin z).$$

Now the P.I. $= \dfrac{1}{\delta^3 - \delta^2 + 2} 10(e^z + e^{-z})$

$$= 5e^z + 10e^{-z} \frac{1}{\delta^3 - 4\delta^2 + 5\delta} 1$$

$$= 5e^z + 10e^{-z} \frac{1}{5\delta}\left(1 + \frac{\delta^2 - 4\delta}{5}\right)^{-1} 1$$

$$= 5e^z + 10e^{-z} \frac{1}{5\delta}\left(1 - \frac{\delta^2 - 4\delta}{5}\right) 1$$

$$= 5e^z + 2ze^{-z}$$

So the general solution is

$$y = c_1 e^{-z} + e^z (c_2 \cos z + c_3 \sin z) + 5e^z + 2ze^{-z}$$

$$= \frac{1}{x}(c_1 + 2\ln x) + x[c_2 \cos(\ln x) + c_3 \sin(\ln x) + 5].$$

where $c_i, i = 1, 2, 3,$ arbitrary constant.

7. Solve $x^3D^3y = 6y$.

Solution

Putting $x = e^z$ in the given equation, we get

$$\delta(\delta - 1)(\delta - 2)y - 6y = 0 \Rightarrow [\delta^3 - 3\delta^2 + 2\delta - 6]y = 0.$$

So the auxiliary equation is $m^3 - 3m^2 + 2m - 6 = 0$

and its solutions are $m = 3, \pm i\sqrt{2}$. So the general solution is

$$y = c_1 e^{3x} + c_2 \cos\sqrt{2}z + c_3 \sin\sqrt{2}z = c_1 x^3 + c_2 \cos\left(\sqrt{2}\ln x\right) + c_3 \sin\left(\sqrt{2}\ln x\right),$$

where $c_i, i = 1, 2, 3,$ arbitrary constant.

8. Solve: $(x^4D^4 + 2x^3D^3 + x^2D^2 - xD + 1)y = \log x$.

Solution

Putting $x = e^z$ in the given equation we get

$$\delta(\delta - 1)(\delta - 2)(\delta - 3)y + 2\delta(\delta - 1)(\delta - 2)y + \delta(\delta - 1)y - \delta y + y = z$$

$$\Rightarrow [\delta^4 - 4\delta^3 + 6\delta^2 - 4\delta + 1]y = z \Rightarrow (\delta - 1)^4 y = z.$$

So the auxiliary equation is

$$(m - 1)^4 = 0$$

and its solutions are $m = 1, 1, 1, 1$. So the C.F. is

$$(c_1 + zc_2 + z^2c_3 + z^3c_4)e^z.$$

Now P.I. $=\dfrac{1}{(1-\delta)^4}z=(1+4\delta+\cdots)z=z+4$. So the general solution is

$$y=(c_1+zc_2+z^2c_3+z^3c_4)e^z+z+4$$

$$=[c_1+c_2\log x+c_3(\log x)^2+c_4(\log x)^3]x+\log x+4.$$

where $c_i, i=1,2,3,$ is an arbitrary constant.

9. Solve $(x^3D^3+2x^2D^2)y=x+\sin(\ln x)$.

Solution

Putting $x=e^z$ in the given equation, we get

$$\delta(\delta-1)(\delta-2)y+2\delta(\delta-1)y=e^z+\sin z \Rightarrow [\delta^3-\delta^2]y=e^z+\sin z.$$

So the auxiliary equation is

$$m^2(m-1)=0$$

and its solutions are $m=0, 0, 1$. So the C.F. is

$$c_1+zc_2+c_3e^z.$$

Now P.I. $=\dfrac{1}{\delta^3-\delta^2}e^z+\dfrac{1}{\delta^3-\delta^2}\sin z=e^z\dfrac{1}{\delta}1+\dfrac{1}{1-\delta}\sin z$

$$=ze^z+\frac{1+\delta}{1-\delta^2}\sin z=ze^z+\frac{1}{2}(\sin z+\cos z).$$

So the general solution is

$$y=c_1+zc_2+c_3e^z+ze^z+\frac{1}{2}(\sin z+\cos z)$$

$$=c_1+c_2\log x+c_3x+x\log x+\frac{1}{2}[\sin(\log x)+\cos(\log x)],$$

where $c_i, i=1,2,3,$ is an arbitrary constant.

10. Solve: $x^2D^2y+4xDy+2y=x+\sin x$.

Solution

Putting $x=e^z$, i.e., $z=\ln x$ in the given equation, we get

$$[\delta(\delta-1)+4\delta+2]y=e^z+\sin e^z,\ \text{where } \delta=\frac{d}{dz}\ \text{or, } [\delta^2+3\delta+2]=e^z+\sin e^z.$$

So the auxiliary equation is $m^2+3m+2=0$ and its solutions are $m=-2,-1$. Hence the C.F. is

$$y=Ae^{-z}+Be^{-2z}$$

and P.I. $=\dfrac{1}{(\delta+1)(\delta+2)}(e^z+\sin e^z)$

$$=\frac{e^z}{6}+\frac{1}{\delta+2}\left[e^{-z}\int e^z\sin e^z dz\right]=\frac{e^z}{6}+e^{-2z}\int e^{2z}e^{-z}(-\cos e^z)dz$$

$$=\frac{e^z}{6}-e^{-2z}\int e^z\cos e^z dz=\frac{e^z}{6}-e^{-2z}\sin e^z.$$

So the general solution is

$$y = Ae^{-z} + Be^{-2z} + \frac{e^z}{6} - e^{-2z}\sin e^z = \frac{A}{x} + \frac{B}{x^2} + \frac{x}{6} - \frac{\sin x}{x^2}.$$

where A, B are arbitrary constants.

11. Solve: $x^3D^3y + 2x^2D^2y - xDy + y = x\ln x$.

Solution

Putting $x = e^z$, i.e., $z = \ln x$ in the given equation, we get

$$[\delta(\delta-1)(\delta-2) + 2\delta(\delta-1) - \delta + 1]y = ze^z, \text{ where } \delta = \frac{d}{dz}$$

or, $[\delta^3 - \delta^2 - \delta + 1]y = ze^z$.

So the auxiliary equation is $m^3 - m^2 - m + 1 = 0$ and its solutions are $m = -1, 1, 1$. Hence the C.F. is

$$y = (A + Bz)e^z + Ce^{-z}.$$

and P.I. $= \dfrac{1}{\delta^3 - \delta^2 - \delta + 1} ze^z$

$$= e^z \frac{1}{\delta^2(\delta + 2)} z$$

$$= e^z \frac{1}{2\delta^2}\left(1 - \frac{\delta}{2} + \cdots\right) z = e^z \frac{1}{2\delta^2}\left(z - \frac{1}{2}\right) = \frac{z^2e^z(2z-3)}{24}.$$

So the general solution is

$$y = (A + Bz)e^z + Ce^{-z} + \frac{z^2(2z-3)}{24} = (A + B\ln x)x + \frac{C}{x} + \frac{x(\ln x)^2(2\ln x - 3)}{24},$$

where A, B, C are arbitrary constants.

12. Solve: $(5 + 2x)^2D^2y - 6(5 + 2x)Dy + 8y = 8(5 + 2x)^2$.

Solution

Putting $5 + 2x = e^z$, i.e., $z = \log(5 + 2x)$ in the given equation we get

$$[4\delta(\delta - 1) - 12\delta + 8]y = 8e^{2z}, \text{ where } \delta = \frac{d}{dz}$$

$$\text{or, } [4\delta^2 - 16\delta + 8]y = 8e^{2z} \quad \text{or, } [\delta^2 - 4\delta + 2]y = 2e^{2z}.$$

So the auxiliary equation is $m^2 - 4m + 2 = 0$ whose roots are $m = 2 \pm \sqrt{2}$. Hence the C.F. is

$$y = e^{2z}\left[Ae^{\sqrt{2}z} + Be^{-\sqrt{2}z}\right].$$

The P.I. $= \dfrac{1}{\delta^2 - 4\delta + 2}2e^{2z} = \dfrac{1}{4 - 8 + 2}2e^{2z} = -e^{2z}$.

Hence the general solution is

$$y = e^{2z}\left[Ae^{\sqrt{2}z} + Be^{-\sqrt{2}z}\right] - e^{2z} = (5+2x)^2\left[A(5+2x)^{\sqrt{2}} + B(5+2x)^{-\sqrt{2}} - 1\right],$$

where A, B are arbitrary constants.

13. Solve: $(1 + 2x)^2D^2y - 6(1 + 2x)Dy + 16y = 8(1 + 2x)^2$.

Solution

Putting $1 + 2x = e^z$, i.e., $z = \ln(1 + 2x)$ in the given equation, we get

$$[4\delta(\delta-1)-12\delta+16]y = 8e^{2z}, \text{ where } \delta = \frac{d}{dz}$$

$$\text{or,} \quad [4\delta^2-16\delta+16]y = 8e^{2z}$$

$$\text{or,} \quad (\delta-2)^2 y = 2e^{2z}.$$

So the auxiliary equation is $(m - 2)^2 = 0$. Solving we get $m = 2, 2$. So, the C.F. is

$$y = (A+Bz)e^{2z}.$$

$$\text{The, P.I.} = \frac{1}{(\delta-2)^2}2e^{2z} = 2e^{2z}\frac{1}{\delta^2}1 = z^2e^{2z}.$$

Hence, the general solution is

$$y = (A+Bz)e^{2z} + z^2e^{2z} = (1+2x)^2[A + B\ln(1+2x) + \{\ln(1+2x)\}^2],$$

where A, B are arbitrary constants.

14. $(1 + x)^2D^2y + (1 + x)Dy + y = 4\cos\ln(1 + x)$.

Solution

Putting $1 + x = e^z$, i.e., $z = \ln(1 + x)$ in the given equation, we get

$$[\delta(\delta-1)+\delta+1]y = 4\cos z, \text{ where } \delta = \frac{d}{dz} \text{ or, } [\delta^2+1]y = 4\cos z.$$

So the auxiliary equation is $m^2 + 1 = 0$, whose roots are $m = \pm i$. Hence the C.F. is

$$y = A\cos z + B\sin z.$$

$$\text{The P.I.} = \frac{1}{\delta^2+1}4\cos z = -4\frac{z\sin z}{2} = -2z\sin z.$$

Hence the general solution is

$$y = A\cos z + B\sin z - 2z\sin z$$
$$= A\cos\ln(1+x) + B\sin\ln(1+x) - 2\ln(1+x)\sin\ln(1+x),$$

where A, B are arbitrary constants.

15. $(a + x)^2D^2y - 4(a + x)Dy + 6y = x$.

Solution

Putting $a + x = e^z$, i.e., $z = \log(a + x)$ in the given equation, we get

$$[\delta(\delta-1)-4\delta+6]y = e^z - a, \text{ where } \delta = \frac{d}{dz} \text{ or, } [\delta^2-5\delta+6]y = e^z - a.$$

So the auxiliary equation is $m^2 - 5m + 6 = 0$, whose roots are $m = 2, 3$. Hence the C.F. is

$$y = Ae^{2z} + Be^{3z}$$

$$\text{and P.I.} = \frac{1}{\delta^2-5\delta+6}e^z - a\frac{1}{\delta^2-5\delta+6}e^0 = \frac{1}{2}e^z - \frac{a}{6} = \frac{3e^z-a}{6}.$$

Hence the general solution is

$$y = Ae^{2z} + Be^{3z} + \frac{3e^z - a}{6} = A(x+a)^2 + B(x+a)^3 + \frac{3x+2a}{6}.$$

where A, B are arbitrary constants.

16. Solve: $x^2y_2 + xy_1 - y = x^2\log x$ by the method of variation of parameter.

Solution

We rewrite the given equation as

$$y_2 + \frac{1}{x}y_1 - \frac{1}{x^2}y = \log x.$$

Putting $x = e^z$, i.e., $z = \log x$ in the equation $x^2y_2 + xy_1 - y = 0$, we get

$$[\delta(\delta-1)+\delta-1]y = 0, \text{ where } \delta = \frac{d}{dz}.$$

So, the auxiliary equation is $m^2 - 1 = 0$, i.e., $m = \pm 1$. Hence, the complementary function of the given equation is

$$y = ae^z + be^{-z} = ax + \frac{b}{x}.$$

We assume the particular integral as

$$y_p = v_1(x)x + \frac{v_2(x)}{x}, \tag{1}$$

where $v_1(x)$, $v_2(x)$ are unknown functions. Differentiating this with respect to x, we get

$$y_p' = v_1(x) - \frac{v_2(x)}{x^2} \tag{2}$$

provided $$v_1'(x)x + \frac{v_2'(x)}{x} = 0. \tag{3}$$

Again differentiating (2) with respect to x, we get

$$y_p'' = v_1'(x) - \frac{v_2'(x)}{x^2} + \frac{2v_2(x)}{x^3}. \tag{4}$$

Using (1), (2), (4) in the given equation, we get

$$v_1'(x) - \frac{v_2'(x)}{x^2} = \log x. \tag{5}$$

Solving (3) and (5) we get $v_2'(x) = -\frac{1}{2}x^2\log x$, $v_1'(x) = \frac{1}{2}\log x$.

Integrating, we get $v_1(x) = \frac{x}{2}(\log x - 1)$, $v_2(x) = -\frac{x^3}{18}(3\log x - 1)$.

Hence, the P.I. is $y = \frac{x^2}{2}(\log x - 1) - \frac{x^2}{18}(3\log x - 1) = \frac{1}{3}x^2\log x - \frac{4}{9}x^2$.

So, the complete solution is $y = ax + \frac{b}{x} + \frac{1}{3}x^2\log x - \frac{4}{9}x^2$.

EXERCISE

- Solve the following equations:

1. $(x^2D^2 - xD + 1)y = 0$
2. $(x^2D^2 + xD + 1)y = 0$
3. $(x^2D^2 + 1)y = 0,$
4. $(x^2D^2 + xD - 1)y = 4,$
5. $(x^2D^2 - 2)y = x^2 + \frac{1}{x},$
6. $(x^2D^2 - 3xD + 4)y = 2x^2,$
7. $(x^2D^2 - 2xD + 2)y = \frac{1}{x},$
8. $(x^2D^2 + xD - 1)y = \sin(\log x) + x\cos(\log x),$
9. $(x^2D^2 - 3xD + 4)y = 2x^2,$
10. $(x^2D^2 + xD - 4)y = x^2,$
11. $(x^2D^2 - 4xD + 6)y = \ln x^2,$
12. $(x^2D^2 - 3xD + 5)y = \sin(\log x),$
13. $(x^2D^2 - 3xD + 5)y = x^2\sin(\log x),$
14. $(x^2D^2 + xD + 1)y = \log x \sin(\log x),$
15. $(x^3D^3 + x^2D^2)y = x,$
16. $(x^3D^3 + 2x^2D^2 - xD + 1)y = \frac{1}{x}$
17. $(x^3D^3 + 2x^2D^2 + 2)y = 10\left(x + \frac{1}{x}\right),$
18. $(x^3D^3 - x^2D^2 + 2xD - 2)y = x^3,$
19. $(x^3D^2 + xD - 1)y = 3x^4,$
20. $(x^3D^3 + 2xD - 2)y = 3x + x^2\log x,$
21. $(x^3D^3 + 3x^2D^2 + xD + 8)y = 65\cos(\log x),$
22. $(x^3D^3 - 4x^2D^2 + 5xD - 2)y = x^3,$
23. $(x^3D^3 + 3x^2D^2 + xD + 1)y = x\log x,$
24. $(x^4D^4 + 6x^3D^3 + 4x^2D^2 - 2xD - 4)y = x^2 + 2\cos(\log x),$
25. $(x^4D^4 + 6x^3D^3 + 9x^2D^2 + 3xD + 1)y = (1 + \log x)^2,$
26. $(x+a)^2D^2y - 4(x+a)Dy + 6y = x,$
27. $(1+x)^2D^2y + (1+x)Dy + y = 4\cos\ln(1+x),$
28. $(1+x)^2D^2y + (1+x)Dy + y = 2\sin\ln(1+x),$
29. $(1+x)^2D^2y - 3(1+x)Dy + 4y = x,$
30. $(1+x)^2D^2y + (1+x)Dy = 4x^2 + 14x + 12,$
31. $(2x-1)^3D^3y + (2x-1)Dy - 2y = 0,$
32. $(2+3x)^2D^2y + 5(2+3x)Dy - 3y = x^2 + x + 1,$
33. $(2+x)^2D^2y + (2+x)Dy + 4y = 2\sin\{2\ln(2+x)\}.$

- Solve the following equations using method of variation of parameter:

34. $x^2y''+xy'-y=x^2e^x$,
35. $x^2y''+xy'-y=x^2e^{2x}$,
36. $x^2y''+xy'-y=x$,
37. $x^2y''-xy'=x^3e^x$,
38. $x^2y''+3xy'+y=\dfrac{1}{(1-x)^2}$,
39. $x^2y''+xy'-y=x^2\log x$,
40. $x^2y''+2xy'=\log x$,
41. $x^2y''-2xy'+2y=x$,
42. $x^2y''+5xy'+4y=x\log x$,
43. $x^2y''-xy'+2y=x\log x$.

ANSWERS

1. $(c_1+c_2\ln x)x$,
2. $c_1\cos(\ln x)+c_2\sin(\ln x)$,
3. $\sqrt{x}\left[c_1\cos\dfrac{\sqrt{3}\ln x}{2}+c_2\sin\dfrac{\sqrt{3}\ln x}{2}\right]$,
4. $c_1x+\dfrac{c_2}{x}-4$,
5. $c_1x^2+\dfrac{c_2}{x}+\dfrac{1}{3}\left(x^2-\dfrac{1}{x}\right)\ln x$,
6. $(c_1+c_2\ln x)x^2+(x\ln x)^2$,
7. $c_1x+c_2x^2+\dfrac{1}{6x}$,
8. $c_1x+\dfrac{c_2}{x}-\dfrac{1}{2}\sin(\ln x)+\dfrac{x}{5}[2\sin(\ln x)-\cos(\ln x)]$,
9. $(c_1+c_2\ln x)x^2+x^2(\ln x)^2$,
10. $c_1x^2+c_2x^{-2}+\dfrac{1}{4}x^2\ln x$,
11. $c_1x^3+c_2x^2+\dfrac{1}{18}(3\ln x^2+5)$,
12. $x^2[c_1\cos(\ln x)+c_2\sin(\ln x)]+\dfrac{1}{8}[\sin(\ln x)+\cos(\ln x)]$,
13. $x^2[A\cos(\ln x)+B\sin(\ln x)]-\dfrac{x^2}{2}\ln x\cos(\ln x)$,
14. $c_1\cos(\ln x)+c_2\sin(\ln x)+\dfrac{\ln x}{4}[\sin(\ln x)-\ln x\cos(\ln x)]$,
15. $c_1+(c_2+c_3\ln x)x+\dfrac{x}{2}(\ln x)^2$,
16. $(c_1+c_2\ln x)x+\dfrac{c^3}{x}+\dfrac{1}{4x}\ln x$,
17. $\dfrac{c_1}{x}+x[c_2\cos(\ln x)+c_3\sin(\ln x)]+\dfrac{2}{x}\ln x+5x$,
18. $(c_1+c_2\ln x)x+c_3x^2+\dfrac{x^3}{4}$,
19. $c_1x+c_2x\ln x+c_3x(\ln x)^2+\dfrac{x^4}{9}$,

20. $x[c_1 + c_2 \cos(\ln x) + c_3 \sin(\ln x) + 3\ln x] + \frac{1}{2}x^2(\ln x - 2)$,

21. $\frac{c_1}{x^2} + x\{c_2 \cos(\sqrt{3}\ln x) + c_3 \sin(\sqrt{3}\ln x)\} + 8\cos(\ln x) - \sin(\ln x)$,

22. $c_1 x^2 + x^{\frac{5}{2}}\left\{c_2 x^{\frac{\sqrt{21}}{2}} + c_3 x^{-\frac{\sqrt{21}}{2}}\right\} - \frac{1}{5}x^3$,

23. $\frac{c_1}{x} + \sqrt{x}\left[c_2 \cos\left(\frac{\sqrt{3}\ln x}{2}\right) + c_3 \sin\left(\frac{\sqrt{3}\ln x}{2}\right)\right] + \frac{x}{2}\left(\ln x - \frac{3}{2}\right)$,

24. $c_1 x^2 + c_2 x^{-2} + c_3 \cos(\ln x) + c_4 \sin(\ln x) + \frac{1}{20}x^2 \ln x - \frac{1}{5}\ln x \sin(\ln x)$,

25. $(c_1 + c_2 \ln x)\sin(\ln x) + (c_3 + c_4 \ln x)\cos(\ln x) + (\ln x)^2 + 2\ln x - 3$,

26. $c_1(x+a)^2 + c_2(x+a)^3 + \frac{(x+a)}{2} - \frac{a}{6}$,

27. $c_1 \cos\{\ln(1+x) - c_2\} + 2\ln(1+x)\sin \ln(1+x)$,

28. $c_1 \cos \ln(1+x) + c_2 \sin \ln(1+x) - \ln(1+x)\cos \ln(1+x)$,

29. $(1+x)^2[c_1 + c_2 \ln(1+x)] + x + \frac{3}{4}$,

30. $c_1 + c_2 \ln(1+x) + \{\ln(1+x)\}^2 + x^2 + 8x$,

31. $c_1(2x-1) + c_2(2x-1)\cosh\left[\frac{\sqrt{3}}{2}\ln(2x-1) + c_3\right]$,

32. $c_1(3x+2)^{-1} + c_2(3x+2)^{\frac{1}{3}} + \frac{1}{405}(3x+2)^2 - \frac{1}{108}(3x+2) - \frac{7}{27}$,

33. $A\cos\{2\ln(2+x)\} + B\sin\{2\ln(2+x)\} - \frac{1}{2}\ln(2+x)\cos\{2\ln(2+x)\}$,

34. $ax + \frac{b}{x} + e^x\left(1 - \frac{1}{x}\right)$,

35. $ax + \frac{b}{x} + \frac{1}{8}e^{2x}\left(2 - \frac{1}{x}\right)$,

36. $ax + \frac{b}{x} + \frac{x^3}{8}$,

37. $a + bx + (x-1)e^x$,

38. $\frac{a}{x} + \frac{b}{x}\ln x + \frac{1}{x}\ln\left(\frac{x}{1-x}\right)$,

39. $ax + \frac{b}{x} + \frac{x^2}{3}\ln x - \frac{4x^2}{9}$,

40. $a + \frac{b}{x} + \frac{1}{2}(\ln x)^2 - \ln x$,

41. $ax + bx^2 - x(1 + \ln x)$,

42. $(a + b\ln x)x^{-2} + \frac{x}{27}(3\ln x - 2)$,

43. $x[a\cos(\ln x) + b\sin(\ln x)] + x\ln x$.

Chapter 8

SECOND ORDER DIFFERENTIAL EQUATION WITH VARIABLE COEFFICIENTS

8.1 Introduction

The general form of a linear second order differential equation is

$$\frac{d^2y}{dx^2}+P\frac{dy}{dx}+Qy=R, \tag{1}$$

where P, Q, R are functions of x only.

In this chapter we will consider the following cases and methods:

- Complete solution of the above equation in terms of known integral,
- Solution of the associated homogeneous equation obtained by inspection,
- Reduction of the above equation to its normal form (i.e., by removing $\frac{dy}{dx}$),
- Relation between two integrals,
- Method of operational factors,
- Change of independent variable.

8.2 Complete Solution of a Second Order Linear Differential Equation in Terms of Known Integral

Let $y = u(x)$ be a known integral of the complementary function of the equation given by (1). Then we have

$$\frac{d^2u}{dx^2}+P\frac{du}{dx}+Qu=0. \tag{2}$$

Let $y = u(x)\,v(x)$ be the complete solution of the equation (1), then using

$$\frac{dy}{dx}=u\frac{dv}{dx}+v\frac{du}{dx} \text{ and } \frac{d^2y}{dx^2}=u\frac{d^2v}{dx^2}+2\frac{dv}{dx}\frac{du}{dx}+v\frac{d^2u}{dx^2}$$

in equation (1) we get

$$u\frac{d^2v}{dx^2}+\left(2\frac{du}{dx}+Pu\right)\frac{dv}{dx}+\left(\frac{d^2u}{dx^2}+P\frac{du}{dx}+Qu\right)v=R.$$

Using the result (2) in the above equation we get

$$\frac{d^2v}{dx^2}+\left(\frac{2}{u}\frac{du}{dx}+P\right)\frac{dv}{dx}=\frac{R}{u}$$

which is a linear differential equation in $\frac{dv}{dx}$ and can be solved, if $u(x)$ be known. Let

$$\frac{dv}{dx} = f(x, C), \text{ where } C \text{ is an arbitrary constant.}$$

Again integrating, we get

$$v = \int f(x, C)dx + D, \text{ where } D \text{ is an arbitrary constant.}$$

Hence, the complete solution of the equation (1) is

$$y = uv = u\int f(x, C)dx + Du.$$

8.3 Solution of the Associated Homogeneous Equation Obtained by Inspection

From the above article, it is clear that complete solution of a second order linear differential equation can be found easily if one of the integrals of its complementary function is available. In order to adopt this technique we must know one of the integrals of the complementary function of the differential equation.

Let $y = e^{mx}$ be a trial solution of the equation,

$$\frac{d^2y}{dx^2} + P\frac{dy}{dx} + Qy = 0. \tag{3}$$

Then the auxiliary equation is $m^2 + Pm + Q = 0$ as $e^{mx} \neq 0$. This shows that e^{mx} will be the complementary function of the equation if $m^2 + Pm + Q = 0$. Using this we see that

a. e^x is a solution of (3), if $1 + P + Q = 0$,

b. e^{-x} is a solution of (3), if $1 - P + Q = 0$,

c. e^{sx}, $s \neq 0$, is a solution of (3), if $1 + \frac{P}{s} + \frac{Q}{s^2} = 0$.

Let $y = x^m$ be a trial solution of the equation (3), then we have

$$m(m-1) + Pmx + Qx^2 = 0.$$

Using this we see that

a. x is a solution of (3), if $P + Qx = 0$,

b. x^s is a solution of (3), if $s(s - 1) + Psx + Qx^2 = 0$.

ILLUSTRATIVE EXAMPLES

1. Solve $x^2\frac{d^2y}{dx^2} + x\frac{dy}{dx} - 9y = 0$, given that $y = x^3$ is a solution.

Solution

Let $y = vx^3$, where v is a function, be the complete integral of the given equation. Here

$$P = \frac{1}{x}, R = 0.$$

So the transformed equation is

$$\frac{d^2v}{dx^2}+\left(\frac{2}{x^3}3x^2+\frac{1}{x}\right)\frac{dv}{dx}=0 \text{ or, } \frac{dp}{dx}+\frac{7}{x}p=0, \text{ where } p=\frac{dv}{dx}.$$

This is a linear differential equation in $p=\frac{dv}{dx}$, whose integrating factor is $e^{7\int\frac{dx}{x}}=e^{7\ln x}=x^7$. Multiplying the above equation by x^7, we get

$$\frac{d}{dx}(px^7)=0 \quad \text{or,} \quad \frac{dv}{dx}=\frac{C}{x^7}.$$

Again integrating, we get

$$v=-\frac{C}{6x^6}+D.$$

So the complete integral of the given equation is

$$y=-\frac{C}{6x^3}+Dx^3, \text{ where } C, D \text{ are arbitrary constants.}$$

2. Solve $x\frac{d^2y}{dx^2}-(2x-1)\frac{dy}{dx}+(x-1)y=0$, given that $y=e^x$ is a solution.

Solution

Let $y=ve^x$, where v is a function, be the complete integral of the given equation. Here

$$P=\frac{1-2x}{x}, R=0.$$

So the transformed equation is

$$\frac{d^2v}{dx^2}+\left(2e^{-x}e^x-2+\frac{1}{x}\right)\frac{dv}{dx}=0 \text{ or, } \frac{d^2v}{dx^2}+\frac{1}{x}\frac{dv}{dx}=0 \text{ or, } \frac{dp}{dx}+\frac{p}{x}=0.$$

This is a linear differential equation in $p=\frac{dv}{dx}$, whose integrating factor is $e^{\int\frac{dx}{x}}=e^{\ln x}=x$. Multiplying the above equation by x, we get

$$\frac{d}{dx}(px)=0, \text{ or, } \frac{dv}{dx}=\frac{C}{x}.$$

Again integrating, we get

$$v=C\ln x+D.$$

So the complete integral of the given equation is

$$y=(C\ln x+D)e^x, \text{ where } C, D \text{ are arbitrary constants.}$$

3. Solve: $(1-x^2)\frac{d^2y}{dx^2}+x\frac{dy}{dx}-y=x(1-x^2)^{3/2}$, in terms of known integral.

Solution

Comparing the given equation with the standard form of second order linear differential equation, we get

$$P=\frac{x}{1-x^2}, Q=\frac{-1}{1-x^2} \text{ and } R=x\sqrt{1-x^2}.$$

Here, we see that $P + xQ = 0$, so $y = x$ is a part of the complementary function, i.e., a known integral. We assume $y = xv$, where v is a function of x only, be the complete integral of the given equation. Using this in the given equation, we get

$$\frac{d^2v}{dx^2}+\left(\frac{2}{x}+\frac{x}{1-x^2}\right)\frac{dv}{dx}=\sqrt{1-x^2}$$

or, $$\frac{dp}{dx}+\left(\frac{2}{x}+\frac{x}{1-x^2}\right)p=\sqrt{1-x^2}, \text{ where } p=\frac{dv}{dx}.$$

This is a linear differential equation whose integrating factor is

$$e^{\int\left(\frac{2}{x}+\frac{x}{1-x^2}\right)dx}=e^{2\ln x-\frac{1}{2}\ln(1-x^2)}=\frac{x^2}{\sqrt{1-x^2}}.$$

Multiplying the above equation by this factor, we get

$$\frac{d}{dx}\left(\frac{px^2}{\sqrt{1-x^2}}\right)=x^2.$$

Integrating, we get

$$p=\frac{dv}{dx}=\frac{\sqrt{1-x^2}}{x^2}\left(\frac{1}{3}x^3+C\right).$$

Again integrating, we get

$$v=-\frac{1}{9}(1-x^2)^{3/2}-C\left[\frac{\sqrt{1-x^2}}{x}+\sin^{-1}x\right]+D.$$

Hence, the complete integral of the above equation is

$$y=-\frac{1}{9}x(1-x^2)^{3/2}-C\left[\sqrt{1-x^2}+x\sin^{-1}x\right]+Dx,$$

where C, D are arbitrary constants.

4. Solve: $x^2\dfrac{d^2y}{dx^2}-2x(1+x)\dfrac{dy}{dx}+2(1+x)y=x^3$, in terms of known integral.

Solution

Proceeding as Ex.3 we see that $P=-2\left(1+\frac{1}{x}\right), Q=2\left(\frac{1}{x}+\frac{1}{x^2}\right)$ and $R = x$. So $P + xQ = 0$, i.e., $y = x$ is a part of the complementary function, i.e., a known integral. We assume $y = xv$, where v is a function of x only, is the complete integral of the given equation. Using this in the given equation, we get

$$\frac{d^2v}{dx^2}+\left(\frac{2}{x}-\frac{2(1+x)}{x}\right)\frac{dv}{dx}=1 \quad \text{or,} \quad \frac{d^2v}{dx^2}-2\frac{dv}{dx}=1 \quad \text{or,} \quad \frac{dp}{2p+1}=dx.$$

Integrating, we get

$$\ln(2p+1)=2x+C$$

or, $$2p+1=Ke^{2x},$$

or, $$2dv=(Ke^{2x}-1)dx.$$

Integrating, we get

$$v = \frac{K}{4}e^{2x} - \frac{x}{2} + D.$$

So the complete integral is

$$y = xv = \frac{K}{4}xe^{2x} - \frac{x^2}{2} + Dx, \text{ where } K, D \text{ are arbitrary constants.}$$

5. Solve $\frac{d^2y}{dx^2} - 2\tan x\frac{dy}{dx} + 3y = 2\sec x$, given that $y = \sin x$ is a part of its complementary function.

Solution

Let $y = v\sin x$, where v is a function, be the complete integral of the given equation. Here

$$P = -2\tan x, R = 2\sec x.$$

So the transformed equation is

$$\frac{d^2v}{dx^2} + \left(\frac{2}{\sin x}\cos x - 2\tan x\right)\frac{dv}{dx} = \frac{2}{\sin x \cos x}$$

or, $$\frac{dp}{dx} + 2(\cot x - \tan x)p = 4\operatorname{cosec} 2x.$$

This is a linear differential equation $p = \frac{dv}{dx}$ whose integrating factor is

$$e^{2\int(\cot x - \tan x)dx} = e^{2(\ln\sin x + \ln\cos x)} = e^{\ln(\sin x\cos x)^2} = \frac{1}{4}\sin^2 2x.$$

Multiplying the above equation by this factor we get

$$\frac{d}{dx}\left(\frac{p}{4}\sin^2 2x\right) = \sin 2x.$$

Integrating, we get

$$p = \frac{dv}{dx} = \frac{4}{\sin^2 2x}\left(-\frac{1}{2}\cos 2x + C\right) = -2\cot 2x\operatorname{cosec} 2x + 4C\operatorname{cosec}^2 2x.$$

Again integrating, we get

$$v = \operatorname{cosec} 2x - 2C\cot 2x + D.$$

So the complete integral of the given equation is

$$y = v\sin x = \frac{1}{2}\sec x + D\sin x - C\frac{\cot^2 x - 1}{\cot x}\sin x$$

$$= \frac{1}{2}\sec x + D\sin x + C(\sec x - 2\cos x)$$

$$= \left(A\cos 2x + B\sin 2x + \frac{1}{2}\right)\sec x, \text{ where } A, B \text{ are two arbitrary constants.}$$

8.4 Reduction to Normal Form

In the article 8.3, we have discussed a few special cases of determining one of the parts of the complementary function of a second order linear differential equation and in such cases the method of solving the equation discussed in the article 8.2 can be applied. In this section, we will discuss a method of solving a second order linear differential equation when no integral solution of its complementary function is available as it is not always possible to have such solution in hand.

In the article 8.2, we have shown that under the substitution $y = uv$, equation (1) becomes

$$u\frac{d^2v}{dx^2}+\left(2\frac{du}{dx}+Pu\right)\frac{dv}{dx}+\left(\frac{d^2u}{dx^2}+P\frac{du}{dx}+Qu\right)v=R.$$

If we set, $2\frac{du}{dx}+Pu=0$, i.e., $u=e^{-\frac{1}{2}\int Pdx}$, then the above equation reduces to

$$u\frac{d^2v}{dx^2}+\left(\frac{d^2u}{dx^2}+P\frac{du}{dx}+Qu\right)v=R.$$

Now, from $u=e^{-\frac{1}{2}\int Pdx}$, we get

$$\frac{du}{dx}=-\frac{1}{2}Pu \text{ and } \frac{d^2u}{dx^2}=-\frac{1}{2}\left[P\frac{du}{dx}+u\frac{dP}{dx}\right]=\frac{1}{4}P^2u-\frac{1}{2}u\frac{dP}{dx}.$$

Using these results in the above equation, we get

$$u\frac{d^2v}{dx^2}+\left(Qu-\frac{1}{2}u\frac{dP}{dx}-\frac{1}{4}P^2u\right)v=R \text{ or, } \frac{d^2v}{dx^2}+Lv=S, \tag{4}$$

where $L=Q-\frac{1}{2}\frac{dP}{dx}-\frac{1}{4}P^2$ and $S=Re^{\frac{1}{2}\int Pdx}$. This is called the *normal form* of the equation (1). For known values of P, Q, R both u, v can be found easily and hence the complete integral of the equation (1) can be determined.

8.5 Relation between Two Integrals

If u, v be known integrals of the equation

$$\frac{d^2y}{dx^2}+P\frac{dy}{dx}+Qy=0,$$

then we have
$$\frac{d^2u}{dx^2}+P\frac{du}{dx}+Qu=0$$

and
$$\frac{d^2v}{dx^2}+P\frac{dv}{dx}+Qv=0.$$

Multiplying the equations by v, u respectively and then subtracting, we obtain

$$v\frac{d^2u}{dx^2}-u\frac{d^2v}{dx^2}+P\left(v\frac{du}{dx}-u\frac{dv}{dx}\right)=0$$

or,
$$\frac{d}{dx}\left(v\frac{du}{dx}-u\frac{dv}{dx}\right)+P\left(v\frac{du}{dx}-u\frac{dv}{dx}\right)=0$$

or, $$\frac{d\left(v\frac{du}{dx}-u\frac{dv}{dx}\right)}{\left(v\frac{du}{dx}-u\frac{dv}{dx}\right)}=-Pdx.$$

Integrating, we get

$$v\frac{du}{dx}-u\frac{dv}{dx}=Ae^{-\int Pdx}. \tag{5}$$

This is the relation between two integrals u, v of the given differential equation.

The above equation is a linear differential in one of the integrals if other is known and can be solved easily. Thus the above relation between the integrals can be applied to get the complete integral of a second order linear differential equation when one of those is known.

WORKED-OUT EXAMPLES

1. Solve $x^2\frac{d^2y}{dx^2}-2x\frac{dy}{dx}+y(x^2a^2+2)=0$, by reducing it to normal form.

Solution

Here $P=-\frac{2}{x}, Q=a^2+\frac{2}{x^2}$ and $R=0$. To reduce it to normal form we choose

$$u=e^{\int\frac{dx}{x}}=e^{\ln x}=x.$$

Now we put $y=uv=xv$, where v is a function of x only, in the given equation and get

$$\frac{d^2v}{dx^2}+Lv=0,$$

where $L=Q-\frac{1}{2}\frac{dP}{dx}-\frac{1}{4}P^2=a^2+\frac{2}{x^2}-\frac{1}{2}\frac{2}{x^2}-\frac{1}{4}\frac{4}{x^2}=a^2.$

Solving the equation $\frac{d^2v}{dx^2}+a^2v=0$, we get

$$v=A\cos ax+B\sin ax.$$

Hence, the complete integral is

$$y=x(A\cos ax+B\sin ax),\ \text{where } A, B \text{ are arbitrary constants.}$$

2. Solve $\frac{d^2y}{dx^2}-\frac{2}{x}\frac{dy}{dx}+y\left(1+\frac{2}{x^2}\right)=xe^x$, by reducing it to normal form.

Solution

Here $P=-\frac{2}{x}, Q=1+\frac{2}{x^2}$ and $R=xe^x$. To reduce it to normal form we choose

$$u=e^{\int\frac{dx}{x}}=e^{\ln x}=x.$$

Now we put $y = uv = xv$, where v is a function of x only, in the given equation and get

$$\frac{d^2v}{dx^2} + Lv = S,$$

where $L = Q - \frac{1}{2}\frac{dP}{dx} - \frac{1}{4}P^2 = 1 + \frac{2}{x^2} - \frac{1}{2}\frac{2}{x^2} - \frac{1}{4}\frac{4}{x^2} = 1$ and $S = \frac{R}{u} = e^x$.

Solving the equation $\frac{d^2v}{dx^2} + v = e^x$, we get

$$v = A\cos x + B\sin x + \frac{1}{D^2+1}e^x = A\cos x + B\sin x + \frac{1}{2}e^x.$$

Hence, the complete integral is

$$y = x\left(A\cos x + B\sin x + \frac{1}{2}e^x\right), \text{ where } A, B \text{ are arbitrary constants.}$$

3. Solve $x^2\frac{d^2y}{dx^2} - 2x(3x-2)\frac{dy}{dx} + 3x(3x-4)y = e^{3x}$, by reducing it to normal form.

Solution

Here $P = -2\left(3 - \frac{2}{x}\right), Q = 3\left(3 - \frac{4}{x}\right)$ and $R = \frac{e^{3x}}{x^2}$. To reduce it to normal form we choose

$$u = e^{\int\left(3-\frac{2}{x}\right)dx} = \frac{e^{3x}}{x^2}.$$

Now we put $y = uv = v\frac{e^{3x}}{x^2}$, where v is a function of x only, in the given equation and get

$$\frac{d^2v}{dx^2} + Lv = S,$$

where $L = Q - \frac{1}{2}\frac{dP}{dx} - \frac{1}{4}P^2 = 3\left(3 - \frac{4}{x}\right) + \frac{2}{x^2} - \left(3 - \frac{2}{x}\right)^2 = -\frac{2}{x^2}$ and $S = \frac{R}{u} = 1$.

So the normal form of the given equation is

$$\frac{d^2v}{dx^2} - \frac{2}{x^2}v = 1.$$

Now, if we put $x = e^z$, then the above equation becomes

$$[\delta(\delta-1) - 2]v = e^{2z} \quad \text{or,} \quad [\delta^2 - \delta - 2]v = e^{2z}.$$

Its solution is

$$v = Ae^{2z} + Be^{-z} + \frac{1}{(\delta-2)(\delta+1)}e^{2z} = Ae^{2z} + Be^{-z} + \frac{1}{3(\delta-2)}e^{2z}$$

$$= Ae^{2z} + Be^{-z} + e^{2z}\frac{1}{3(\delta+2-2)}1 = Ax^2 + \frac{B}{x} + \frac{1}{3}x^2\ln x.$$

So, the complete solution of the given equation is

$$y=\left(Ax^2+\frac{B}{x}+\frac{1}{3}x^2\ln x\right)\frac{e^{3x}}{x^2}.$$

4. Solve $x^2\dfrac{d^2y}{dx^2}+x\dfrac{dy}{dx}-y=0$, given that $x+\dfrac{1}{x}$ is one of its solutions.

Solution

Here $P=\dfrac{1}{x}$. Taking $v=x+\dfrac{1}{x}$ and using (5), we get

$$\left(x+\frac{1}{x}\right)\frac{du}{dx}-u\left(1-\frac{1}{x^2}\right)=Ae^{-\int\frac{dx}{x}} \quad \text{or,} \quad \frac{du}{dx}-\frac{x^2-1}{x^2+1}\frac{1}{x}u=\frac{A}{x^2+1}.$$

This is a linear differential equation in u, whose integrating factor is

$$e^{-\int\frac{x^2-1}{x^2+1}\frac{1}{x}dx}=e^{-\ln\left(x+\frac{1}{x}\right)}=\frac{x}{x^2+1}.$$

Multiplying the above equation by this factor, we get

$$\frac{d}{dx}\left(\frac{xu}{x^2+1}\right)=\frac{Ax}{(x^2+1)^2}.$$

Integrating, we get

$$\frac{xu}{x^2+1}=-\frac{1}{2}\frac{A}{x^2+1}+B \quad \text{or,} \quad u=-\frac{1}{2}\frac{A}{x}+B\left(x+\frac{1}{x}\right).$$

Hence, the complete integral of the given equation is

$$y=\frac{C}{x}+D\left(x+\frac{1}{x}\right),$$ where C, D are two arbitrary constants.

5. Solve $x^2\dfrac{d^2y}{dx^2}+x\dfrac{dy}{dx}-9y=0$, given that x^3 is one of its solutions.

Solution

Here $P=\dfrac{1}{x}$. Taking $v=x^3$ and using (5), we get

$$x^3\frac{du}{dx}-3x^2u=Ae^{-\int\frac{dx}{x}}=Ae^{-\ln x} \quad \text{or,} \quad \frac{du}{dx}-\frac{3}{x}u=\frac{A}{x^4}.$$

This is a linear differential equation in u whose integrating factor is $e^{-3\int\frac{dx}{x}}=e^{-3\ln x}=\dfrac{1}{x^3}$. Multiplying the above equation by this factor, we get

$$\frac{d}{dx}\left(\frac{u}{x^3}\right)=\frac{A}{x^7}.$$

Integrating, we get

$$\frac{u}{x^3}=-\frac{A}{6x^6}+B \quad \text{or,} \quad u=-\frac{A}{6}\frac{1}{x^3}+Bx^3.$$

So the complete integral of the given equation is

$$y = \frac{C}{x^3} + Dx^3, \text{ where } C, D \text{ are two arbitrary constants.}$$

6. Solve $x\frac{d^2y}{dx^2} + (1-2x)\frac{dy}{dx} + (x-1)y = 0$, given that e^x is one of its solutions.

Solution

Here $P = \frac{1}{x} - 2$. Taking $v = e^x$ and using (5), we get

$$e^x \frac{du}{dx} - ue^x = Ae^{-\int\left(\frac{1}{x}-2\right)dx} = A\frac{e^{2x}}{x} \quad \text{or,} \quad \frac{du}{dx} - u = A\frac{e^x}{x}.$$

This is a linear differential equation in u, whose integrating factor is $e^{-\int dx} = e^{-x}$. Multiplying the above equation by e^{-x}, we get

$$\frac{d}{dx}\left(ue^{-x}\right) = \frac{A}{x}.$$

Integrating, we get

$$ue^{-x} = A\ln x + B \quad \text{or,} \quad u = (A\ln x + B)e^x.$$

Hence, the complete integral of the given equation is

$$y = (C\ln x + D)e^x, \text{ where } C, D \text{ are two arbitrary constants.}$$

8.6 Method of Operational Factors

The equation (1) can be written as

$$f(D)y = R$$

where $f(D) = D^2 + PD + Q$. Now if $f(D)$ can be factorised as $f(D) = f_1(D)f_2(D)$, where $f_1(D), f_2(D)$ are two linear factors of $f(D)$, then above equation can be written as

$$f_1(D)f_2(D)y = R.$$

where $f_2(D)y = v$, and $f_1(D)v = R$. Thus solution of a second order linear differential equation is confined in finding the solutions of two linear equations. To get it we first solve the linear equation in v given by $f_1(D)v = R$. Then using the value of v in $f_2(D)y = v$, we get a linear equation in y. This can be solved for y, i.e., for the solution of the given equation.

It is to be mentioned that the factors $f_1(D)$, $f_2(D)$ of $f(D)$ are not commutative if at least one of those contains the variable x, that is why the factors must be put in right order.

WORKED-OUT EXAMPLES

1. Solve $xD^2y+(x-2)Dy-2y=x^3$, where $D=\frac{d}{dx}$, by the method of operational factors.

Solution

Here we see that $xD^2+(x-2)D-2=xD(D+1)-2(D+1)=(xD-2)(D+1)$. So the given equation can be written as

$$(xD-2)(D+1)y=x^3.$$

First we solve the linear differential equation

$$(xD-2)v=x^3 \text{ or, } Dv-\frac{2}{x}v=x^2,$$

whose integrating factor is $e^{-2\int\frac{dx}{x}}=e^{-2\ln x}=\frac{1}{x^2}$. Multiplying the above equation by this factor, we get

$$\frac{d}{dx}\left(\frac{v}{x^2}\right)=1.$$

Integrating, we get $$\frac{v}{x^2}=x+A$$

or, $$v=x^3+Ax^2.$$

Using this in the given equation we get $(D+1)y=x^3+Ax^2$. This is again a linear differential equation whose integrating factor is e^x. Multiplying the equation by e^x, we get

$$\frac{d}{dx}(ye^x)=(x^3+Ax^2)e^x.$$

Integrating we get

$$ye^x=x^3e^x+(A-3)\int x^2e^x dx+B=x^3e^x+C(x^2-2x+2)e^x+B$$

or, $y=x^3+C(x^2-2x+2)+Be^{-x}$, where B, C are arbitrary constants.

2. Solve $x^2D^2y-3xDy+3y=2x^3-x^2$, where $D=\frac{d}{dx}$, by the method of operational factors.

Solution

Here we see that

$$x^2D^2-3xD+3=x(xD^2+D)-xD-3(xD-1)$$
$$=xD(xD-1)-3(xD-1)=(xD-3)(xD-1).$$

So the given equation can be written as

$$(xD-3)(xD-1)y=2x^3-x^2.$$

First we solve the linear differential equation,

$$(xD-3)v=2x^3-x^2 \text{ or, } Dv-\frac{3}{x}v=2x^2-x.$$

This is a linear differential equation whose integrating factor is

$e^{-3\int\frac{dx}{x}}=e^{-3\ln x}=\frac{1}{x^3}$. Multiplying the above equation by this factor, we get

$$\frac{d}{dx}\left(\frac{v}{x^3}\right) = \frac{2}{x} - \frac{1}{x^2}.$$

Integrating, we get $\frac{v}{x^3} = 2\ln x + \frac{1}{x} + A$ or, $v = 2x^3 \ln x + x^2 + Ax^3$.

Using this in the given equation, we get

$$(xD - 1)y = 2x^3 \ln x + x^2 + Ax^3 \text{ or, } Dy - \frac{1}{x}y = 2x^2 \ln x + x + Ax^2.$$

This is again a linear differential equation, whose integrating factor is $e^{-\int \frac{dx}{x}} = e^{-\ln x} = \frac{1}{x}$. Multiplying the equation by $\frac{1}{x}$, we get

$$\frac{d}{dx}\left(\frac{y}{x}\right) = 2x \ln x + 1 + Ax.$$

Integrating, we get

$$\frac{y}{x} = x^2\left(\ln x - \frac{1}{2}\right) + x + A\frac{x^2}{2} + B$$

or, $y = x^2(x \ln x + 1) + Cx^3 + Bx$, where B, C are two arbitrary constants.

8.7 Change of Independent Variable

If we put $z = f(x)$ in the differential equation (1), then it becomes

$$\frac{d^2y}{dz^2}\left(\frac{dz}{dx}\right)^2 + \frac{dy}{dz}\frac{d^2z}{dx^2} + P\frac{dy}{dz}\frac{dz}{dx} + Qy = R$$

or,
$$\frac{d^2y}{dz^2}\left(\frac{dz}{dx}\right)^2 + \left(P\frac{dz}{dx} + \frac{d^2z}{dx^2}\right)\frac{dy}{dz} + Qy = R$$

or,
$$\frac{d^2y}{dz^2} + P_1\frac{dy}{dz} + Q_1 y = R_1,$$

where $P_1 = \frac{\left(P\frac{dz}{dx} + \frac{d^2z}{dx^2}\right)}{\left(\frac{dz}{dx}\right)^2}$, $Q_1 = \frac{Q}{\left(\frac{dz}{dx}\right)^2}$ and $R_1 = \frac{R}{\left(\frac{dz}{dx}\right)^2}$.

If we choose z in such a way that $P\frac{dz}{dx} + \frac{d^2z}{dx^2} = 0$, then the above equation is transformed to

$$\frac{d^2y}{dz^2} + Q_1 y = R_1.$$

Thus, if we choose $\frac{dz}{dx} = e^{-\int P dx}$, i.e., $z = \int e^{-\int P dx} dx$, then equation (1) reduces to the above form which can be solved, if $Q_1 = K$ or, $\frac{K}{z^2}$.

Further, if we choose z in such a way that both P_1, Q_1 become constant, then the transformed equation can be solved easily.

WORKED-OUT EXAMPLES

1. Solve $\frac{d^2y}{dx^2}+\frac{2}{x}\frac{dy}{dx}+\frac{a^2}{x^4}y=0$ by changing the independent variables.

Solution

Here $P=\frac{2}{x}, Q=\frac{a^2}{x^4}$ and $R=0$. Now, we put

$$z=\int e^{-\int P dx}dx=\int e^{-2\int\frac{dx}{x}}dx=\int\frac{1}{x^2}dx=-\frac{1}{x}.$$

Then the given equation reduces to

$$\frac{d^2y}{dz^2}+Q_1y=R_1,$$

where $Q_1=\frac{Q}{\left(\frac{dz}{dx}\right)^2}=a^2$ and $R_1=\frac{R}{\left(\frac{dz}{dx}\right)^2}=0.$

i.e., reduces to

$$\frac{d^2y}{dz^2}+a^2y=0.$$

Solving it, we get

$$y=A\cos az+B\sin az=A\cos\left(\frac{a}{x}\right)-B\sin\left(\frac{a}{x}\right).$$

2. Solve $\frac{d^2y}{dx^2}-\frac{1}{x}\frac{dy}{dx}-4x^2y=8x^2\sin x^2$ by changing the independent variable.

Solution

Here $P=-\frac{1}{x}, Q=-4x^2$ and $R=8x^2\sin x^2$. Now, we put

$$z=\int e^{-\int P dx}dx=\int e^{\int\frac{dx}{x}}dx=\int x\,dx=\frac{x^2}{2}.$$

Then the given equation reduces to

$$\frac{d^2y}{dz^2}+Q_1y=R_1,$$

where $Q_1=\frac{Q}{\left(\frac{dz}{dx}\right)^2}=-4$ and $R_1=\frac{R}{\left(\frac{dz}{dx}\right)^2}=8\sin x^2.$

i.e., reduces to

$$\frac{d^2y}{dz^2}-4y=8\sin 2z.$$

Solving it, we get

$$y = Ae^{2z} + Be^{-2z} + 8\frac{1}{D^2-4}\sin 2z = Ae^{2z} + Be^{-2z} + 8\frac{1}{-4-4}\sin 2z$$

$$= Ae^{x^2} + Be^{-x^2} - \sin x^2.$$

3. Solve $\dfrac{d^2y}{dx^2} + \left(4x - \dfrac{1}{x}\right)\dfrac{dy}{dx} + 4x^2y = 3xe^{-x^2}$ by changing the independent variable.

Solution

Here $P = 4x - \dfrac{1}{x}$, $Q = 4x^2$ and $R = 3xe^{-x^2}$. Now, we choose

$$Q_1 = \frac{Q}{\left(\dfrac{dz}{dx}\right)^2} = 4 \text{ or, } \frac{dz}{dx} = x \text{ or, } z = \frac{x^2}{2}.$$

Then the given equation reduces to

$$\frac{d^2y}{dz^2} + P_1\frac{dy}{dx} + Q_1y = R_1$$

where $P_1 = \dfrac{\left(4x - \dfrac{1}{x}\right)x + 1}{x^2} = 4$ and $R_1 = \dfrac{3xe^{-x^2}}{x^2} = \dfrac{3}{x}e^{-x^2}$, i.e., reduces to

$$\frac{d^2y}{dz^2} + 4\frac{dy}{dz} + 4y = \frac{3}{\sqrt{2z}}e^{-2z}.$$

Solving it we get

$$y = (A + 2Bz)e^{-2z} + \frac{1}{(D+2)^2}\frac{3}{\sqrt{2z}}e^{-2z}$$

$$= (A + 2Bz)e^{-2z} + 3e^{-2z}\frac{1}{\sqrt{2}}\frac{1}{(D-2+2)^2}\frac{1}{\sqrt{z}}$$

$$= (A + 2Bz)e^{-2z} + 3e^{-2z}\frac{1}{\sqrt{2}}2\left(\frac{2}{3}\right)z^{3/2}$$

$$= (A + Bx^2)e^{-x^2} + x^3e^{-x^2}.$$

4. Solve $x^6\dfrac{d^2y}{dx^2} + 3x^5\dfrac{dy}{dx} + a^2y = x^{-2}$ by changing the independent variable.

Solution

Here $P = \dfrac{3}{x}$, $Q = \dfrac{a^2}{x^6}$ and $R = \dfrac{1}{x^8}$. Now, we choose

$$Q_1 = \frac{Q}{\left(\dfrac{dz}{dx}\right)^2} = a^2 \text{ or, } \frac{dz}{dx} = \frac{1}{x^3} \text{ or, } z = -\frac{1}{2x^2}.$$

Then the given equation reduces to

$$\frac{d^2y}{dz^2} + P_1\frac{dy}{dx} + Q_1y = R_1,$$

where $P_1 = 0$ and $R_1 = \frac{1}{x^2}$, i.,e., reduces to

$$\frac{d^2y}{dz^2} + a^2 y = -2z.$$

Solving it, we get

$$y = A\sin(az) + B\cos(az) + \frac{1}{D^2 + a^2}(-2z) = A\sin(az) + B\cos(az) - \frac{2z}{a^2}$$

$$= -A\sin\left(\frac{a}{2x^2}\right) + B\cos\left(\frac{a}{2x^2}\right) + \frac{1}{a^2x^2}.$$

EXERCISE

- Solve the following equations in terms of known integral:

1. $x^2\frac{d^2y}{dx^2} - 2y = 0$, given that $y = x^2$ is a solution,
2. $x\frac{d^2y}{dx^2} - (2x-1)\frac{dy}{dx} + (x-1)y = 0$, given that $y = e^x$ is a solution,
3. $x^2\frac{d^2y}{dx^2} + x\frac{dy}{dx} - y = 0$, given that $y = x$ is a solution,
4. $x^2\frac{d^2y}{dx^2} + x\frac{dy}{dx} - 9y = 0$, given that $y = x^3$ is a solution,
5. $x^2\frac{d^2y}{dx^2} - x(x+2)\frac{dy}{dx} + (x+2)y = 0$, given that $y = x$ is a solution,
6. $x^2\frac{d^2y}{dx^2} - 3x\frac{dy}{dx} + 4y = 0$, given that $y = x^2$ is a solution,
7. $x^2\frac{d^2y}{dx^2} + x\frac{dy}{dx} - y = 0$, given that $y = x + \frac{1}{x}$ is a solution,
8. $\frac{d^2y}{dx^2} - 2\tan x\frac{dy}{dx} + 3y = 2\sec x$, given that $y = \sec x$ is a solution,
9. $(1-x^2)\frac{d^2y}{dx^2} - x\frac{dy}{dx} - y = 0$, given that $y = e^{\sin^{-1}x}$ is a solution,
10. $x^2\frac{d^2y}{dx^2} - 2x(x+1)\frac{dy}{dx} + 2(x+1)y = x^3$, given that $y = x$ is a solution,
11. $(x\sin x + \cos x)\frac{d^2y}{dx^2} - x\cos x\frac{dy}{dx} + y\cos x = 0$, given that $y = x$ is a solution,
12. $(x-3)\frac{d^2y}{dx^2} - (4x-9)\frac{dy}{dx} + 3(x-2)y = 0$, given that $y = e^x$ is a solution,

13. $x^2\dfrac{d^2y}{dx^2}+x\dfrac{dy}{dx}-y=0,$

14. $x\dfrac{d^2y}{dx^2}-(2x-1)\dfrac{dy}{dx}+(x-1)y=0,$

15. $x\dfrac{d^2y}{dx^2}+(1-x)\dfrac{dy}{dx}-y=e^x,$

16. $x^2\dfrac{d^2y}{dx^2}-x(x+2)\dfrac{dy}{dx}+(x+2)y=x^3,$

17. $x^2\dfrac{d^2y}{dx^2}-x(x+2)\dfrac{dy}{dx}+(x+2)y=x^3e^x,$

18. $\dfrac{d^2y}{dx^2}-\cot x\dfrac{dy}{dx}-(1-\cot x)y=e^x\sin x,$

19. $\dfrac{d^2y}{dx^2}-\dfrac{x}{x-1}\dfrac{dy}{dx}+\dfrac{1}{x-1}y=x-1.$

- **Solve the following equations after reducing to normal form:**

20. $\dfrac{d^2y}{dx^2}-\dfrac{2}{x}\dfrac{dy}{dx}+\left(a^2+\dfrac{2}{x^2}\right)y=0,$

21. $\cos x\dfrac{d^2y}{dx^2}-2\sin x\dfrac{dy}{dx}+y\cos x=0,$

22. $\dfrac{d^2y}{dx^2}-2bx\dfrac{dy}{dx}+b^2x^2y=0,$

23. $\dfrac{d^2y}{dx^2}-\dfrac{1}{\sqrt{x}}\dfrac{dy}{dx}+\dfrac{x+\sqrt{x}-8}{4x^2}y=0,$

24. $\dfrac{d^2y}{dx^2}-4x\dfrac{dy}{dx}+(4x^2-1)y=-3e^{x^2}\sin 2x,$

25. $\dfrac{d^2y}{dx^2}-4x\dfrac{dy}{dx}+(4x^2-1)y=e^{x^2}.$

26. $\dfrac{d^2y}{dx^2}-4x\dfrac{dy}{dx}+4x^2y=xe^{x^2},$

27. $\dfrac{d^2y}{dx^2}+2\tan x\dfrac{dy}{dx}+(\tan^2 x+\sec^2 x)y=\sec x\tan x,$

28. $\dfrac{d^2y}{dx^2}-2x\dfrac{dy}{dx}+(x^2+2)y=e^{\frac{x(x+2)}{2}}.$

- Solve the following equations using the relation between its two integrals:

29. $x^2\frac{d^2y}{dx^2}+x\frac{dy}{dx}-9y=0$, given that $y=x^3$ is a solution,

30. $(\sin x - x\cos x)\frac{d^2y}{dx^2}-x\sin x\frac{dy}{dx}+y\sin x=0$,, given that $y=\sin x$ is a solution,

31. $x^2\frac{d^2y}{dx^2}-3x\frac{dy}{dx}+4y=0$, given that $y=x^2$ is a solution,

32. $x(x\cos x-2\sin x)\frac{d^2y}{dx^2}+(x^2+2)\sin x\frac{dy}{dx}-2y(x\sin x+\cos x)=0$, given that $y=x^2$ is a solution.

- Solve the following equations using the method of operational factors:

33. $x\frac{d^2y}{dx^2}+(1-x)\frac{dy}{dx}-y=e^x$,

34. $x\frac{d^2y}{dx^2}-(1-x)\frac{dy}{dx}-y=x^2$,

35. $(x+2)\frac{d^2y}{dx^2}-(2x+5)\frac{dy}{dx}+2y=(1+x)e^x$,

36. $3x^2\frac{d^2y}{dx^2}+2(1-3x^2)\frac{dy}{dx}-4y=0$,

37. $x\frac{d^2y}{dx^2}+(x-2)\frac{dy}{dx}-2y=x^3$,

38. $(x+1)\frac{d^2y}{dx^2}+(x-1)\frac{dy}{dx}-2y=0$.

- Solve the following equations by changing the independent variable:

39. $\frac{d^2y}{dx^2}+\tan x\frac{dy}{dx}+y\cos^2 x=0$,

40. $\frac{d^2y}{dx^2}+\cot x\frac{dy}{dx}+4y\,\mathrm{cosec}^2 x=0$,

41. $\frac{d^2y}{dx^2}-\frac{1}{x}\frac{dy}{dx}+4x^2y=x^4$,

42. $(1+x)^2\frac{d^2y}{dx^2}+(1+x)\frac{dy}{dx}+y=4\cos\{\ln(1+x)\}$,

43. $\frac{d^2y}{dx^2}-(1+4e^x)\frac{dy}{dx}+3e^{2x}y=e^{2(x+e^x)}$,

44. $\frac{d^2y}{dx^2}-\frac{2}{x}\frac{dy}{dx}+\left(1+\frac{2}{x^2}\right)y=xe^x$,

ANSWERS

1. $Ax^2 + \frac{B}{x}$, 2. $Ae^x \ln x + Be^x$,

3. $Ax + \frac{B}{x}$, 4. $Ax^3 + \frac{B}{x^3}$,

5. $(A + Be^x)x$, 6. $(A + B\ln x)x^2$,

7. $A\left(x + \frac{1}{x}\right) + \frac{B}{x}$, 8. $\left(A\cos 2x + B\sin 2x + \frac{1}{2}\right)\sec x$,

9. $\left(A + Be^{-2\sin^{-1}x}\right)e^{\sin^{-1}x}$, 10. $\left(Ae^{2x} + B - \frac{x}{2}\right)x$,

11. $Ax + B\cos x$, 12. $Ae^x + B(4x^3 - 42x^2 + 150x - 183)e^{3x}$,

13. $Ax + \frac{B}{x}$, 14. $(A + B\ln x)e^x$,

15. $e^x \ln x + Ae^x \int \frac{dx}{xe^x} + Be^x$, 16. $(A + Be^x)x - x^2$,

17. $x\{(x-1)e^x + Ae^x + B\}$, 18. $-\frac{1}{2}e^x \cos x - Ae^{-x}(2\sin x + \cos x) + Be^x$,

19. $Ae^x + Bx - x^2 - 1$, 20. $x(A\cos ax + B\sin ax)$,

21. $(A\cos\sqrt{2}x + B\sin\sqrt{2}x)\sec x$, 22. $(A\cos\sqrt{b}x + B\sin\sqrt{b}x)e^{\frac{bx^2}{2}}$,

23. $\left(Ax^2 + \frac{B}{x}\right)e^{\sqrt{x}}$, 24. $(A\cos x + B\sin x + \sin 2x)e^{x^2}$,

25. $(Ae^x + Be^{-x} - 1)e^{x^2}$, 26. $(A\cos\sqrt{2}x + B\sin\sqrt{2}x)e^{x^2} + \frac{1}{2}xe^{x^2}$,

27. $\left\{\frac{1}{2}(\tan x - x) + Ax + B\right\}\cos x$, 28. $e^{\frac{x^2}{2}}(A\cos\sqrt{3}x + B\sin\sqrt{3}x) + \frac{1}{4}e^{\frac{x(x+2)}{2}}$,

29. $Ax^3 + \frac{B}{x^3}$, 30. $Ax + B\sin x$,

31. $(A + B\ln x)x^2$, 32. $Ax^2 + B\sin x$,

33. $e^x \ln x + Ae^x \int \frac{dx}{xe^x} + Be^x$, 34. $A(x-1) + Be^x + x^2 - 2x + 2$,

35. $A(2x+5) + Be^{2x} - e^x$, 36. $Ae^{2x} + Be^{2x}\int e^{\left(\frac{2}{3x} - 2x\right)}dx$,

37. $A(x^2 - 2x + 2) + Be^{-x} + x^3$, 38. $A(x^2 + 1) + Be^{-x}$,

39. $A\cos(\sin x)+B\sin(\sin x)$, **40.** $A\cos\left(2\ln\tan\frac{x}{2}\right)+B\sin\left(2\ln\tan\frac{x}{2}\right)$,

41. $A\cos x^2+B\sin x^2+\frac{1}{4}x^2$,

42. $A\cos\{\ln(1+x)\}+B\sin\{\ln(1+x)\}+2\ln(1+x)\sin\{\ln(1+x)\}$,

43. $Ae^{e^x}+Be^{3e^x}-e^{2e^x}$, **44.** $\left(A\cos x+B\sin x+\frac{1}{2}e^x\right)x$.

Chapter 9

SIMULTANEOUS DIFFERENTIAL EQUATION WITH CONSTANT COEFFICIENTS

9.1 Introduction

The differential equations, so far we have discussed, involved $\frac{dy}{dx}$ and higher order derivatives of y, i.e., where the number of dependent variable was one. In this section we have considered a system of differential equations involving more than one dependent variables, depending on one independent variable. To get solutions of such system of equations we must have the number of equations to be equal to the number of dependent variables.

Suppose we have the following system of linear differential equations:

$$\begin{aligned} \frac{dx_1}{dt} &= a_{11}x_1 + a_{12}x_2 + \ldots + a_{1n}x_n \\ \frac{dx_2}{dt} &= a_{21}x_1 + a_{22}x_2 + \ldots + a_{2n}x_n \\ &\cdots\cdots\cdots\cdots\cdots\cdots \\ &\cdots\cdots\cdots\cdots\cdots\cdots \\ \frac{dx_n}{dt} &= a_{n1}x_1 + a_{n2}x_2 + \ldots + a_{nn}x_n \end{aligned} \tag{1}$$

where the coefficients a_{ij} are constants. Here t is the argument and $x_1(t), x_2(t), \ldots, x_n(t)$ are unknown functions. The system of equations (1) is a system of homogeneous linear differential equations with constant coefficients.

9.2 Method of Solution

The basic procedure for solving such system is to eliminate all dependent variables except one by differentiating. The equation resulting by the elimination is then solved for the variable. Each of the dependent variables is obtained in a similar way.

If we rewrite the system of equations as

$$F_1(D)x + G_1(D)y = H_1(t),$$

and

$$F_2(D)x + G_2(D)y = H_2(t),$$

where $D = \frac{d}{dt}$ and $H_1(t)$, $H_2(t)$ are functions of t only, then any one of x, y, say x, can be eliminated to get a differential equation in y and the equation can be solved using the methods discussed earlier for ordinary differential equation. Using this any one of the above equations we get another differential equation in x, which gives the value of x.

Again considering the above equations as simultaneous equations in x, y, those can be solved using Crammer's rule. This gives two separate equations in x, y, which can be solved. Finally, using these values of x, y in the equations, we get relations among the arbitrary constants appeared in the solutions. Using these we express the arbitrary constants appeared in any of the variables with those appeared in other variable. Thus we get the values of x, y in which the number of arbitrary constants is equal to the order of differential equations.

WORKED-OUT EXAMPLES

1. Consider the system of differential equations:

$$\frac{dx}{dt}-7x+y=0; \quad \frac{dy}{dt}-2x-5y=0.$$

Solution

Method 1. Differentiating the first equation, w.r.t. t, we get

$$\frac{d^2x}{dt^2}-7\frac{dx}{dt}+\frac{dy}{dt}=0,$$

Using the second equation, we get

$$\frac{d^2x}{dt^2}-7\frac{dx}{dt}+2x+5y=0.$$

Again using the first equation, we get

$$\frac{d^2x}{dt^2}-12\frac{dx}{dt}+37x=0.$$

Putting $x=e^{mt}$, we get the auxiliary equation $m^2-12m+37=0$, which gives $m=6\pm i$.

So $$x=e^{6t}[A\cos t+B\sin t].$$

Differentiating this with respect to t, we get

$$\frac{dx}{dt}=6x+e^{6t}[B\cos t-A\sin t].$$

Putting these values of $x, \frac{dx}{dt}$ in the first equation, we get

$$y=e^{6t}[(A-B)\cos t+(A+B)\sin t].$$

Hence, $x=e^{6t}[A\cos t+B\sin t]$ and $y=e^{6t}[(A-B)\cos t+(A+B)\sin t]$ are the required solution.

Method 2. We rewrite the system of equations as

$$(D-7)x+y=0; \ -2x+(D-5)y=0, \text{ where } D=\frac{d}{dt}.$$

Eliminating y, we get

$$[(D-7)(D-5)+2]x=0 \Rightarrow [D^2-12D+37]x=0.$$

Auxiliary equation of the above equation is $m^2-12m+37=0$, which gives $m = 6 \pm i$.

So $\quad x=e^{6t}[A\cos t+B\sin t]$.

Putting this in the first equation, we get

$$y=e^{6t}[(A-B)\cos t+(A+B)\sin t].$$

Method 3. Rewriting the system of equation as

$$(D-7)x+y=0; \; -2x+(D-5)y=0, \text{ where } D=\frac{d}{dt}$$

and using Crammer's rule we get

$$\begin{vmatrix} D-7 & 1 \\ -2 & D-5 \end{vmatrix} x = \begin{vmatrix} 0 & 1 \\ 0 & D-5 \end{vmatrix} \text{ and } \begin{vmatrix} D-7 & 1 \\ -2 & D-5 \end{vmatrix} y = \begin{vmatrix} D-7 & 0 \\ -2 & 0 \end{vmatrix},$$

i.e., $[D^2-12D+37]x=0$ and $[D^2-12D+37]y=0$, which give

$$x=e^{6t}[A\cos t+B\sin t] \text{ and } y=e^{6t}[C\cos t+D\sin t].$$

Putting these values in $\frac{dx}{dt}-7x+y=0$, we get

$$[B-A+C]\cos t-[A+B-D]\sin t=0,$$

which is true for all values of t. This implies

$$B - A + C = 0 \text{ and } A + B - D = 0.$$

Hence the solutions are

$$x=e^{6t}[A\cos t+B\sin t] \text{ and } y=e^{6t}[(A-B)\cos t+(A+B)\sin t].$$

2. Solve $\frac{dx}{dt}=4x-2y; \frac{dy}{dt}=x+y.$

Solution

Differentiating the first equation with respect to t, we get

$$\frac{d^2x}{dt^2}=4\frac{dx}{dt}-2\frac{dy}{dt}=4\frac{dx}{dt}-2(x+y), \text{ using the second equation}$$

$$=4\frac{dx}{dt}-2x-4x+\frac{dx}{dt}, \text{ using the first equation}$$

or, $$\frac{d^2x}{dt^2}-5\frac{dx}{dt}+6x=0.$$

Its auxiliary equation is $m^2 - 5m + 6 = 0$, whose solutions are $m = 2, 3$. So the general solution of the above equation in x is

$$x = Ae^{2t} + Be^{3t}, \text{ where } A, B \text{ are arbitrary constants.}$$

Using this value of x in second equation we get

$$\frac{dy}{dt} - y = Ae^{2t} + Be^{3t}$$

which is a linear differential equation whose integrating factor is e^{-t}. Multiplying the above equation by e^{-t}, we get

$$\frac{d}{dt}\left(ye^{-t}\right) = Ae^{t} + Be^{2t}.$$

Integrating, we get

$$ye^{-t} = Ae^{t} + \frac{1}{2}Be^{2t} \text{ or, } y = Ae^{2t} + \frac{1}{2}Be^{3t}.$$

Hence $x = Ae^{2t} + Be^{3t}$ and $y = Ae^{2t} + \frac{1}{2}Be^{3t}$ are the required solutions.

3. Solve $\frac{dx}{dt} = y + e^{t}; \frac{dy}{dt} = -x + 1.$

Solution

From the first equation, we get

$$y = \frac{dx}{dt} - e^{t}.$$

Differentiating first equation with respect to t, we get

$$\frac{d^2x}{dt^2} = \frac{dy}{dt} + e^{t} = -x + 1 + e^{t} \text{ or, } (D^2+1)x = 1 + e^{t},\ D = \frac{d}{dt}.$$

The roots of the auxiliary equation of the corresponding homogeneous equation are $\pm i$. So the complementary function is

$$x = A\cos t + B\sin t$$

and P.I. $= \frac{1}{D^2+1}(1+e^{t}) = 1 + \frac{1}{2}e^{t}$. So the general solution of the above differential equation in x is

$$x = A\cos t + B\sin t + 1 + \frac{1}{2}e^{t}.$$

Using this in $y = \frac{dx}{dt} - e^{t}$, we get

$$y = -A\sin t + B\cos t - \frac{1}{2}e^{t}.$$

Hence, $x = A\cos t + B\sin t + 1 + \frac{1}{2}e^{t}$ and $y = -A\sin t + B\cos t - \frac{1}{2}e^{t}$ are the required solutions.

4. Solve the following system of equations: $\frac{dx}{dt} + 4x + 3y = t; \frac{dy}{dt} + 2x + 5y = e^{t}.$

Solution

Rewriting the system of equations as

$$(D+4)x + 3y = t;\ 2x + (D+5)y = e^{t}, \text{ where } D = \frac{d}{dt}.$$

Eliminating y, we get

$$(D+5)(D+4)x-6x=(D+5)t-3e^t,$$

i.e., $$[D^2+9D+14]x=1+5t-3e^t.$$

Solution of this equation is

$$x=Ae^{-7t}+Be^{-2t}+\frac{1}{D^2+9D+14}\left(1+5t-3e^t\right)$$

$$=Ae^{-7t}+Be^{-2t}+\frac{1}{14}\left(1+5t-\frac{45}{14}\right)-\frac{1}{8}e^t$$

$$=Ae^{-7t}+Be^{-2t}+\frac{1}{14}\left(5t-\frac{31}{14}\right)-\frac{1}{8}e^t.$$

Differentiating this with respect to t, we get

$$\frac{dx}{dt}=-7Ae^{-7t}-2Be^{-2t}+\frac{5}{14}-\frac{1}{8}e^t.$$

Using the values of $x, \frac{dx}{dt}$ in the first equation, we get

$$y=\frac{1}{3}\left(3Ae^{-7t}-2Be^{-2t}-\frac{3t}{7}+\frac{5}{8}e^t+\frac{27}{98}\right).$$

Hence, $$x=Ae^{-7t}+Be^{-2t}+\frac{1}{14}\left(5t-\frac{31}{14}\right)-\frac{1}{8}e^t$$

and $$y=\frac{1}{3}\left(3Ae^{-7t}-2Be^{-2t}-\frac{3t}{7}+\frac{5}{8}e^t+\frac{27}{98}\right)$$

are the required solutions.

5. Solve the following system of differential equation:

$$\frac{dx}{dt}+5x+y=e^t;\ \frac{dy}{dt}-x+3y=e^{2t}.$$

Solution

Rewriting the system of equation as

$$(D+5)x+y=e^t;\ -x+(D+3)y=e^{2t},\text{ where } D=\frac{d}{dt}.$$

Eliminating y, we get

$$[(D+5)(D+3)+1]x=4e^t-e^{2t},$$

i.e., $[D^2+8D+16]x=4e^t-e^{2t}$, which gives

$$x=(A+Bt)e^{-4t}+\frac{1}{(D+4)^2}(4e^t-e^{2t})=(A+Bt)e^{-4t}+\frac{4}{25}e^t-\frac{1}{36}e^{2t}.$$

Differentiating it with respect to t, we get

$$\frac{dx}{dt}=(B-4A-4Bt)e^{-4t}+\frac{4}{25}e^t-\frac{1}{18}e^{2t}.$$

Using the values of $x, \frac{dx}{dt}$ in the first equation we get

$$y = e^t - \frac{dx}{dt} - 5x$$

$$= e^t - (B - 4A - 4Bt)e^{-4t} - \frac{4}{25}e^t + \frac{1}{18}e^{2t} - 5\left[(A + Bt)e^{-4t} + \frac{4}{25}e^t - \frac{1}{36}e^{2t}\right]$$

$$= -(A + B + Bt)e^{-4t} + \frac{1}{25}e^t + \frac{7}{36}e^{2t}.$$

Hence, $\quad x = (A + Bt)e^{-4t} + \frac{4}{25}e^t - \frac{1}{36}e^{2t}$

and $\quad y = -(A + B + Bt)e^{-4t} + \frac{1}{25}e^t + \frac{7}{36}e^{2t}$ are the required solutions.

6. Solve the following system of differential equations:

$$2\frac{dx}{dt} + \frac{dy}{dt} - 4x - y = e^t;\ \frac{dx}{dt} + 3x + y = 0.$$

Solution

Rewriting the system of equations as

$$2(D - 2)x + (D - 1)y = e^t;\ (D + 3)x + y = 0, \text{ where } D = \frac{d}{dt}.$$

Eliminating y we get

$$[2(D - 2) - (D - 1)(D + 3)]x = e^t, \text{ i.e., } [D^2 + 1]x = -e^t,$$

which gives

$$x = A\cos t + B\sin t - \frac{1}{D^2 + 1}e^t = A\cos t + B\sin t - \frac{1}{2}e^t.$$

Putting this value of x in the second equation, we get

$$y = -(D + 3)x = (A - 3B)\sin t - (B + 3A)\cos t + 2e^t.$$

Hence,

$$x = A\cos t + B\sin t - \frac{1}{2}e^t \text{ and } y = (A - 3B)\sin t - (B + 3A)\cot t + 2e^t$$

are the required solutions.

7. Solve the following system of differential equations:

$$\frac{dx}{dt} + \frac{dy}{dt} - 2y = 2\cos t - 7\sin t;\ \frac{dx}{dt} - \frac{dy}{dt} + 2x = 4\cos t - 3\sin t.$$

Solution

We rewrite the equation as

$$Dx + (D - 2)y = 2\cos t - 7\sin t; (D + 2)x - Dy = 4\cos t - 3\sin t.$$

Eliminating y, we get

$$[-D^2 - (D^2 - 4)]x = -D(2\cos t - 7\sin t) - (D - 2)(4\cos t - 3\sin t)$$

$$(D^2 - 2)x = -9\cos t, \text{ which gives } x = Ae^{\sqrt{2}t} + Be^{-\sqrt{2}t} + 3\cos t.$$

Adding the given equations, we get

$$2(D+1)x - 2y = 6\cos t - 10\sin t$$

or, $$y = (D+1)[Ae^{\sqrt{2}t} + Be^{-\sqrt{2}t} + 3\cos t] - 3\cos t + 5\sin t$$

$$= (1+\sqrt{2})Ae^{\sqrt{2}t} + (1-\sqrt{2})Be^{-\sqrt{2}t} + 2\sin t.$$

Hence, $$x = Ae^{\sqrt{2}t} + Be^{-\sqrt{2}t} + 3\cos t$$

and $$y = (1+\sqrt{2})Ae^{\sqrt{2}t} + (1-\sqrt{2})Be^{-\sqrt{2}t} + 2\sin t$$

are the required solutions.

8. Solve the following system of differential equations:

$$\frac{dx}{dt} + \frac{dy}{dt} + 2x + y = 0; \quad \frac{dy}{dt} + 5x + 3y = 0.$$

Solution

The given equations can be written as

$$(D+2)x + (D+1)y = 0,$$
$$5x + (D+3)y = 0.$$

Eliminating y, we get $(D^2+1)x = 0.$

Auxiliary equation of the above differential equation in x is $m^2 + 1 = 0$, which gives $m = \pm i$. So, the general solution of the above equation is

$$x = A\cos t + B\sin t, \text{ where } A, B \text{ are arbitrary constants.}$$

Differentiating it, we get

$\frac{dx}{dt} = B\cos t - A\sin t.$ Substracting the given equation, we get

$$\frac{dx}{dt} = 3x + 2y$$

or, $$2y = B\cos t - A\sin t - 3(A\cos t + B\sin t)$$

or, $$y = \frac{B-3A}{2}\cos t - \frac{A+3B}{2}\sin t.$$

Hence $x = A\cos t + B\sin t$ and $y = \frac{B-3A}{2}\cos t - \frac{A+3B}{2}\sin t$ are the required solutions.

9. Solve the following system of differential equations:

$$\frac{d^2x}{dt^2} - 2x - 3y = e^{2t}; \quad \frac{d^2y}{dt^2} + x + 2y = 0.$$

Solution

The given equations can be written as

$$[D^2 - 2]x - 3y = e^{2t}; \; x + [D^2+2]y = 0.$$

Solving the above system for x, we get

$$[(D^2-2)(D^2+2)+3]x=(D^2+2)e^{2t} \qquad [D^4-1]x=6e^{2t},$$

which gives $x=A\cos t+B\sin t+Ee^t+Fe^{-t}+\frac{2}{5}e^{2t}$.

Putting this value of x in the first equation, we get

$$y=-(A\cos t+B\sin t)-\frac{1}{3}(Ee^t+Fe^{-t})-\frac{1}{15}e^{2t}.$$

Hence, $\quad x=A\cos t+B\sin t+Ee^t+Fe^{-t}+\frac{2}{5}e^{2t}$ and

$$y=-(A\cos t+B\sin t)-\frac{1}{3}(Ee^t+Fe^{-t})-\frac{1}{15}e^{2t}$$

are the required solutions.

10. Solve the following system of differential equations:

$$\frac{d^2x}{dt^2}-3x-4y=0;\ \frac{d^2y}{dt^2}+x+y=0.$$

Solution

The given equations can be written as

$$[D^2-3]x-4y=0;\ x+[D^2+1]y=0.$$

Eliminating y, we get

$$[(D^2-3)(D^2+1)+4]x=0,\ \text{i.e.,}\ (D^2-1)^2x=0.$$

Auxiliary equation of the above differential equation in x is

$$(m^2-1)^2=0.$$

which gives $m = 1, -1, 1, -1$.

Hence $\quad x=(A+Bt)e^t+(C+Dt)e^{-t}$.

Putting this value of x in the first equation we get

$$y=\frac{1}{2}[B(1-t)-A]e^t-\frac{1}{2}[C+(1+t)D]e^{-t}.$$

Hence, $\quad x=(A+Bt)e^t+(C+Dt)e^{-t}$

and $\quad y=\frac{1}{2}[B(1-t)-A]e^t-\frac{1}{2}[C+(1+t)D]e^{-t}$ are the required solutions.

11. Solve the following system of differential equations:

$$\frac{dx}{dt}=2y;\ \frac{dy}{dt}=2z;\ \frac{dz}{dt}=2x.$$

Solution

From the given equations, we get

$$\frac{d^3x}{dt^3}=2\frac{d^2y}{dt^2}=4\frac{dz}{dt}=8x.$$

Solution of the equation $\frac{d^3x}{dt^3} - 8x = 0$ is $x = Ae^{2t} + e^{-t}\left[B\cos\sqrt{3}t + C\sin\sqrt{3}t\right]$.

So, $y = \frac{1}{2}\frac{dx}{dt} = Ae^{2t} + \frac{e^{-t}}{2}\left[-\sqrt{3}B\sin\sqrt{3}t + \sqrt{3}C\cos\sqrt{3}t - B\cos\sqrt{3}t - C\sin\sqrt{3}t\right]$

$$= Ae^{2t} + \frac{e^{-t}}{2}\left[(\sqrt{3}C - B)\cos\sqrt{3}t - (\sqrt{3}B + C)\sin\sqrt{3}t\right]$$

and $z = \frac{1}{2}\frac{dy}{dt} = Ae^{2t} + \frac{e^{-t}}{4}\left[-(\sqrt{3}C - B)\sqrt{3}\sin\sqrt{3}t - (\sqrt{3}B + C)\sqrt{3}\cos\sqrt{3}t - (\sqrt{3}C - B)\cos\sqrt{3}t + (\sqrt{3}B + C)\sin\sqrt{3}t\right]$

$$= Ae^{2t} + \frac{e^{-t}}{2}\left[(\sqrt{3}B - C)\sin\sqrt{3}t - (\sqrt{3}C + B)\cos\sqrt{3}t\right].$$

Hence $x = Ae^{2t} + e^{-t}\left[B\cos\sqrt{3}t + C\sin\sqrt{3}t\right]$,

$$y = Ae^{2t} + \frac{e^{-t}}{2}\left[(\sqrt{3}C - B)\cos\sqrt{3}t - (\sqrt{3}B + C)\sin\sqrt{3}t\right]$$

and $z = Ae^{2t} + \frac{e^{-t}}{2}\left[(\sqrt{3}B - C)\sin\sqrt{3}t - (\sqrt{3}C + B)\cos\sqrt{3}t\right]$

are the required solutions.

12. Solve the following system of differential equations:

$$\frac{dx}{dt} = z + y; \quad \frac{dy}{dt} = x + z; \quad \frac{dz}{dt} = y + x.$$

Solution

Differentiating the first one and using all the equations in that we get

$$\frac{d^2x}{dt^2} - \frac{dx}{dt} - 2x = 0.$$

Solving this, we get

$$x = Ae^{-t} + Be^{2t}.$$

From the second and third equations, we get

$$\frac{d}{dt}(z - y) + z - y = 0.$$

Solving this, we get

$$z - y = Ce^{-t}.$$

Again we have $\quad z + y = \frac{dx}{dt} = -Ae^{-t} + 2Be^{2t}$.

Adding and subtracting we get

$$y = -\frac{A + C}{2}e^{-t} + Be^{2t}$$

and $$z = \frac{C - A}{2}e^{-t} + Be^{2t}.$$

Hence, $x = Ae^{-t} + Be^{2t}$, $y = -\frac{A + C}{2}e^{-t} + Be^{2t}$, $z = \frac{C - A}{2}e^{-t} + Be^{2t}$

are the required solutions.

EXERCISE

- Solve the following system of differential equations:

1. $\frac{dx}{dt} = -\omega y; \frac{dy}{dt} = \omega x,$

2. $\frac{dx}{dt} + 2x - y = 0, \frac{dy}{dt} - x + 2y = 0,$

3. $\frac{dx}{dt} - 3x - 2y = 0; \frac{dy}{dt} + 5x + 3y = 0,$

4. $\frac{dx}{dt} + 2x + 3y = 0; \frac{dy}{dt} + 3x + 2y = 2e^{2t},$

5. $\frac{dx}{dt} + 5x + y = e^{t}; \frac{dy}{dt} - x + 3y = e^{2t},$

6. $\frac{dx}{dt} + x + 2y = e^{t}; \frac{dy}{dt} + 2x + y = 3e^{t},$

7. $\frac{dx}{dt} + 2\frac{dy}{dt} - 2x + 2y = 3e^{t}; 3\frac{dx}{dt} + \frac{dy}{dt} + 2x + y = 4e^{2t},$

8. $\frac{dx}{dt} = ax + by; \frac{dy}{dt} = bx + ay,$

9. $\frac{dx}{dt} = ax + by; \frac{dy}{dt} = cx + dy,$

10. $\frac{dx}{dt} - 7x + y = e^{t}, \frac{dy}{dt} - 2x - 5y = 2,$

11. $\frac{dx}{dt} + 2x - 3y = t; \frac{dy}{dt} - 3x + 2y = e^{2t},$

12. $\frac{dy}{dx} + y = z + e^{x}; \frac{dz}{dx} + z = y + e^{x},$

13. $\frac{dx}{dt} + \frac{dy}{dt} + 2y = 0, \frac{dx}{dt} - 3x - 2y = 0,$

14. $\frac{dx}{dt} + \frac{dy}{dt} + 2x + y = 0, \frac{dy}{dy} + 5x + 3y = 0,$

15. $4\frac{dx}{dt} + 9\frac{dy}{dt} + 2x + 31y = e^{t}; 3\frac{dx}{dt} + 7\frac{dy}{dt} + x + 24y = 3,$

16. $\frac{dx}{dt} - \frac{dy}{dt} - y = -e^{t}; \frac{dy}{dt} + x - y = e^{2t},$

17. $\frac{dx}{dt} + \frac{dy}{dt} - x + 3y = e^{-t} - 1; \frac{dx}{dt} + \frac{dy}{dt} + 2x + y = e^{2t} + t,$

18. $\frac{dx}{dt} + \frac{dy}{dt} - x = 2t + 1; 2\frac{dx}{dt} + 2\frac{dy}{dt} + x = t,$

19. $3\frac{dy}{dt} - \frac{dx}{dt} + y = e^{t}, 3\frac{dx}{dt} - \frac{dy}{dt} + x = 0,$

20. $\frac{d^2x}{dt^2} - 3x - 4y = 0; \frac{d^2y}{dt^2} + x + y = 0,$

21. $\frac{d^2x}{dt^2} - 2\frac{dy}{dt} - x = e^{t}\cos t, \frac{d^2y}{dt^2} + 2\frac{dx}{dt} - y = e^{t}\sin t,$

22. $\frac{d^2x}{dt^2} + \frac{d^2y}{dt^2} + \frac{dx}{dt} + x + y = e^{t}, \frac{d^2x}{dt^2} + \frac{d^2y}{dt^2} + \frac{dx}{dt} = e^{-t},$

23. $\frac{d^2x}{dt^2} + a^2y = 0, \frac{d^2y}{dt^2} - a^2x = 0,$

24. $\frac{dx}{dt}+\frac{dy}{dt}+x-y=e^t; \frac{d^2x}{dt^2}+\frac{d^2y}{dt^2}+\frac{dx}{dt}-\frac{dy}{dt}+x+y=t^2,$

25. $\frac{dx}{dt}=x^2+xy; \frac{dy}{dt}=y^2+xy,$

26. $\frac{dx}{dt}+2x-7y=0, \frac{dy}{dt}-3x-2y=0,$ given that $x(0)=9$, $y(0)=-1$,

27. $\frac{dx}{dt}+2x+8y=0, \frac{dy}{dt}-x-6y=0,$ given that $x(0)=4$, $y(0)=1$,

28. $\frac{dx}{dt}-6x+4y=0, \frac{dy}{dt}-x-2y=0,$ given that $x(0)=2$, $y(0)=3$,

29. $\frac{d^2x}{dt^2}+\frac{dy}{dt}+x+y=t, \frac{dy}{dt}+2x+y=0,$ given that $x(0)=y(0)=0$,

30. $\frac{d^2y}{dt^2}-16x=t, \frac{d^2x}{dt^2}-y=1,$ given that $x=y=0$, $\frac{dx}{dt}=1, \frac{dy}{dt}=\frac{1}{4}$ at $t=0$.

31. $t\frac{dx}{dt}+y=0, t\frac{dy}{dt}+x=0,$

32. $t\frac{dx}{dt}+2x-2y=t; t\frac{dy}{dt}+x+5y=t^2,$

33. Show that solutions of the equations $\frac{dx}{dt}=-2y, \frac{dy}{dt}=x$ is given by $x^2+2(y^2+C)=0$.

ANSWERS

1. $x=A\cos\omega t+B\sin\omega t; y=A\sin\omega t-B\cos\omega t,$
2. $x=Ae^{-t}+Be^{-3t}, y=Ae^{-t}-Be^{-3t},$
3. $x=A\cos t+B\sin t; y=\frac{1}{2}(B-3A)\cos t-\frac{1}{2}(A+3B)\sin t,$
4. $x=Ae^t+Be^{-5t}-\frac{6}{7}e^{2t}; y=-Ae^t+Be^{-5t}+\frac{8}{7}e^{2t},$
5. $x=(A+Bt)e^{-4t}+\frac{4}{25}e^t-\frac{1}{36}e^{2t}; y=-[A+(1+t)B]e^{-4t}+\frac{1}{25}e^t+\frac{7}{36}e^{2t}.$
6. $x=Ae^t+Be^{-3t}-te^t, y=-Ae^t+Be^{-3t}+e^t+te^t,$
7. $x=\frac{1}{2}e^{2t}-\frac{3}{11}e^t+Ae^{-\frac{6t}{5}}; y=\frac{15}{22}e^t+Be^{-t}-8Ae^{-\frac{6t}{5}},$
8. $x=Ae^{(a+b)t}+Be^{(a-b)t}; y=Ae^{(a+b)t}-Be^{(a-b)t},$
9. $x=Ae^{m_1t}+Be^{m_2t}; y=\frac{1}{b}[(m_1-a)Ae^{m_1t}+(m_2-a)Be^{m_2t}],$ where

   $m_{2,1}=\frac{1}{2}\left\{a+d\pm\sqrt{(a-d)^2+4bc}\right\},$

10. $x=(A\cos t+B\sin t)e^{6t}-\frac{2}{13}e^{t}-\frac{2}{37},$

$y=\left[(A-B)\cos t+(A+B)\sin t\right]e^{6t}+\frac{1}{13}e^{t}-\frac{14}{37},$

11. $x=Ae^{-5t}+Be^{t}+\frac{3}{7}e^{2t}-\frac{2t}{5}-\frac{13}{25};\ y=-Ae^{-5t}+Be^{t}+\frac{4}{7}e^{2t}-\frac{3t}{5}-\frac{12}{25},$

12. $y=A+Be^{-2x}+e^{x};\ z=A-Be^{-2x}+e^{x},$

13. $x=Ae^{-3t}+Be^{2t},\ y=-3Ae^{-3t}-\frac{1}{2}Be^{2t},$

14. $x=A\cos t+B\sin t;\ y=\frac{1}{2}(B-3A)\cos t-\frac{1}{2}(A+3B)\sin t,$

15. $x=(A\sin t+B\cos t)e^{-4t}+\frac{31}{26}e^{t}-\frac{93}{17};$

$y=[(A-B)\sin t-(A+B)\cos t]e^{-4t}-\frac{2}{13}e^{t}+\frac{6}{17},$

16. $x=(A-B)\cos t+(A+B)\sin t+\frac{3}{5}e^{2t};\ y=A\cos t+B\sin t+\frac{2}{5}e^{2t}+\frac{1}{2}e^{t},$

17. $x=2Ae^{-\frac{7t}{5}}+\frac{5}{17}e^{2t}+\frac{3t}{7}-\frac{1}{49};\ y=3Ae^{-\frac{7t}{5}}-\frac{1}{17}e^{2t}+\frac{1}{2}e^{-t}+\frac{t}{7}-\frac{26}{49},$

18. $x=-t-\frac{2}{3},\ y=\frac{t^{2}}{2}+\frac{4}{3}t+A,$

19. $x=Ae^{-\frac{1}{2}t}+Be^{-\frac{1}{4}t}+\frac{1}{15}e^{t},\ y=Ae^{-\frac{1}{2}t}-Be^{-\frac{1}{4}t}+\frac{4}{15}e^{t},$

20. $x=(A+Bt)e^{t}+(C+Dt)e^{-t};\ y=\frac{1}{2}[B(1-t)-A]e^{t}-\frac{1}{2}[D(1+t)+C]e^{-t},$

21. $x=(c_1+c_2t)\cos t+(c_3+c_4t)\sin t+\frac{e^{t}}{25}(4\sin t-3\cos t);$

$y=-(c_1+c_2t)\sin t+(c_3+c_4t)\cos t+\frac{e^{t}}{25}(3\sin t+4\cos t),$

22. $x=-e^{t}-2e^{-t}-A;\ y=2e^{t}+e^{-t}+A,$

23. $x=e^{\frac{at}{\sqrt{2}}}\left(A\cos\frac{at}{\sqrt{2}}+B\sin\frac{at}{\sqrt{2}}\right)+e^{-\frac{at}{\sqrt{2}}}\left(C\cos\frac{at}{\sqrt{2}}+D\sin\frac{at}{\sqrt{2}}\right)$

$y=e^{\frac{at}{\sqrt{2}}}\left(A\sin\frac{at}{\sqrt{2}}-B\cos\frac{at}{\sqrt{2}}\right)+e^{-\frac{at}{\sqrt{2}}}\left(D\cos\frac{at}{\sqrt{2}}-C\sin\frac{at}{\sqrt{2}}\right).$

24. $x=\frac{1}{2}(e^{t}-2t+t^{2});\ y=\frac{1}{2}(2t+t^{2}-3e^{t}),$

25. $x = -\dfrac{1}{(1+c_1)t + c_2}; \; y = -\dfrac{c_1}{(1+c_1)t + c_2},$

26. $x = 2e^{5t} + 7e^{-5t}, \; y = 2e^{5t} - 3e^{-5t},$

27. $x = 4e^{4t}[\cos 2t - 2\sin 2t], \; y = e^{4t}[\cos 2t + 3\sin 2t],$

28. $x = 2e^{4t} - 8te^{4t}, \; y = 3e^{4t} - 4te^{4t},$

29. $x = -2e^{t} + 2e^{-t} - t, \; y = 2(e^{t} - 2et^{-1} + t - 1),$

30. $x = \dfrac{1}{16}(3e^{2t} - e^{-2t}) + \dfrac{9}{32}\sin 2t - \dfrac{1}{8}\cos 2t - \dfrac{t}{16},$

$y = \dfrac{1}{4}(3e^{2t} - e^{-2t}) - \dfrac{9}{8}\sin 2t + \dfrac{1}{2}\cos 2t - 1,$

31. $x = c_1 t + \dfrac{c_2}{t}; \; y = -c_1 t + \dfrac{c_2}{t},$

32. $x = \dfrac{A}{t^3} + \dfrac{B}{t^4} + \dfrac{3}{10}t + \dfrac{1}{15}t^2; \; y = -\dfrac{A}{2t^3} - \dfrac{B}{t^4} - \dfrac{1}{20}t + \dfrac{2}{15}t^2.$

9.3 Simultaneous Equations—Another Form

If the equations are given as

$$P_1 dx + Q_1 dy + R_1 dz = 0,$$

and

$$P_2 dx + Q_2 dy + R_2 dz = 0,$$

where $P_k, Q_k, R_k, k = 1, 2, 3,$ are functions of x, y, z only. Then solving, we get

$$\frac{dx}{Q_1 R_2 - Q_2 R_1} = \frac{dy}{R_1 P_2 - R_2 P_1} = \frac{dz}{P_1 Q_2 - P_2 Q_1} \quad \text{or,} \quad \frac{dx}{P} = \frac{dy}{Q} = \frac{dz}{R}$$

where P, Q, R are functions of x, y, z.

9.4 Method of Solution of the Equation of the Form $\frac{dx}{P} = \frac{dy}{Q} = \frac{dz}{R}$

Choose any set of two ratios of the given three ratios. Integrating this, we get a relation in x, y, z, say, $u(x, y, z) = 0$. Next choose another set of two ratios and integrate it to get another relation in x, y, z, say, $v(x, y, z) = 0$. These two relations give the complete solution.

Again choose multipliers p, q, r so that $pP + qQ + rR = 0$ and $pdx + qdy + rdz$ becomes an exact differential. Then from the given equation we see that

$$\frac{dx}{P} = \frac{dy}{Q} = \frac{dz}{R} = \frac{pdx + qdy + rdz}{0},$$

i.e., $pdx + qdy + rdz = 0$, which gives another equation on integration.

So, using any combination of these methods discussed so far, we get two independent relations in x, y, z, which constitute the general solution.

9.5 Geometrical Interpretation of These Equations $\frac{dx}{P}=\frac{dy}{Q}=\frac{dz}{R}$

From solid geometry, we know that the direction cosines of a tangent to a curve are $\left(\frac{dx}{ds}, \frac{dy}{ds}, \frac{dz}{ds}\right)$, i.e., the direction cosines of a tangent to a curve are proportional to dx, dy, dz. Again, from the given equations, we see that dx, dy, dz are proportional to P, Q, R. So, geometrically the given equations represent a system of curves in space, such that direction cosines of the tangent to any point on the curves are proportional to P, Q, R.

WORKED-OUT EXAMPLES

1. Solve: $\frac{dx}{yz}=\frac{dy}{xz}=\frac{dz}{xy}$.

Solution

From $\frac{dx}{yz}=\frac{dy}{xz}$, we get $x^2-y^2=A$.

Similarly, from $\frac{dy}{xz}=\frac{dz}{xy}$ we get $y^2-z^2=B$.

So the general solution is given by $x^2-y^2=A$ and $y^2-z^2=B$.

2. Solve: $\frac{dx}{xz}=\frac{dy}{yz}=\frac{dz}{xy}$.

Solution

From $\frac{dx}{xz}=\frac{dy}{yz}$, we get $\ln x=\ln y+\ln A \quad A=\frac{x}{y}$.

Again, from the given equations, we get

$$ydx = xdy = zdx$$

$$\Rightarrow ydx + xdy = 2zdz \Rightarrow d(xy) = d(z^2)$$

$$\Rightarrow d(xy - z^2) = 0 \Rightarrow xy - z^2 = B.$$

So the general solution is given by $\frac{x}{y}=A$ and $xy-z^2=B$.

3. Solve: $\frac{dx}{x^2+2y^2}=\frac{dy}{-xy}=\frac{dz}{xz}$.

Solution

From $\frac{dy}{-xy}=\frac{dz}{xz}$, we get $\frac{dy}{y}+\frac{dz}{z}=0$.

Integrating, we get $yz=A$.

Again, from the given equation, we get

$$\frac{xdx+ydy}{x(x^2+y^2)}=\frac{dz}{xz} \quad \text{or,} \quad \frac{d(x^2+y^2)}{x^2+y^2}=2\frac{dz}{z}.$$

Integrating we get

$$\ln(x^2+y^2) = \ln z^2 + \ln B \quad \text{or,}\ x^2+y^2 = z^2 B.$$

So the general solution is given by $yz = A$ and $x^2+y^2 = z^2B$.

4. Solve: $\dfrac{dx}{2xz} = \dfrac{dy}{2yz} = \dfrac{dz}{z^2-x^2-y^2}$.

Solution

From $\dfrac{dx}{2xz} = \dfrac{dy}{2yz}$, we get $\dfrac{dx}{x} = \dfrac{dy}{y}$.

Integrating, we get $x = Ay$.

Again, from the given equation, we get

$$\frac{dy}{2yz} = \frac{xdx+ydy+zdz}{z(z^2+x^2+y^2)} \quad \text{or,} \quad \frac{dy}{y} = \frac{d(z^2+x^2+y^2)}{z^2+x^2+y^2}.$$

Integrating, we get $\quad x^2+y^2+z^2 = By.$

So the general solution is given by $x = Ay$ and $x^2+y^2+z^2 = By$.

5. Solve: $\dfrac{dx}{1} = \dfrac{dy}{-2} = \dfrac{dz}{3x^2\sin(y+2x)}$.

Solution

From $\dfrac{dx}{1} = \dfrac{dy}{-2}$, we get $2x+y = A$.

Using this in the given equation, we get

$$\frac{dx}{1} = \frac{dz}{3x^2\sin A} \quad \text{or,}\ 3x^2dx = \operatorname{cosec} A\, dz.$$

Integrating, we get

$$x^3\sin(y+2x) - z = B.$$

So the general solution is given by $y+2x = A$ and $x^3\sin(y+2x)-z = B$.

6. Solve: $\dfrac{dx}{y} = \dfrac{dy}{x} = \dfrac{dz}{xyz^2(x^2-y^2)}$.

Solution

From the first two ratios, we get $xdx - ydy = 0$.

Integrating, we get $x^2-y^2 = A$.

Using the above result in $\dfrac{dy}{x} = \dfrac{dz}{xyz^2(x^2-y^2)}$, we get $Aydy = \dfrac{dz}{z^2}$.

Integrating, we get $Ay^2 = -\dfrac{2}{z} + B$ or, $(x^2-y^2)y^2 + \dfrac{2}{z} = B$.

So the general solution is given by $x^2-y^2 = A$ and $(x^2-y^2)y^2 + \dfrac{2}{z} = B$.

7. Solve: $\dfrac{dx}{x+y}=\dfrac{dy}{x+y}=\dfrac{dz}{-(x+y+2z)}$.

Solution

From $\dfrac{dx}{x+y}=\dfrac{dy}{x+y}$, we get $dx = dy$.
Integrating, we get $x - y = A$.
Again, from the given equation, we get

$$\frac{dx+dy}{2(x+y)}=\frac{dz}{-(x+y+2z)} \quad \text{or,} \quad \frac{dz}{du}+\frac{z}{u}=-\frac{1}{2}, \text{ where } u = x + y.$$

This is a linear differential equation in z, whose integrating factor is $e^{\int \frac{du}{u}} = u$.
Multiplying the above equation by this factor, we get

$$\frac{d}{du}(zu)=-\frac{1}{2}u.$$

Integrating, we get

$$zu=-\frac{1}{4}u^2+B \quad \text{or,} \quad z(x+y)=-\frac{1}{4}(x+y)^2+B$$

$$\text{or,} \quad z(x+y)=-\frac{1}{4}(x-y)^2 \quad xy+B \quad \text{or,} \quad xy+yz+xz=C.$$

So the general solution is given by $x - y = A$ and $xy + yz + xz = C$.

8. Solve: $\dfrac{dx}{-x(x+y)}=\dfrac{dy}{y(x+y)}=\dfrac{dz}{(x-y)(2x+2y+z)}$.

Solution

From the first two ratios, we get $\dfrac{dx}{x}+\dfrac{dy}{y}=0$.
Integrating, we get $xy = A$.
Again from the given equation, we get

$$\frac{dx+dy}{-(x-y)(x+y)}=\frac{dz}{(x-y)(2x+2y+z)} \quad \text{or,} \quad \frac{dz}{du}+\frac{z}{u}=-2, \text{ where } u = x + y.$$

This is a linear differential equation in z, whose integrating factor is $e^{\int \frac{du}{u}} = u$.
Multiplying the above equation by this factor we get

$$\frac{d}{du}(zu)=-2u.$$

Integrating we get $\quad z(x+y)+(x+y)^2 = B.$

So the general solution is given by $xy = A$ and $z(x+y)+x^2+y^2 = C$.

9. Solve: $\dfrac{dx}{z(x+y)}=\dfrac{dy}{z(x-y)}=\dfrac{dz}{x^2+y^2}$.

Solution

From the equation we see that

$$\frac{xdx-ydy}{xz(x+y)-yz(x-y)}=\frac{dz}{x^2+y^2}$$

or, $\dfrac{xdx - ydy}{z(x^2 + y^2)} = \dfrac{dz}{x^2 + y^2}$

or, $xdx - ydy = zdz.$

Integrating, we get $x^2 - y^2 - z^2 = A.$

Again, from the equation, we get

$$\frac{ydx + xdy}{yz(x + y) + xz(x - y)} = \frac{dz}{x^2 + y^2}.$$

or, $\dfrac{d(xy)}{z(x^2 + y^2)} = \dfrac{dz}{x^2 + y^2}$

or, $d(xy) = zdz.$

Integrating, we get $2xy - z^2 = B.$

So the general solution is given by $x^2 - y^2 - z^2 = A$ and $2xy - z^2 = B.$

10. Solve: $\dfrac{xdx}{z^2 - 2yz - y^2} = \dfrac{dy}{y + z} = \dfrac{dz}{y - z}.$

Solution

Each ratio is equal to $\dfrac{xdx + ydy + zdz}{z^2 - 2yz - y^2 + y(y + z) + z(y - z)} = \dfrac{d(x^2 + y^2 + z^2)}{0}.$

So, $x^2 + y^2 + z^2 = A.$

Again, from $\dfrac{dy}{y + z} = \dfrac{dz}{y - z}$, we get

$$\frac{ydy - zdz}{y(y + z) - z(y - z)} = \frac{ydz + zdy}{y(y - z) + z(y + z)}$$

or, $\dfrac{ydy - zdz}{y^2 + z^2} = \dfrac{ydz + zdy}{y^2 + z^2}$

or, $\dfrac{1}{2}d(y^2 - z^2) = d(yz).$

Integrating we get $y^2 - z^2 - 2yz = B.$

So the general solution is given by $x^2 + y^2 + z^2 = A$ and $y^2 - z^2 - 2yz = B.$

11. Solve $\dfrac{dx}{y^2 + yz - x^2} = \dfrac{dy}{y^2 + xz - x^2} = \dfrac{dz}{z(y - x)}.$

Solution

From the equation, we get

$\dfrac{dx - dy}{yz - xz} = \dfrac{dz}{z(y - x)}$ or, $dx - dy - dz = 0.$

Integrating, we get $x - y - z = A.$

Again, from the equation, we get

$$\frac{xdx - ydy}{x(y^2 + yz - x^2) - y(y^2 + xz - x^2)} = \frac{dz}{z(y - x)}$$

or, $$\frac{xdx - ydy}{(x^2 - y^2)(y - x)} = \frac{dz}{z(y - x)}$$

or, $$\frac{d(x^2 - y^2)}{x^2 - y^2} = 2\frac{dz}{z},$$

Integrating, we get

$$\ln(x^2 - y^2) = \ln z^2 + \ln B$$

or, $$x^2 - y^2 = Bz^2.$$

So the general solution is given by $x - y - z = A$ and $x^2 - y^2 = Bz^2$.

12. Solve: $\dfrac{dx}{y + z} = \dfrac{dy}{x + z} = \dfrac{dz}{x + y}$.

Solution

From the equation, we get

$$\frac{dx - dy}{y - x} = \frac{dy - dz}{z - y}.$$

Integrating, we get $\quad x - y = A(y - z)$.

Again, from the equation, we get

$$\frac{dx - dy}{y - x} = \frac{dx + dy + dz}{2(x + y + z)}$$

Integrating we get $\quad (x + y + z)(x - y)^2 = B$.

So the general solution is given by $x - y = A(y - z)$ and $(x + y + z)(x - y)^2 = B$.

13. Solve: $\dfrac{dx}{\cos(x + y)} = \dfrac{dy}{\sin(x + y)} = \dfrac{dz}{z}$.

Solution

From the given equation, we get

$$\frac{dx + dy}{\cos(x + y) + \sin(x + y)} = \frac{dx - dy}{\cos(x + y) - \sin(x + y)} = \frac{dz}{z}.$$

From the first two ratios of the equation, we get

$$\frac{\cos(x + y) - \sin(x + y)}{\cos(x + y) + \sin(x + y)} d(x + y) = d(x - y).$$

Integrating, we get

$$\ln\{\cos(x + y) + \sin(x + y)\} = x - y + k$$

or, $$[\cos(x + y) + \sin(x + y)]e^{y - x} = A.$$

Again, from $\quad \dfrac{dx + dy}{\cos(x + y) + \sin(x + y)} = \dfrac{dz}{z},$

we get $$\frac{d(x + y)}{\sin\left(x + y + \frac{\pi}{4}\right)} = \sqrt{2}\frac{dz}{z}.$$

Integrating, we get

$$\sqrt{2}\ln z = \ln\tan\frac{1}{2}\left(x+y+\frac{\pi}{4}\right)+\ln B$$

or, $$z^{\sqrt{2}}\cot\frac{1}{2}\left(x+y+\frac{\pi}{4}\right)=B.$$

So the general solution is given by

$$[\cos(x+y)+\sin(x+y)]e^{y-x}=A \text{ and } z^{\sqrt{2}}\cot\frac{1}{2}\left(x+y+\frac{\pi}{4}\right)=B.$$

EXERCISE

- Solve the following equations:

1. $\frac{dx}{yz}=\frac{dy}{xz}=\frac{dz}{xy}$,

2. $\frac{dx}{y^2z}=\frac{dy}{x^2z}=\frac{dz}{xy^2}$,

3. $\frac{dx}{0}=\frac{dy}{z}=\frac{dz}{-y}$,

4. $\frac{dx}{y}=\frac{dy}{x}=\frac{dz}{z}$,

5. $\frac{dx}{y^2}=\frac{dy}{x^2}=\frac{dz}{x^2y^2z^2}$,

6. $\frac{dx}{x^2+2y^2}=\frac{dy}{-xy}=\frac{dz}{xz}$,

7. $\frac{dx}{xy}=\frac{dy}{y^2}=\frac{dz}{xyz-zx^2}$,

8. $\frac{dx}{1}=\frac{dy}{3}=\frac{dz}{5z+\tan(y-3x)}$,

[*Hints*: From first two ratios we get $y-3x=c$ and use this in the third ratio.]

9. $\frac{dx}{1}=\frac{dy}{-2}=\frac{dz}{3x^2\sin(y+2x)}$,

[*Hints*: From first two ratios we get $y+2x=c$ and use this in the third ratio.]

10. $\frac{dx}{z}=\frac{dy}{-z}=\frac{dz}{z^2+(x+y)^2}$,

[*Hints*: From first two ratios we get $y+x=c$ and use this in the third ratio.]

11. $dx=-dy=(x+y)\frac{dz}{z}$,

12. $\frac{dx}{xz}=\frac{dy}{-yz}=\frac{dz}{z+xy}$,

13. $\frac{yzdx}{y-z}=\frac{xzdy}{z-x}=\frac{xydz}{x-y}$,

14. $\frac{dx}{mz-ny}=\frac{dy}{nx-lz}=\frac{dz}{ly-mx}$,

15. $\frac{ldx}{mn(y-z)}=\frac{mdy}{nl(z-x)}=\frac{ndz}{lm(x-y)}$,

16. $\frac{dx}{y}=\frac{dy}{-x}=\frac{dz}{bx-ay}$,

17. $\frac{dx}{xz(z^2+xy)}=\frac{dy}{-yz(z^2+xy)}=\frac{dz}{x^4}$,

18. $\frac{dx}{xy}=\frac{dy}{y^2}=\frac{ydz}{xy^2z-2x^3}$,

19. $\frac{dx}{4y-3z}=\frac{dy}{4x-2z}=\frac{dz}{2y-3x}$,

20. $\frac{adx}{(b-c)yz}=\frac{bdy}{(c-a)xz}=\frac{cdz}{(a-b)xy}$,

21. $\dfrac{dx}{x(y-z)}=\dfrac{dy}{y(z-x)}=\dfrac{dz}{z(x-y)}$,

22. $\dfrac{dx}{x^2(y-z)}=\dfrac{dy}{y^2(z-x)}=\dfrac{dz}{z^2(x-y)}$,

23. $\dfrac{dx}{x(y^2+z)}=\dfrac{dy}{-y(x^2+z)}=\dfrac{dz}{z(x^2-y^2)}$,

24. $\dfrac{dx}{x(y^2-z^2)}=\dfrac{dy}{-y(x^2+z^2)}=\dfrac{dz}{z(x^2+y^2)}$,

25. $\dfrac{dx}{x(y^2-z^2)}=\dfrac{dy}{y(z^2-x^2)}=\dfrac{dz}{z(x^2-y^2)}$,

26. $\dfrac{dx}{y-xz}=\dfrac{dy}{x+yz}=\dfrac{dz}{x^2+y^2}$,

27. $\dfrac{dx}{x^2+y^2}=\dfrac{dy}{2xy}=\dfrac{dz}{(x+y)z}$,

28. $\dfrac{dx}{z^2+y^2}=\dfrac{dy}{-xy}=\dfrac{dz}{-xz}$,

29. $\dfrac{dx}{x}=\dfrac{dy}{z}=\dfrac{dz}{-y}$,

[*Hints*: From the given ratios, we get $ydy+zdz=0$ and $\dfrac{dx}{x}=\dfrac{zdy-ydz}{y^2+z^2}=\dfrac{d\left(\frac{y}{z}\right)}{1+\left(\frac{y}{z}\right)^2}$]

30. $\dfrac{dx}{x+2z}=\dfrac{dy}{4xz-y}=\dfrac{dz}{y+2x^2}$,

31. $\dfrac{dx}{y+1}=\dfrac{dy}{x+1}=\dfrac{dz}{z}$,

[*Hints*: From the given ratios, we get $\dfrac{d(x+y)}{x+y+2}=\dfrac{dz}{z}$ and $\dfrac{d(x-y)}{y-x}=\dfrac{dz}{z}$]

32. $dx=dy=\dfrac{dz}{x+y+z}$,

33. $\dfrac{dx}{x^2-yz}=\dfrac{dy}{y^2-xz}=\dfrac{dz}{z^2-yx}$,

34. $\dfrac{dx}{x-y}=\dfrac{dy}{x+y}=\dfrac{dz}{2xz}$.

[*Hints*: From the given ratios, we get $d(x+y)=\dfrac{dz}{z}$ and $\dfrac{xdx+ydy}{x^2+y^2}=\dfrac{xdy-ydx}{x^2+y^2}$]

ANSWERS

1. $x^2-y^2=A,\ y^2-z^2=B$,
2. $x^2-z^2=A,\ x^3-y^3=B$,
3. $x=A,\ y^2+z^2=B$,
4. $x^2-y^2=A,\ x+y=Bz$,
5. $x^3-y^3=A,\ x^3+\dfrac{3}{z}=B$,
6. $yz=A,\ y^2(x^2+y^2)=B$,
7. $x=Ay,\ x=\ln\left(z-\dfrac{2x}{y}\right)+B$,
8. $y-3x=A,\ x=\dfrac{1}{5}\ln[5z+\tan(y-3x)]+B$,

9. $y+2x=A,\ x^3\sin(y+2x)-z=B,$

10. $x+y=A,\ (x+y)^2+z^2=Be^{2x},$

11. $x+y=A,\ x-(x+y)\ln z=B,$

12. $xy=A,\ \ln x+xy\ln(z+xy)-z=B,$

13. $xyz=A,\ x+y+z=B,$

14. $lx+my+nz=A,\ x^2+y^2+z^2=B,$

15. $l^2x+m^2y+n^2z=A,\ l^2x^2+m^2y^2+n^2z^2=B,$

16. $ax+by+z=A,\ x^2+y^2=B,$

17. $xy=A,\ x^4-z^4-2xyz^2=B,$

18. $\dfrac{x}{y}=A,\ x-\ln\left(z-2\dfrac{x^2}{y^2}\right)=B,$

19. $2x-3y-4z=A,\ x^2-y^2-z^2=B,$

20. $ax^2+by^2+cz^2=A,\ a(a-c)x^2+b(b-c)y^2=B,$

21. $x+y+z=A,\ xyz=B,$

22. $\dfrac{1}{x}+\dfrac{1}{y}+\dfrac{1}{z}=A,\ xyz=B,$

23. $xyz=A,\ x^2+y^2-2z=B,$

24. $\dfrac{x}{yz}=A,\ x^2+y^2+z^2=B,$

25. $x^2+y^2+z^2=A,\ xyz=B,$

26. $xy-z=A,\ x^2-y^2+z^2=B,$

27. $x+y=Az,\ x^2-y^2=By,$

28. $\dfrac{y}{z}=A,\ x^2+y^2+z^2=B,$

29. $y^2+z^2=A,\ xe^{-\tan^{-1}\left(\frac{y}{z}\right)}=B,$

30. $xy-z^2=A,\ x^2-y-z=B,$

31. $z(x-y)=A,\ \dfrac{z}{x+y+2}=B,$

32. $x-y=A,\ x+y+z+2=Be^x,$

33. $x-y=A(y-z),\ y-z=B(z-x),$

34. $x+y-\ln z=A,\ x^2+y^2=Be^{2\tan^{-1}\left(\frac{y}{x}\right)}.$

Chapter 10

TOTAL DIFFERENTIAL EQUATION

10.1 Introduction

If an expression of the form $f(x, y, z) = C$ is given, then we have

$$\frac{\partial f}{\partial x}dx+\frac{\partial f}{\partial y}dy+\frac{\partial f}{\partial z}dz=0.$$

This is of the form $Pdx + Qdy + Rdz = 0$, (1)

where P, Q, R are functions of x, y, z.

A differential equation of the form (1) is called a *total differential equation*. From above it is clear that an expression of the form $f(x, y, z) = C$ gives rise to the total differential equation, but the converse of the above is not always true. In other words, a total differential equation may or may not have solution. So it is necessary to check its integrability at first.

10.2 Condition of Integrability of a Total Differential Equation

The necessary and sufficient condition for integrability of a total differential equation of the form

$$Pdx + Qdy + Rdz = 0$$

is
$$P\left(\frac{\partial Q}{\partial z}-\frac{\partial R}{\partial y}\right)+Q\left(\frac{\partial R}{\partial x}-\frac{\partial P}{\partial z}\right)+R\left(\frac{\partial P}{\partial y}-\frac{\partial Q}{\partial x}\right)=0. \tag{2}$$

Necessary condition: Let $f(x, y, z) = C$ be a solution of the given differential equation, then we have

$$k(Pdx + Qdy + Rdz) = df = 0.$$

This gives $\quad \frac{\partial f}{\partial x}=kP, \frac{\partial f}{\partial y}=kQ, \frac{\partial f}{\partial z}=kR,$ where k is a function of x, y, z.

Now using the above results in

$$\frac{\partial}{\partial x}\left(\frac{\partial f}{\partial y}\right)=\frac{\partial}{\partial y}\left(\frac{\partial f}{\partial x}\right)$$

We get $\quad \frac{\partial}{\partial x}(kQ)=\frac{\partial}{\partial y}(kP)$ or, $k\left(\frac{\partial P}{\partial y}-\frac{\partial Q}{\partial x}\right)=Q\frac{\partial k}{\partial x}-P\frac{\partial k}{\partial y}.$

Similarly, we get two other results

$$k\left(\frac{\partial Q}{\partial z}-\frac{\partial R}{\partial y}\right)=R\frac{\partial k}{\partial y}-Q\frac{\partial k}{\partial z}$$

and
$$k\left(\frac{\partial R}{\partial x}-\frac{\partial P}{\partial z}\right)=P\frac{\partial k}{\partial z}-R\frac{\partial k}{\partial x}.$$

Multiplying the above results by R, P, Q respectively and adding, we get

$$P\left(\frac{\partial Q}{\partial z}-\frac{\partial R}{\partial y}\right)+Q\left(\frac{\partial R}{\partial x}-\frac{\partial P}{\partial z}\right)+R\left(\frac{\partial P}{\partial y}-\frac{\partial Q}{\partial x}\right)=0.$$

This is the necessary conditions for integrability.

Sufficient condition: Conversely, let the condition (2) be satisfied by the functions P, Q, R of the equation

$$Pdx + Qdy + Rdz = 0.$$

Then a similar relation holds for the coefficients of

$$kPdx + kQdy + kRdz = 0, \tag{3}$$

where k is a function of x, y, z.

If $Pdx + Qdy$ be not exact, then it can be made so by multiplying it by an integrating factor and hence equation (3) can be considered as the equation of interest. Considering $Pdx + Qdy$ as an exact equation we have

$$\frac{\partial Q}{\partial x}=\frac{\partial P}{\partial y}$$

and
$Pdx + Qdy = dV$, where V is a function of x, y, z.

This gives
$$\frac{\partial V}{\partial x}=P, \frac{\partial V}{\partial y}=Q.$$

Hence
$$\frac{\partial P}{\partial z}=\frac{\partial^2 V}{\partial z\partial x} \text{ and } \frac{\partial Q}{\partial z}=\frac{\partial^2 V}{\partial z\partial y}.$$

Putting these values in (2), we get

$$\frac{\partial V}{\partial x}\left(\frac{\partial^2 V}{\partial z\partial y}-\frac{\partial R}{\partial y}\right)-\frac{\partial V}{\partial y}\left(\frac{\partial^2 V}{\partial z\partial x}-\frac{\partial R}{\partial x}\right)=0$$

or,
$$\frac{\partial V}{\partial x}\frac{\partial}{\partial y}\left(\frac{\partial V}{\partial z}-R\right)+\frac{\partial V}{\partial y}\frac{\partial}{\partial x}\left(R-\frac{\partial V}{\partial z}\right)=0.$$

This shows that there exists a relation between $\left(\frac{\partial V}{\partial z}-R\right)$ and V, which is independent of x, y. In other words, $\frac{\partial V}{\partial z}-R$ is a function of z, V only. Let

$$\frac{\partial V}{\partial z}-R=\phi(z, V).$$

Then
$$Pdx + Qdy + Rdz$$

$$=\frac{\partial V}{\partial x}dx+\frac{\partial V}{\partial y}dy+\left(\frac{\partial V}{\partial z}-\phi(z, V)\right)dz = dV-\phi(z, V)dz.$$

This shows that the equation $Pdx + Qdy + Rdz = 0$ becomes $dV - \phi(z, V)dz = 0$ and this is integrable. Hence if the condition is satisfied by P, Q, R, then the equation $Pdx + Qdy + Rdz = 0$ is integrable.

Note: If $Pdx + Qdy + Rdz$ is an exact differential, then we can write

$$Pdx + Qdy + Rdz = df = \frac{\partial f}{\partial x}dx + \frac{\partial f}{\partial y}dy + \frac{\partial f}{\partial z}dz.$$

So, $$\frac{\partial f}{\partial x} = P, \frac{\partial f}{\partial y} = Q, \frac{\partial f}{\partial z} = R.$$

From these we get

$$\frac{\partial P}{\partial y} = \frac{\partial^2 f}{\partial x \partial y} = \frac{\partial^2 f}{\partial y \partial x} = \frac{\partial Q}{\partial x}, \frac{\partial Q}{\partial z} = \frac{\partial R}{\partial y} \text{ and } \frac{\partial R}{\partial x} = \frac{\partial P}{\partial z}.$$

Thus a total differential equation $Pdx + Qdy + Rdz = 0$, where P, Q, R are functions of x, y, z is said to be exact, if the following conditions are satisfied:

$$\frac{\partial P}{\partial y} = \frac{\partial Q}{\partial x}, \frac{\partial Q}{\partial z} = \frac{\partial R}{\partial y}, \frac{\partial R}{\partial x} = \frac{\partial P}{\partial z}.$$

10.3 Method of Solution

If the condition of integrability is satisfied by P, Q, R of the total differential equation $Pdx + Qdy + Rdz = 0$, then we can use any one of the following methods/cases for solving it.

Solution by Inspection: In many cases, it is possible to rearrange the terms of a total differential equation in such a way that equation becomes exact. So the solution can be obtained by just integrating after reducing it to the exact form.

WORKED-OUT EXAMPLES

1. Solve: $yz \ln z dx - xz \ln z dy + xy dz = 0.$

Solution

Here $P = yz \ln z, Q = -xz \ln z, R = xy.$ So

$$P\left(\frac{\partial Q}{\partial z} - \frac{\partial R}{\partial y}\right) + Q\left(\frac{\partial R}{\partial x} - \frac{\partial P}{\partial z}\right) + R\left(\frac{\partial P}{\partial y} - \frac{\partial Q}{\partial x}\right)$$

$$= yz \ln z(-x - x\ln z - x) - xz \ln z(y - y\ln z - y) + xy(z\ln z + z\ln z) = 0$$

Thus the condition of integrability is satisfied.

Now $$yz \ln z dx - xz \ln z dy + xy dz = 0$$

$$\Rightarrow \frac{dx}{x} - \frac{dy}{y} + \frac{dz}{z \ln z} = 0$$

$$\Rightarrow d(\ln x) - d(\ln y) + d\{\ln(\ln z)\} = 0 \Rightarrow d\left[\ln\left(\frac{x \ln z}{y}\right)\right] = 0.$$

Integrating, we get $\ln\left(\frac{x \ln z}{y}\right) = \ln A$ or, $\frac{x \ln z}{y} = A.$

2. Solve: $(yz + xyz)dx + (xz + xyz)dy + (xy + xyz)dz = 0$.

Solution

Here $P = yz + xyz$, $Q = xz + xyz$, $R = xy + xyz$. So

$$P\left(\frac{\partial Q}{\partial z}-\frac{\partial R}{\partial y}\right)+Q\left(\frac{\partial R}{\partial x}-\frac{\partial P}{\partial z}\right)+R\left(\frac{\partial P}{\partial y}-\frac{\partial Q}{\partial x}\right)$$

$$= (yz + xyz)(xy - xz) + (xz + xyz)(yz - xy) + (xy + xyz)(xz - yz)$$

$$= xyz\,(y - z + z - x + x - y + xy - xz + yz - xy + xz - yz) = 0.$$

Thus the condition of integrability is satisfied.

Now $$(yz + xyz)dx + (xz + xyz)dy + (xy + xyz)dz = 0$$

$$\Rightarrow \frac{dx}{x}+\frac{dy}{y}+\frac{dz}{z}+dx+dy+dz=0$$

$$\Rightarrow d[\ln(xyz)+x+y+z]=0$$

Integrating, we get $\ln(xyz)+x+y+z=A$.

3. Solve: $(y+z)dx+(x+z)dy+(x+y)dz=0$.

Solution

Here $P=y+z$, $Q=x+z$, $R=x+y$. So

$$P\left(\frac{\partial Q}{\partial z}-\frac{\partial R}{\partial y}\right)+Q\left(\frac{\partial R}{\partial x}-\frac{\partial P}{\partial z}\right)+R\left(\frac{\partial P}{\partial y}-\frac{\partial Q}{\partial x}\right)$$

$$=(y+z)(1-1)+(x+z)(1-1)+(x+y)(1-1)=0.$$

Thus the condition of integrability is satisfied.

Now $$(y+z)dx+(x+z)dy+(x+y)dz=0$$

$$\Rightarrow xdy + ydx + zdx + xdz + ydz + zdy = 0$$

$$\Rightarrow d(xy+yz+xz)=0.$$

Integrating, we get $xy + yz + xz = A$.

4. Solve: $(ydx+xdy)(a-z)+xydz=0$.

Solution

Here $P=y(a-z)$, $Q=x(a-z)$, $R=xy$. So

$$P\left(\frac{\partial Q}{\partial z}-\frac{\partial R}{\partial y}\right)+Q\left(\frac{\partial R}{\partial x}-\frac{\partial P}{\partial z}\right)+R\left(\frac{\partial P}{\partial y}-\frac{\partial Q}{\partial x}\right)$$

$$=y(a-z)(-x-x)+x(a-z)(y+y)+xy(a-z-a+z)=0.$$

Thus the condition of integrability is satisfied.

Now $$(ydx+xdy)(a-z)+xydz=0$$

$$\Rightarrow \frac{d(xy)}{xy}+\frac{dz}{a-z}=0 \Rightarrow d[\ln xy-\ln(a-z)]=0.$$

Integrating, we get $\quad \ln xy - \ln(a-z) = \ln C$

or, $\quad xy = C(a-z)$.

Considering one variable as constant: If taking one variable, say z, as constant a total differential equation becomes integrable, then the solution of the equation $Pdx + Qdy = 0$ can be taken as $f(x, y) = u$, where u is independent of x, y and so can be taken as a function of z.

Now taking total differential of $f(x, y) = u(z)$ and comparing it with

$$Pdx + Qdy + Rdz = 0,$$

we get a linear differential equation in u with the aid of $f(x, y) = u(z)$ if the coefficients of du or dz are functions of x, y. Solving that we get $u(z)$ and hence we get the complete solution of the given equation.

WORKED-OUT EXAMPLES

1. Solve: $(yz+z^2)dx - xzdy + xydz = 0$.

Solution

Here $P = yz + z^2, Q = -xz, R = xy$. So

$$P\left(\frac{\partial Q}{\partial z}-\frac{\partial R}{\partial y}\right)+Q\left(\frac{\partial R}{\partial x}-\frac{\partial P}{\partial z}\right)+R\left(\frac{\partial P}{\partial y}-\frac{\partial Q}{\partial x}\right)$$

$$=(yz+z^2)(-x-x)-xz(y-y-2z)+xy(z+z)$$

$$=-2xyz-2xz^2+2xz^2+2xyz=0.$$

Thus the condition of integrability is satistified.

Now taking x as constant, we get

$$\frac{dy}{y}=\frac{dz}{z}.$$

Integrating, we get $\quad y = u(x)z$.

Taking differential of $u(x)=\dfrac{y}{z}$ we get

$$xz^2u'(x)dx - xzdy + xydz = 0.$$

Comparing this with the given equation, we get

$$xz^2u'(x) = yz + z^2 = z^2(1+u)$$

or, $$\frac{du}{u+1}=\frac{dx}{x}.$$

Integrating we get $\quad u + 1 = Cx$

or, $\quad y + z = Cxz$.

2. Solve: $(y^2+z^2-x^2)dx-2xydy-2xzdz=0.$

Solution

Here $P=y^2+z^2-x^2, Q=-2xy, R=-2xz.$ So

$$P\left(\frac{\partial Q}{\partial z}-\frac{\partial R}{\partial y}\right)+Q\left(\frac{\partial R}{\partial x}-\frac{\partial P}{\partial z}\right)+R\left(\frac{\partial P}{\partial y}-\frac{\partial Q}{\partial x}\right)$$

$$=(y^2+z^2-x^2)(0-0)-2xy(-2z-2z)-2xz(2y+2y)=0.$$

Thus the condition of integrability is satisfied. Now taking x as constant, we get

$$ydy+zdz=0.$$

Integrating, we get $\quad y^2+z^2=u(x).$

Taking differential of it, we get

$$u'(x)dx-2ydy-2zdz=0.$$

Comparing it with given equation, we get

$$xu'(x)=y^2+z^2-x^2=u(x)-x^2$$

or, $\quad \dfrac{xu'(x)-u(x)}{x^2}=-1$

or, $\quad \dfrac{d}{dx}\left(\dfrac{u(x)}{x}\right)=-1$

or, $\quad \dfrac{u(x)}{x}=-x+A$

or, $\quad u(x)=-x^2+Ax$

or, $\quad y^2+z^2+x^2=Ax.$

3. Solve: $3x^2dx+3y^2dy-(x^3+y^3+e^{2z})dz=0.$

Solution

Here $P=3x^2, Q=3y^2, R=-(x^3+y^3+e^{2z}).$ So

$$P\left(\frac{\partial Q}{\partial z}-\frac{\partial R}{\partial y}\right)+Q\left(\frac{\partial R}{\partial x}-\frac{\partial P}{\partial z}\right)+R\left(\frac{\partial P}{\partial y}-\frac{\partial Q}{\partial x}\right)$$

$$=3x^2(0+3y^2)+3y^2(-3x^2-0)-(x^3+y^3+e^{2z})(0-0)=0.$$

Thus the condition of integrability is satisfied. Now taking z as constant, we get

$$3x^2dx+3y^2dy=0.$$

Integrating, we get $\quad x^3+y^3=u(z).$

Taking differential of it, we get

$$3x^2dx+3y^2dy-u'(z)dz=0.$$

Comparing it with the given equation we get

$$u'(z) = x^3 + y^3 + e^{2z} = u(z) + e^{2z}.$$

or, $$u'(z) - u(z) = e^{2z}.$$

This is a linear differential equation in $u(z)$, whose integrating factor is $e^{-\int dz} = e^{-z}$. So

$$\frac{d}{dz}(ue^{-z}) = e^z$$

or, $$ue^{-z} = e^z + A$$

or, $$x^3 + y^3 - e^{2z} = Ae^z.$$

4. Solve: $2yzdx - 2xzdy - (x^2 - y^2)(z-1)dz = 0.$

Solution

Here $P = 2yz, Q = -2xz, R = -(x^2 - y^2)(z-1)$. So

$$P\left(\frac{\partial Q}{\partial z} - \frac{\partial R}{\partial y}\right) + Q\left(\frac{\partial R}{\partial x} - \frac{\partial P}{\partial z}\right) + R\left(\frac{\partial P}{\partial y} - \frac{\partial Q}{\partial x}\right)$$

$$= 2yz(-2x - 2yz + 2y) - 2xz(-2xz + 2x - 2y) - (x^2 - y^2)(z-1)(2z + 2z)$$

$$= -4xyz - 4y^2z^2 + 4y^2z + 4x^2z^2 - 4x^2z + 4xyz - (4z^2 - 4z)(x^2 - y^2) = 0.$$

Thus the condition of integrability is satisfied. Now taking z as constant, we get

$$\frac{dx}{x} - \frac{dy}{y} = 0.$$

Integrating, we get $$\frac{x}{y} = u(z).$$

Taking its differential and comparing the result with the given equation, we get

$$\frac{u'(z)}{u^2(z) - 1} = \frac{z-1}{2z}.$$

Integrating, we get $$\ln\frac{u(z)-1}{u(z)+1} = z - \ln z + \ln A$$

or, $$\frac{u(z)-1}{u(z)+1}\frac{z}{A} = e^z$$

or, $$z(x-y) = A(x+y)e^z.$$

Homogeneous Equation: A total differential equation $Pdx + Qdy + Rdz = 0$ is called a homogeneous equation, if the functions P, Q, R are homogeneous functions of same degree. In such cases, one variable, say z, may be separated from the other two by the following substitutions:

$$x = zu, \; y = zv,$$

so that $$dx = zdu + udz, \; dy = zdv + vdz.$$

Use of above results in the given equation will reduce it to one which can be integrated easily.

WORKED-OUT EXAMPLES

1. Solve: $(z+y)zdx - xzdy + xydz = 0$.

Solution

Here $P = (z+y)z, Q = -xz, R = xy$.

So
$$P\left(\frac{\partial Q}{\partial z} - \frac{\partial R}{\partial y}\right) + Q\left(\frac{\partial R}{\partial x} - \frac{\partial P}{\partial z}\right) + R\left(\frac{\partial P}{\partial y} - \frac{\partial Q}{\partial x}\right)$$
$$= (z+y)z(-x-x) - xz(y-2z-y) + xy(z+z)$$
$$= -2xz^2 - 2xyz + 2xz^2 + 2xyz = 0.$$

Thus the condition of integrability is satisfied. Since the given equation is homogeneous equation, we put

$$x = zu,\ y = zv,$$

so that
$$dx = zdu + udz,\ dy = zdv + vdz.$$

Using these in the given equation, we get

$$(v+1)z^2(zdu + udz) - uz^2(zdv + vdz) + uvz^2dz = 0$$

or,
$$(v+1)zdu - uzdv + [u(v+1) - uv + uv]dz = 0$$

or,
$$\frac{du}{u} - \frac{dv}{v+1} + \frac{dz}{z} = 0$$

Integrating, we get

$$\frac{uz}{v+1} = A$$

or,
$$xz = A(y+z).$$

2. Solve: $(yz + y^2)dx + (xz + z^2)dy + (y^2 - xy)dz = 0$.

Solution

Here $P = yz + y^2, Q = xz + z^2, R = y^2 - xy$. So

$$P\left(\frac{\partial Q}{\partial z} - \frac{\partial R}{\partial y}\right) + Q\left(\frac{\partial R}{\partial x} - \frac{\partial P}{\partial z}\right) + R\left(\frac{\partial P}{\partial y} - \frac{\partial Q}{\partial x}\right)$$

$$= (yz + y^2)(x + 2z + x - 2y) + (xz + z^2)(-y - y) + (y^2 - xy)(z + 2y - z)$$
$$= 2xyz + 2xy^2 + 2yz^2 + 2zy^2 - 2zy^2 - 2y^3 - 2xyz - 2yz^2 + 2y^3 - 2xy^2$$
$$= 0.$$

Thus the condition of integrability is satisfied.

Since the given equation is homogeneous equation, we put

$$x = zu,\ y = zv,$$

so that
$$dx = zdu + udz,\ dy = zdv + vdz.$$

Using these in the given equation we get

$$(vz^2 + z^2v^2)(zdu + udz) + (uz^2 + z^2)(zdv + vdz) + (z^2v^2 - z^2uv)dz = 0$$

or, $$v(v+1)(zdu + udz) + (u+1)(zdv + vdz) + v(v-u)dz = 0$$

or, $$v(v+1)du + (u+1)dv + [uv(v+1) + v(u+1) + v(v-u)]\frac{dz}{z} = 0$$

or, $$\frac{du}{u+1} + \frac{dv}{v(v+1)} + \frac{dz}{z} = 0.$$

Integrating, we get

$$\frac{(u+1)vz}{v+1} = A \quad \text{or,} \quad y(x+z) = A(y+z).$$

3. Solve: $yz(y+z)dx + xz(x+z)dy + xy(x+y)dz = 0.$

Solution

Here $P = yz(y+z), Q = xz(x+z), R = xy(x+y)$. So

$$P\left(\frac{\partial Q}{\partial z} - \frac{\partial R}{\partial y}\right) + Q\left(\frac{\partial R}{\partial x} - \frac{\partial P}{\partial z}\right) + R\left(\frac{\partial P}{\partial y} - \frac{\partial Q}{\partial x}\right)$$

$$= yz(y+z)2x(z-y) + xz(x+z)2y(x-z) + xy(x+y)2z(y-x)$$

$$= 2xyz[z^2 - y^2 + x^2 - z^2 + y^2 - x^2] = 0.$$

Thus the condition of integrability is satisfied. Since the given equation is homogeneous equation, we put

$$x = zu, \; y = zv,$$

so that $$dx = zdu + udz, \; dy = zdv + vdz.$$

Using these in the given equation, we get

$$vz^3(v+1)(zdu + udz) + uz^3(u+1)(zdv + vdz) + z^3uv(u+v)dz = 0$$

or, $$v(v+1)(zdu + udz) + u(u+1)(zdv + vdz) + uv(u+v)dz = 0$$

or, $$v(v+1)du + u(u+1)dv + 2uv[1+u+v]\frac{dz}{z} = 0$$

or, $$\frac{(v+1)du}{u(u+v+1)} + \frac{(u+1)dv}{v(u+v+1)} + 2\frac{dz}{z} = 0$$

or, $$\frac{du}{u} + \frac{dv}{v} - \frac{d(u+v+1)}{u+v+1} + 2\frac{dz}{z} = 0.$$

Integrating, we get

$$\frac{uvz^2}{(u+v+1)} = A$$

or, $$xyz = A(x+y+z).$$

4. Solve: $(2x-y)zdx+(2y-x)zdy-(x^2-xy+y^2)dz=0$.

Solution

Here $P=(2x-y)z, Q=(2y-x)z, R=-(x^2-xy+y^2)$. So

$$P\left(\frac{\partial Q}{\partial z}-\frac{\partial R}{\partial y}\right)+Q\left(\frac{\partial R}{\partial x}-\frac{\partial P}{\partial z}\right)+R\left(\frac{\partial P}{\partial y}-\frac{\partial Q}{\partial x}\right)$$
$$=(2x-y)z(2y-x+2y-x)+(2y-x)z(y-2x-2x+y)$$
$$-(x^2-xy+y^2)(-z+z)$$
$$=0.$$

Thus the condition of integrability is satisfied. Since the given equation is homogeneous equation, we put

$$x=zu, y=zv,$$

so that $$dx=zdu+udz, dy=zdv+vdz.$$

Using these in the given equation, we get

$$(2u-v)z^2(zdu+udz)+(2v-u)z^2(zdv+vdz)-(u^2-uv+v^2)z^2dz=0$$

or, $$(2u-v)(zdu+udz)+(2v-u)(zdv+vdz)-(u^2-uv+v^2)dz=0$$

or, $$(2u-v)zdu+(2v-u)zdv+[(2u-v)u+(2v-u)v-(u^2-uv+v^2)]dz=0$$

or, $$(2u-v)zdu+(2v-u)zdv+(v^2-uv+v^2)dz=0$$

or, $$\frac{(2u-v)du+(2v-u)dv}{u^2-uv+v^2}+\frac{dz}{z}=0$$

or, $$\frac{d(u^2-uv+v^2)}{u^2-uv+v^2}+\frac{dz}{z}=0.$$

Integrating, we get

$$(u^2-uv+v^2)z=A \quad \text{or,}\quad x^2-xy+y^2=Az.$$

Auxiliary Equations: If the equation $Pdx + Qdy + Rdz = 0$ be integrable, then functions P, Q, R satisfy the equation, known as condition of integrability,

$$P\left(\frac{\partial Q}{\partial z}-\frac{\partial R}{\partial y}\right)+Q\left(\frac{\partial R}{\partial x}-\frac{\partial P}{\partial z}\right)+R\left(\frac{\partial P}{\partial y}-\frac{\partial Q}{\partial x}\right).$$

Comparing these equations, we get

$$\frac{dx}{\dfrac{\partial Q}{\partial z}-\dfrac{\partial R}{\partial y}}=\frac{dy}{\dfrac{\partial R}{\partial x}-\dfrac{\partial P}{\partial z}}=\frac{dz}{\dfrac{\partial P}{\partial y}-\dfrac{\partial Q}{\partial x}}.$$

This is called auxiliary equations of the given equation and can be solved by the methods discussed for solviñg simultaneous equations, if $\frac{\partial Q}{\partial z}\neq\frac{\partial R}{\partial y}, \frac{\partial R}{\partial x}\neq\frac{\partial P}{\partial z}, \frac{\partial P}{\partial y}\neq\frac{\partial Q}{\partial x}$, i.e., if the given equation is not exact.

Let $u = a$, $v = b$ be two integrals. Now we try to find two functions ϕ, ψ so that the given equation can be written as

$$\phi du + \psi dv = 0.$$

Using $u = a$, $v = b$ in the above equation and comparing it with the given equation we get the values of ϕ, ψ in terms of u, v and hence the equation $\phi du + \psi dv = 0$ can be integrated to get the required integral.

WORKED-OUT EXAMPLES

1. **Solve:** $(z+y)ydx+(x+z)zdy+y(y-x)dz=0.$

Solution

Here $P = yz + y^2, Q = xz + z^2, R = y^2 - xy.$ So

$$P\left(\frac{\partial Q}{\partial z}-\frac{\partial R}{\partial y}\right)+Q\left(\frac{\partial R}{\partial x}-\frac{\partial P}{\partial z}\right)+R\left(\frac{\partial P}{\partial y}-\frac{\partial Q}{\partial x}\right)$$

$$=(yz+y^2)(x+2z+x-2y)+(xz+z^2)(-y-y)+(y^2-xy)(z+2y-z)$$

$$=2xyz+2xy^2+2yz^2+2zy^2-2zy^2-2y^3-2xyz-2yz^2+2y^3-2xy^2=0.$$

Thus the condition of integrability is satisfied. The auxiliary equation of the given equation is

$$\frac{dx}{x-y+z}=\frac{dy}{-y}=\frac{dz}{y}.$$

From the second and third ratios we get $dy + dz = 0$. Integrating, we get

$$y+z=A=u \quad \text{(say)}.$$

From the auxiliary equation, we also have

$$\frac{dx+dz}{x+z}=\frac{dy}{-y}.$$

Integrating we get $\quad (x+z)y=B=v \quad$ (say).

Now we proceed to determine ϕ, ψ in such a way that

$$\phi du+\psi dv=0$$

i.e., $\quad \phi(dy+dz)+\psi[(x+z)dy+y(dx+dz)]=0$

i.e., $\quad y\psi dx+\{\phi+(x+z)\psi\}dy+\{\phi+y\psi\}dz=0$

becomes identical with the given equation. Comparing, we get

$$y\psi=y(y+z),$$

$$\phi+(x+z)\psi=z(x+z),$$

$$\phi+y\psi=y(y-x).$$

Solving, we get $\psi = y + z = u, \phi = -y(x+z) = -v.$

Hence $-vdu + udv = 0.$

Integrating, we get $u = Cv$ or, $y + z = Cy(y+z).$

2. Solve: $z(z-y)dx + (z+x)zdy + x(x+y)dz = 0.$

Solution

Here $P = z(z-y), Q = (z+x)z, R = x(x+y).$ So

$$P\left(\frac{\partial Q}{\partial z} - \frac{\partial R}{\partial y}\right) + Q\left(\frac{\partial R}{\partial x} - \frac{\partial P}{\partial z}\right) + R\left(\frac{\partial P}{\partial y} - \frac{\partial Q}{\partial x}\right)$$

$$= z(z-y)(2z+x-x) + (z+x)z(2x+y+y-2z) + x(x+y)(-z-z)$$

$$= 2z^3 - 2yz^2 + 2z(x^2 - z^2) + 2yz^2 + 2xzy - 2zx^2 - 2xyz = 0.$$

Thus the condition of integrability is satisfied. The auxiliary equation of the given equation is

$$\frac{dx}{2z} = \frac{dy}{2(x+y-z)} = \frac{dz}{-2z}.$$

From the first and third ratios, we get $dx + dz = 0.$

Integrating, we get $x + z = A = u$ (say).

Again from the auxiliary equation, we get

$$\frac{d(x+y)}{2(x+y)} = \frac{dz}{-2z}.$$

Integrating, we get $z(x+y) = B = v$ (say).

Now we proceed to determine ϕ, ψ in such a way that

$$\phi du + \psi dv = 0$$

i.e., $\phi(dx+dz) + \psi[(x+y)dz + z(dx+dy)] = 0$

i.e., $(\phi + z\psi)dx + \psi z dy + \{\phi + \psi(x+y)\}dz = 0$

becomes identical with the given equation. Comparing, we get

$$\phi + z\psi = z(z-y),$$

$$\psi z = z(z+x)$$

and $\phi + \psi(x+y) = x(x+y).$

Solving, we get $\psi = x + z = u,$ and $\phi = -z(x+y) = -v.$

Hence, $-vdu + udv = 0.$

Integrating, we get $u = Cv$ or, $x + z = Cz(x+y).$

3. Solve: $(y^2+yz+z^2)dx+(x^2+xz+z^2)dy+(x^2+xy+y^2)dz=0.$

Solution

Here $P=y^2+yz+z^2, Q=x^2+xz+z^2, R=x^2+xy+y^2$.

So
$$P\left(\frac{\partial Q}{\partial z}-\frac{\partial R}{\partial y}\right)+Q\left(\frac{\partial R}{\partial x}-\frac{\partial P}{\partial z}\right)+R\left(\frac{\partial P}{\partial y}-\frac{\partial Q}{\partial x}\right)$$
$$=P(2z+x-x-2y)+Q(2x+y-y-2z)+R(2y+z-2x-z)$$
$$=2[z^3-y^3+x^3-z^3+y^3-x^3]=0.$$

Thus the condition of integrability is satisfied. The auxiliary equation of the given equation is

$$\frac{dx}{z-y}=\frac{dy}{x-z}=\frac{dz}{y-x}.$$

From this we get each of the ratios being equal to $\dfrac{d(x+y+z)}{0}$. So

$$d(x+y+z)=0.$$

Integrating, we get $\quad x+y+z=A=u$ (say).

Also we see that each of the ratios is equal to $\dfrac{(z+y)dx+(x+z)dy+(y+x)dz}{0}$. So

$$(z+y)dx+(x+z)dy+(y+x)dz=0.$$

Integrating, we get $\quad xy+yz+xz=B=v$ (say).

Now we proceed to determine ϕ, ψ in such a way that

$$\phi du+\psi dv=0$$

i.e., $\quad \phi(dx+dy+dz)+\psi[(z+y)dx+(x+z)dy+(y+x)dz]=0$

i.e., $\quad [\phi+(y+z)\psi]dx+[\phi+(x+z)\psi]dy+[\phi+(y+x)\psi]dz=0$

becomes identical with the given equation. Comparing we get

$$\phi+(y+z)\psi=y^2+yz+z^2,$$

$$\phi+(x+z)\psi=x^2+xz+z^2$$

and $\quad \phi+(x+y)\psi=y^2+yx+x^2.$

Solving we get $\quad \psi=x+y+z=u,$

$$\phi=-(xy+yz+xz)=-v.$$

Hence, $\quad -vdu+udv=0.$

Integrating, we get $\quad u=Cv$ or, $x+y+z=C(xy+yz+xz).$

EXERCISE

- Solve the following equations:

1. $(y+z)dx+dy+dz=0,$
2. $(yz+2x)dx+(xz-2z)dy+(xy-2y)dz=0,$
3. $(yz+2x)dx+(xz+2y)dy+(xy+2z)dz=0,$
4. $(z^2+2xy)dx+(x^2+2yz)dy+(y^2+2xz)dz=0,$
5. $yz\ln z dx - xz\ln z dy + xy dz=0,$
6. $yz(1+x)dx+xz(1+y)dy+xy(1+z)dz=0,$
7. $yzdx+2xzdy-3xydz=0,$
8. $(y^2+z^2-x^2)dx-2xydy-2xzdz=0,$
9. $\dfrac{yzdx}{x^2+y^2}-\dfrac{xzdy}{x^2+y^2}-\tan^{-1}\left(\dfrac{y}{x}\right)dz=0,$
10. $(x-3y-z)dx+(2y-3x)dy+(z-x)dz=0,$
11. $(2x^2+2xy+2xz^2+1)dx+dy+2zdz=0,$
12. $(e^x y+\cos x)dx+(e^x+e^y z)dy+e^y dz=0,$
13. $(y-z)(y+z-2x)dx+(z-x)(z+x-2y)dy+(x-y)(x+y-2z)dz=0,$
14. $(2x^3-z)zdx+2x^2yzdy+x(x+z)dz=0,$
15. $x(y^2-a^2)dx+y(x^2-z^2)dy-z(y^2-a^2)dz=0,$
16. $(mz-ny)dx+(nx-lz)dy+(ly-mx)dz=0,$
17. $(2x^2+2xy+2xz^2+1)dx+dy+2zdz=0,$
18. $2yzdx+xzdy-xy(1+z)dz=0,$
19. $xz^3dx-zdy+2ydz=0,$
20. $(y^2+z^2+x^2)dx-2xydy-2xzdz=0,$
21. $(y+b)(z+c)dx+(x+a)(z+c)dy+(x+a)(y+b)dz=0,$
22. $(x-y)dx-xdy+zdz=0,$
23. $(a+z)ydx+(a+z)xdy-xydz=0,$
24. $z(y+z)dx-xzdy+xydz=0,$
25. $(2x-y)zdx+(2y-x)zdy-(x^2-xy+y^2)dz=0,$
26. $2(y+z)dx-(x+z)dy+(2y-x+z)dz=0,$
27. $yz^2(x^2-yz)dx+zx^2(y^2-xz)dy+xy^2(z^2-yx)dz=0,$

28. $xz^3dx - zdy + 2ydz = 0,$

29. $y(y+z)dx + z(x+z)dy - y(x-y)dz = 0,$

30. $z(z-y)dx + z(z+x)dy + x(x+y)dz = 0,$

31. $yz(y+z)dx + xz(x+z)dy + xy(x+y)dz = 0,$

32. $(x^2+y^2+z^2)dx - 2xydy - 2xzdz = 0,$

33. $(2xy+z)zdx + x^2zdy + (xz+1)dz = 0,$

34. $zydx + x(xy-z)dy + x(xz-y)dz = 0,$

35. $x(y^2-z^2)dx - y(x^2-z^2)dy - z(y^2-x^2)dz = 0,$

36. $zy(1+4xz)dx - xz(1+2xz)dy - xydz = 0,$

37. $z^2dx + (z^2-2yz)dy + (2y^2-yz-xz)dz = 0,$

38. $z(z+y^2)dx + z(z+x^2)dy - xy(x+y)dz = 0,$

39. Find $f(y)$, if $f(y)dx - xzdy - xy \ln y\, dz = 0$ be integrable. Find the corresponding integral.

40. Find $f(z)$, if $(y^2 + z^2 - x^2)dx - 2xydy - 2xf(z)dz = 0$ be integrable. Find the corresponding integral.

41. Find $f(y)$, if $z(1+y)dx - xzdy - xf(y)dz = 0$ be integrable. Find the corresponding integral.

ANSWERS

1. $y + z = Ae^{-x},$

2. $xyz + x^2 - 2yz = A,$

3. $x^2 + y^2 + z^2 + xyz = A,$

4. $x^2y + y^2z + z^2x = A,$

5. $x \ln z = Ay,$

6. $\ln xyz + x + y + z = A,$

7. $xy^2 = Az^3,$

8. $x^2 + y^2 + z^2 = Ax,$

9. $y = x \tan\left(\frac{A}{z}\right),$

10. $x^2 + 2y^2 + z^2 - 6xy - 2xz = A,$

11. $x + y + z^2 = Ae^{-x^2},$

12. $e^x y + ze^y + \sin x = A,$

13. $x(y^2-z^2) + y(z^2-x^2) + z(x^2-y^2) = A,$

14. $x^2 + y^2 + \ln z + \frac{z}{x} = A,$

15. $(x^2 - z^2)(y^2 - a^2) = A,$

16. $nx - lz = A(mz - ny),$

17. $(x + y + z^2)e^{x^2} = A,$

18. $x^2y = Aze^z,$

19. $2y = z^2(x^2 + A),$

20. $y^2 + z^2 = x(x + A),$

21. $(x+a)(y+b)(z+c) = A,$

22. $x^2 - 2xy + z^2 = A$,

23. $xy = A(a + z)$,

24. $xz = A(y + z)$,

25. $x^2 + y^2 - xy = Az$,

26. $(x + z)^2 = A(y + z)$,

27. $x^2 z + y^2 x + z^2 y = Axyz$,

28. $x^2 z^2 - 2y = Az^2$,

29. $(x + z)y = A(y + z)$,

30. $(x + y)z = A(x + z)$,

31. $xyz = c(x + y + z)$,

32. $y^2 + z^2 = x^2 + Ax$,

33. $x^2 y + xz + \ln z = A$,

34. $x^2(y^2 + z^2 - 2A) = 2xyz$,

35. $A(z^2 - x^2) = (A - 1)(y^2 - x^2)$,

36. $x(1 + 2xz) = Cyz$,

37. $(x + y)z - y^2 = Az^2$,

38. $x(y^2 + z) = z(x + y)(1 - Cy)$,

39. $f(y) = ky, x^k = Ay^z$,

40. $f(z) = z, x^2 + y^2 + z^2 = Ax$,

41. $f(y) = k(1 + y), xz^k = A(1 + y)$.

Chapter 11

EXACT DIFFERENTIAL EQUATIONS AND SOME SPECIAL FORMS

11.1 Exact Equation

An equation of the form

$$f(y_n, y_{n-1}, \cdots, y_2, y_1, y) = X, \tag{1}$$

where $y_k = \dfrac{d^k y}{dx^k}$, $k = 1, 2, \cdots, n$ and X is a function of x only, is said to be an exact equation if it can be derived by differentiating once only from an equation of the next lower degree of the form

$$g(y_{n-1}, \ldots, y_2, y_1, y) = \int X dx + C. \tag{2}$$

Equation (2) is said to be the first integral of (1). If the equation (2) is also exact, then we can determine the first integral of (2), i.e., the second integral of the equation (1) and the process can be continued if the integrals obtained successively are found exact. Thus an nth degree equation can have n integrals.

11.2 Condition of Exactness

A linear differential equation of order n of the form

$$P_0 y_n + P_1 y_{n-1} + P_2 y_{n-2} + \ldots + P_{n-1} y_1 + P_n y = X, \tag{3}$$

where P_0, P_1, $\cdots$, P_n, X are all functions of x will be exact if the successive derivatives of the functions P_0, P_1, $\cdots$, P_n satisfy the condition

$$P_n - P'_{n-1} + P''_{n-2} + \ldots + (-1)^n P_0^{(n)} = 0. \tag{4}$$

To prove it, we assume the equation (3) is exact. Then it can be derived by simply differentiating an equation of next lower order, which is called the first integral of the equation (3). Let it be

$$Q_0 y_{n-1} + Q_1 y_{n-2} + \ldots + Q_{n-2} y_1 + Q_{n-1} y = \int X dx + C. \tag{5}$$

Differentiating it and comparing the result with the equation (3), we get

$$P_0 = Q_0 \Rightarrow P_0^{(n)} = Q_0^{(n)},$$

$$P_1 = Q_0' + Q_1 \Rightarrow P_1^{(n-1)} = Q_0^{(n)} + Q_1^{(n-1)},$$

$$P_2 = Q_1' + Q_2 \Rightarrow P_2^{(n-2)} = Q_1^{(n-1)} + Q_2^{(n-2)},$$

$$\cdots\cdots\cdots\cdots\cdots\cdots\cdots\cdots\cdots$$

$$\cdots\cdots\cdots\cdots\cdots\cdots\cdots\cdots\cdots$$

$$P_{n-1} = Q_{n-2}' + Q_{n-1} \Rightarrow P_{n-1}' = Q_{n-2}'' + Q_{n-1}',$$

$$P_n = Q_{n-1}'.$$

Now

$$\begin{aligned} & P_n - P_{n-1}' + P_{n-2}'' + \cdots\ldots + (-1)^n P_0^{(n)} \\ &= Q_{n-1}' - (Q_{n-2}'' + Q_{n-1}') + \cdots + (-1)^{n-2}(Q_1^{(n-1)} + Q_2^{(n-2)}) \\ &\qquad + (-1)^{n-1}(Q_0^{(n)} + Q_1^{(n-1)}) + (-1)^n Q_0^{(n)} \\ &= 0. \end{aligned}$$

Hence, the condition is satisfied.

Again, we see that

$$Q_0 = P_0,$$

$$Q_1 = P_1 - Q_0' = P_1 - P_0',$$

$$Q_2 = P_2 - Q_1' = P_2 - P_1' + P_0'',$$

$$\cdots\cdots\cdots\cdots\cdots\cdots\cdots$$

$$\cdots\cdots\cdots\cdots\cdots\cdots\cdots$$

$$Q_{n-1} = P_{n-1} - P_{n-2}' + P_{n-3}'' - \cdots + (-1)^{n-1} P_0^{(n-1)}.$$

So, if the equation (3) be exact, then its first integral is

$$P_0 y_{n-1} + (P_1 - P_0')\, y_{n-2} + (P_2 - P_1' + P_0'')\, y_{n-3} + \cdots$$

$$+ \left(P_{n-1} - P_{n-2}' + P_{n-3}'' - \cdots + (-1)^{n-1} P_0^{(n-1)}\right) y = \int X\,dx + C. \tag{6}$$

An equation, which is not in the form of an exact equation, can be made exact by multiplying the equation by an integrating factor. The integrating factors depend on the coefficients of the equation. For example, if the coefficients are polynomial functions, then the integrating factor will be of the form x^m. On the other hand, if the coefficients are trigonometric functions, then the integrating factor will be also a trigonometric function.

WORKED-OUT EXAMPLES

1. Solve: $x\dfrac{d^2y}{dx^2} + (1-x)\dfrac{dy}{dx} - y = e^x$.

Solution

Here $P_0 = x$, $P_1 = 1 - x$ and $P_2 = -1$. So $P_2 - P_1' + P_0'' = -1 + 1 + 0 = 0$. Thus the condition of exactness is satisfied.

The first integral is given by

$$P_0\frac{dy}{dx} + (P_1 - P_0')y = \int e^x dx + C \quad \text{or,} \quad x\frac{dy}{dx} - xy = e^x + C \quad \text{or,} \quad \frac{dy}{dx} - y = \frac{1}{x}e^x + \frac{C}{x}.$$

This is a linear differential equation, whose integrating factor is e^{-x}. Multiplying both sides of the equation by e^{-x}, we get

$$\frac{d}{dx}(ye^{-x})=\frac{1}{x}+\frac{Ce^{-x}}{x}.$$

Integrating, we get

$$ye^{-x}=\ln x+C\int\frac{e^{-x}}{x}dx+D,$$

or, $y=e^{x}\ln x+Ce^{x}\int\frac{e^{-x}}{x}dx+De^{x}$, where C, D are arbitrary constants.

2. Solve: $(2x^2+3x)\frac{d^2y}{dx^2}+(6x+3)\frac{dy}{dx}+2y=(x+1)e^{x}$.

Solution

Here $P_0 = 2x^2 + 3x$, $P_1 = 6x + 3$ and $P_2 = 2$. So $P_2-P_1'+P_0''=2-6+4=0$. Thus the condition of exactness is satisfied.

The first integral is given by

$$P_0\frac{dy}{dx}+(P_1-P_0')y=\int(x+1)e^{x}dx+C$$

or, $$(2x^2+3x)\frac{dy}{dx}+2xy=xe^{x}+C$$

or, $$\frac{dy}{dx}+\frac{2}{2x+3}y=\frac{e^{x}}{2x+3}+\frac{C}{x(2x+3)}.$$

This is a linear differential equation whose integrating factor is $2x + 3$. Multiplying the above equation by this factor, we get

$$\frac{d}{dx}[y(2x+3)]=e^{x}+\frac{C}{x}.$$

Integrating, we get

$y(2x+3)=e^{x}+C\ln x+D$, where C, D are arbitrary constants.

3. Solve: $\frac{d^2y}{dx^2}+2e^{x}\frac{dy}{dx}+2e^{x}y=x^2$.

Solution

Here $P_0 = 1$, $P_1 = 2e^x$ and $P_2 = 2e^x$. So $P_2-P_1'+P_0''=2e^{x}-2e^{x}+0=0$. Thus the condition of exactness is satisfied.

The first integral is given by

$$P_0\frac{dy}{dx}+(P_1-P_0')y=\int x^2dx+C \qquad \text{or,} \quad \frac{dy}{dx}+2e^{x}y=\frac{x^3}{3}+C.$$

This is a linear differential equation whose integrating factor is $e^{2e^{x}}$. Multiplying the equation by $e^{2e^{x}}$, we get

$$\frac{d}{dx}(ye^{2e^{x}})=\frac{x^3}{3}e^{2e^{x}}+Ce^{2e^{x}}.$$

Integrating, we get

$$ye^{2e^x} = \frac{1}{3}\int x^3 e^{2e^x} dx + C\int e^{2e^x} dx + D$$

or, $\quad y = \frac{1}{3}e^{-2e^x}\int x^3 e^{2e^x} dx + Ce^{-2e^x}\int e^{2e^x} dx + De^{-2e^x}$, where C, D are arbitrary constants.

4. Solve: $(x^3 - 4x)\frac{d^3y}{dx^3} + (9x^2 - 12)\frac{d^2y}{dx^2} + 18x\frac{dy}{dx} + 6y = 0.$

Solution

Here $P_0 = x^3 - 4x$, $P_1 = 9x^2 - 12$, $P_2 = 18x$ and $P_3 = 6$.

So $P_3 - P_2' + P_1'' - P_0''' = 6 - 18 + 18 - 6 = 0$. Thus the condition of exactness is satisfied.

The first integral is given by

$$P_0\frac{d^2y}{dx^2} + (P_1 - P_0')\frac{dy}{dx} + (P_2 - P_1' + P_0'')y = C$$

or, $\quad (x^3 - 4x)\frac{d^2y}{dx^2} + (6x^2 - 8)\frac{dy}{dx} + 6xy = C.$

Again, taking $P_0 = x^3 - 4x$, $P_1 = 6x^2 - 8$, $P_2 = 6x$, we see that $P_2 - P_1' + P_0'' = 6x - 12x + 6x = 0$, i.e., the first integral of the given equation is also exact. So the first integral of the above equation, i.e., the second integral of the given equation is

$$P_0\frac{dy}{dx} + (P_1 - P_0')y = Cx + D$$

or, $\quad (x^3 - 4x)\frac{dy}{dx} + (3x^2 - 4)\,y = Cx + D.$

Again, taking $P_0 = x^3 - 4x$, $P_1 = 3x^2 - 4$ we see that $P_1 - P_0' = 0$, i.e., the second integral of the given equation is also exact. So the third integral of the given equation is

$$P_0 y = \frac{C}{2}x^2 + Dx + E$$

or, $\quad y = \frac{1}{x^3 - 4x}\left(\frac{C}{2}x^2 + Dx + E\right)$, where C, D, E are arbitrary constants.

5. Solve: $(1 + x + x^2)\frac{d^3y}{dx^3} + (3 + 6x)\frac{d^2y}{dx^2} + 6\frac{dy}{dx} = 0.$

Solution

Here $P_0 = 1 + x + x^2$, $P_1 = 3 + 6x$, $P_2 = 6$ and $P_3 = 0$.

So $P_3 - P_2' + P_1'' - P_0''' = 0 - 0 + 0 - 0 = 0$. Thus the condition of exactness is satisfied.

The first integral is given by

$$P_0\frac{d^2y}{dx^2}+(P_1-P_0')\frac{dy}{dx}+(P_2-P_1'+P_0'')y=C$$

or $$(1+x+x^2)\frac{d^2y}{dx^2}+(2+4x)\frac{dy}{dx}+2y=C.$$

Again, taking $P_0=1+x+x^2, P_1=2+4x, P_2=2,$ we see that $P_2-P_1'+P_0''=$ $2-4+2=0$, i.e., the first integral of the given equation is also exact. So the first integral of the above equation, i.e., the second integral of the given equation is

$$P_0\frac{dy}{dx}+(P_1-P_0')y=Cx+D$$

or, $$(1+x+x^2)\frac{dy}{dx}+(1+2x)y=Cx+D.$$

Again taking $P_0 = 1 + x + x^2$, $P_1 = 1 + 2x$, we see that $P_1-P_0'=0$, i.e., the second integral of the given equation is also exact. So the third integral of the given equation is

$$P_0y=\frac{C}{2}x^2+Dx+E$$

or, $y=\dfrac{1}{1+x+x^2}\left(\dfrac{C}{2}x^2+Dx+E\right)$, where C, D, E are arbitrary constants.

6. Show that the equation $x^3\dfrac{d^3y}{dx^3}+5x^2\dfrac{d^2y}{dx^2}+(2x-x^3)\dfrac{dy}{dx}-(2+x^2)y=40x^3-4x^5$ can be put as an exact equation and find its first integral.

Solution

Here $P_0 = x^3$; $P_1 = 5x^2$, $P_2 = 2x - x^3$ and $P_3 = -(2 + x^2)$.

So $P_3-P_2'+P_1''-P_0'''=-(2+x^2)-2+3x^2+10-6=2x^2\neq 0$. Thus the condition of exactness is not satisfied.

To make it exact, we multiply the equation by x^m. Now we choose m in such a way that the equation

$$x^{m+3}\frac{d^3y}{dx^3}+5x^{m+2}\frac{d^2y}{dx^2}+x^m(2x-x^3)\frac{dy}{dx}-x^m(2+x^2)y=40x^{m+3}-4x^{m+5}$$

becomes exact. Now using $P_3-P_2'+P_1''-P_0'''=0$, we get $m = -2$. Hence, x^{-2} is an integrating factor. Putting $m = -2$, we get

$$x\frac{d^3y}{dx^3}+5\frac{d^2y}{dx^2}+\left(\frac{2}{x}-x\right)\frac{dy}{dx}-\left(\frac{2}{x^2}+1\right)y=40x-4x^3.$$

The first integral of this equation is

$$P_0\frac{d^2y}{dx^2}+(P_1-P_0')\frac{dy}{dx}+(P_2-P_1'+P_0'')y=\int(40x-4x^3)dx+C$$

or, $$x\frac{d^2y}{dx^2}+4\frac{dy}{dx}+\left(\frac{2}{x}-x\right)y=20x^2-x^4+C.$$

7. Show that the equation $x^3\frac{d^3y}{dx^3}+4x^2\frac{d^2y}{dx^2}+x(2+x^2)\frac{dy}{dx}+3x^2y=2x$ can be put as an exact equation and find its first integral.

Solution

Here $P_0=x^3, P_1=4x^2, P_2=x(2+x^2)$ and $P_3=3x^2$.

So $P_3-P_2'+P_1''-P_0'''=3x^2-2-3x^2+8-6=0$. Thus the condition of exactness is satisfied. The first integral of this equation is

$$P_0\frac{d^2y}{dx^2}+(P_1-P_0')\frac{dy}{dx}+(P_2-P_1'+P_0'')y=\int 2x\,dx+C$$

$$\text{or,}\quad x^3\frac{d^2y}{dx^2}+x^2\frac{dy}{dx}+x^3y=x^2+C.$$

8. Solve: $2x^2\frac{d^2y}{dx^2}+15x\frac{dy}{dx}-7y=3x^2.$

Solution

Here $P_0=2x^2, P_1=15x$ and $P_2=-7$. So $P_2-P_1'+P_0''=-7-15+4\neq 0$. Thus the condition of exactness is not satisfied. To make it exact, we multiply the equation by x^m. Now we choose m in such a way that the equation

$$2x^{m+2}\frac{d^2y}{dx^2}+15x^{m+1}\frac{dy}{dx}-7x^my=3x^{m+2}$$

becomes exact,

i.e., $P_2-P_1'+P_0'=-7x^m-15(m+1)x^m+2(m+2)(m+1)x^m=0.$

This gives $2m^2-9m-18=0$. Solving, we get $m=-\frac{3}{2}, 6$. Taking x^6 as an integrating factor, we see that the first integral is given by

$$P_0\frac{dy}{dx}+(P_1-P_0')y=3\int x^8dx+C$$

$$\text{or,}\quad 2x^8\frac{dy}{dx}-x^7y=\frac{1}{3}x^9+C$$

$$\text{or,}\quad \frac{dy}{dx}-\frac{1}{2x}y=\frac{1}{6}x+\frac{C}{2x^8}.$$

This is a linear differential equation whose integrating factor is $\frac{1}{\sqrt{x}}$. Multiplying the above equation by this factor, we get

$$\frac{d}{dx}\left(\frac{y}{\sqrt{x}}\right)=\frac{1}{6}\sqrt{x}+\frac{C}{2x^{\frac{17}{2}}}.$$

Integrating, we get

$$\frac{y}{\sqrt{x}}=\frac{1}{9}x^{\frac{3}{2}}-\frac{C}{15}\frac{1}{x^{\frac{15}{2}}}+D\quad\text{or,}\quad y=\frac{1}{9}x^2-\frac{C}{15}\frac{1}{x^7}+D\sqrt{x}.$$

9. Solve: $y'' + 2\tan xy' + 3y = \tan^2 x \sec x$.

Solution

Clearly, the given equation is not exact. To make it exact we multiply the given equation by $\cos x$ and get

$$y'' \cos x + 2\sin xy' + 3y\cos x = \tan^2 x.$$

Here $P_0 = \cos x$, $P_1 = 2\sin x$ and $P_2 = 3\cos x$.

So, $P_2 - P_1' + P_0'' = 3\cos x - 2\cos x - \cos x = 0$, i.e., the condition of exactness is satisfied.

The first integral is given by

$$P_0 \frac{dy}{dx} + (P_1 - P_0')y = \int \tan^2 xdx + C$$

or, $$\cos x\frac{dy}{dx} + 3\sin xy = \tan x - x + C$$

or, $$\frac{dy}{dx} + 3\tan xy = \tan x\sec x - x\sec x + C\sec x.$$

This is a linear differential equation whose integrating factor is $e^{\int 3\tan xdx} = \sec^3 x$. Multiplying the above equation by $\sec^3 x$, we get

$$\frac{d}{dx}(y\sec^3 x) = \tan x\sec^4 x - x\sec^4 x + +C\sec^4 x$$

or, $y\sec^3 x = \int(\tan x\sec^4 x - x\sec^4 x + C\sec^4 x)dx + D$

$$= \int[\tan x(\tan^2 x+1) - x(\tan^2 x+1) + C(\tan^2 x+1)]\sec^2 xdx + D$$

$$= \frac{\tan^4 x}{4} + \frac{\tan^2 x}{2} + (C-x)\left(\frac{\tan^3 x}{3} + \tan x\right) + \frac{\tan^2 x}{6} + \frac{2}{3}\ln\sec x + D$$

$$= \frac{\tan^4 x}{4} + \frac{2\tan^2 x}{3} + (C-x)\left(\frac{\tan^3 x}{3} + \tan x\right) + \frac{2}{3}\ln\sec x + D.$$

EXERCISE

- Solve the following equations:

1. $(1+x^2)\dfrac{d^2y}{dx^2} + 3x\dfrac{dy}{dx} + y = 0,$

2. $(ax - bx^2)\dfrac{d^2y}{dx^2} + 2a\dfrac{dy}{dx} + 2by = x,$

3. $x\dfrac{d^2y}{dx^2} + 2x\dfrac{dy}{dx} + 2y = 0,$

4. $(1+x^2)\dfrac{d^2y}{dx^2} + 4x\dfrac{dy}{dx} + 2y = \sec^2 x,$

5. $x^2\dfrac{d^2y}{dx^2} + 3x\dfrac{dy}{dx} + y = \dfrac{1}{(1-x)^2},$

6. $\sin x\dfrac{d^2y}{dx^2} - \cos x\dfrac{dy}{dx} + 2y\sin x = 0,$

7. $x\dfrac{d^3y}{dx^3} + (x^2-3)\dfrac{d^2y}{dx^2} + 4x\dfrac{dy}{dx} + 2y = 0,$

8. $(x^3 - x)\dfrac{d^3y}{dx^3} + (8x^2 - 3)\dfrac{d^2y}{dx^2} + 14x\dfrac{dy}{dx} + 4y = \dfrac{2}{x^3}$,

9. $(1+x^2)\dfrac{d^2y}{dx^2} + 3x\dfrac{dy}{dx} + y = 1 + 3x^2$,

10. $x^2(1+x)\dfrac{d^2y}{dx^2} + 2x(2+3x)\dfrac{dy}{dx} + 2(1+3x)y = 0$,

11. $x\dfrac{d^3y}{dx^3} + (x^2 + x + 3)\dfrac{d^2y}{dx^2} + (4x+2)\dfrac{dy}{dx} + 2y = 0$.

- **Show that the following equations can be transformed to exact equations and hence solve these:**

12. $\sqrt{x}\dfrac{d^2y}{dx^2} + 2x\dfrac{dy}{dx} + 3y = x$,

13. $\sin^2 x\dfrac{d^2y}{dx^2} = 2y$,

14. $\dfrac{d^2y}{dx^2} - \cot x\dfrac{dy}{dx} + 2y = 0$,

15. $x^3\dfrac{d^2y}{dx^2} + 2x\dfrac{dy}{dx} - 2y = 1$,

16. $2x^2(x+1)\dfrac{d^2y}{dx^2} + x(7x+3)\dfrac{dy}{dx} - 3y = x^2$.

ANSWERS

1. $y\sqrt{1+x^2} = C\ln(x + \sqrt{1+x^2}) + D$,

2. $xy = \dfrac{x^3}{6a} + \dfrac{1}{3b}C + D(a - bx)^3$,

3. $y = xe^{-2x}\left[A\int x^{-2}e^{2x}dx + B\right]$,

4. $y(1+x^2) = \ln\sec x + Cx + D$,

5. $xy = \ln\left(\dfrac{x}{1-x}\right) + C\ln x + D$,

6. $y\text{cosec}^2 x = \dfrac{C}{2}\left[\ln\tan\dfrac{x}{2} - \text{cosec}\, x\cot x\right] + D$,

7. $yx^{-5}e^{\frac{x2}{2}} = \int\left(C + \dfrac{D}{x}\right)x^{-5}e^{\frac{x2}{2}}dx + E$,

8. $xy\sqrt{x^2-1} = \sec^{-1}x + C\sqrt{x^2-1} + D\ln(x + \sqrt{x^2-1}) + E$,

9. $y\sqrt{1+x^2} = \dfrac{1}{3}(1+x^2)^{\frac{3}{2}} + A\ln[x + \sqrt{1+x^2}] + B$,

10. $yx^2(x+1) = Ax + B$,

11. $xye^{\frac{x(x+2)}{2}} = C\int xe^{\frac{x(x+2)}{2}}dx + D\int e^{\frac{x(x+2)}{2}}dx + E$,

12. $yx^{-1}e^{\frac{4x\sqrt{x}}{3}} = \dfrac{1}{5}e^{\frac{4x\sqrt{4}}{3}} + C\int\dfrac{1}{x^2}e^{\frac{4x\sqrt{x}}{3}}dx + D$,

13. $y = C(1 - x\cot x) + D\cot x$,

14. $y\text{cosec}^2 x = \dfrac{C}{2}\left[\ln\tan\dfrac{x}{2} - \text{cosec}\, x\cot x\right] + D$,

15. $y = -\frac{1}{4}(x+2) + \frac{C}{2}x + Dxe^{\frac{2}{x}}$,

16. $y(x+1) = \frac{1}{7}x^2 + \frac{C}{5}x + Dx^{-\frac{3}{2}}$.

11.3 Some Special Forms

(i) *Equations of the form:* $\frac{d^n y}{dx^n} = f(x)$.

Solution of this form of the equation can be obtained by integrating it successively as follows:

Integrating both sides, we get the first integral as

$$\frac{d^{n-1}y}{dx^{n-1}} = \int f(x)\,dx + C_1.$$

Again integrating it, we get

$$\frac{d^{n-2}y}{dx^{n-2}} = \int\left(\int f(x)\,dx\right)dx + C_1x + C_2 = \iint f(x)(dx)^2 + C_1x + C_2.$$

Proceeding in this way, we get

$$y = \iiint \ldots\ldots \int\iint f(x)(dx)^n + B_0 + B_1x + \ldots\ldots + B_{n-1}x^{n-1}.$$

WORKED-OUT EXAMPLES

1. Solve: $\frac{d^2y}{dx^2} = xe^x$.

Solution

Integrating the given equation, we get

$$\frac{dy}{dx} = (x-1)e^x + C.$$

Again integrating, we get

$$y = \int (x-1)e^x dx + Cx + D = (x-2)e^x + Cx + D.$$

2. Solve: $\frac{d^2y}{dx^2} = x^2 \sin x$.

Solution

Integrating the given equation, we get

$$\frac{dy}{dx} = \int x^2 \sin x dx = -x^2 \cos x + 2x \sin x + 2\cos x + C.$$

Again integrating we get

$$y = -\int x^2 \cos dx + 2\int x \sin x\, dx + 2\int \cos x\, dx + Cx + D$$

$$= -x^2 \sin x - 4x \cos x + 6 \sin x + Cx + D.$$

3. Solve: $\dfrac{d^2y}{dx^2} = e^x \sin x.$

Solution

Integrating the given equation, we get

$$\frac{dy}{dx} = \int e^x \sin x\, dx = \frac{1}{2} e^x (\sin x - \cos x) + C.$$

Again, integrating we get

$$y = \frac{1}{2}\int e^x \sin x dx - \frac{1}{2}\int e^x \cos x dx + Cx + D$$

$$= \frac{1}{4} e^x (\sin x - \cos x) - \frac{1}{4} e^x (\sin x + \cos x) + Cx + D$$

$$= -\frac{1}{2} e^x \cos x + Cx + D.$$

(ii) *Equations of the form:* $\dfrac{d^2y}{dx^2} = f(y).$

Multiplying the equation by $2\dfrac{dy}{dx}$ and using $\dfrac{d}{dx}\left(\dfrac{dy}{dx}\right)^2 = 2\dfrac{dy}{dx}\dfrac{d^2y}{dx^2}$, we get

$$\frac{d}{dx}\left(\frac{dy}{dx}\right)^2 = 2f(y)\frac{dy}{dx}.$$

Integrating both sides, we get

$$\left(\frac{dy}{dx}\right)^2 = 2\int f(y) dy + C$$

or, $$\frac{dy}{\sqrt{2\int f(y) dy + C}} = \pm dx.$$

Again, integrating we get

$$\int \frac{dy}{\sqrt{2\int f(y) dy + C}} = \pm \int dx + D.$$

WORKED-OUT EXAMPLES

1. Solve: $\dfrac{d^2y}{dx^2}=\dfrac{a}{y^3}$.

Solution

Multiplying both sides by $2\dfrac{dy}{dx}$, we get $\dfrac{d}{dx}\left(\dfrac{dy}{dx}\right)^2=\dfrac{2a}{y^3}\dfrac{dy}{dx}$.

Integrating, we get $\left(\dfrac{dy}{dx}\right)^2=C-\dfrac{a}{y^2}$

or, $\dfrac{dy}{dx}=\dfrac{\sqrt{Cy^2-a}}{y}$.

or, $\dfrac{ydy}{\sqrt{Cy^2-a}}=dx$.

Integrating, we get $\sqrt{Cy^2-a}=Cx+D$

or, $y^2=\dfrac{1}{C}[a+(Cx+D)^2]$.

2. Solve: $\cos^3 y\dfrac{d^2y}{dx^2}=\sin y$.

Solution

Multiplying both sides by $2\dfrac{dy}{dx}$, we get

$$\frac{d}{dx}\left(\frac{dy}{dx}\right)^2=2\tan y\sec^2 y\frac{dy}{dx}=\frac{d}{dx}(\tan^2 y).$$

Integrating, we get $\left(\dfrac{dy}{dx}\right)^2=C+\tan^2 y$

or, $\dfrac{dy}{dx}=\pm\sqrt{C+\tan^2 y}$

or, $\dfrac{dy}{\sqrt{C+\tan^2 y}}=\pm dx$

or, $\dfrac{\cos ydy}{\sqrt{C\cos^2 y+\sin^2 y}}=\pm dx$

or, $\dfrac{d(\sin y)}{\sqrt{C+(1-C)\sin^2 y}}=\pm dx$, assuming $C<1$.

Integrating, we get $\ln\left[\sin y+\sqrt{\dfrac{C}{1-C}+\sin^2 y}\right]=\pm\sqrt{1-C}x+D$.

3. Solve: $\sin^3 y \dfrac{d^2y}{dx^2} = \cos y.$

Solution

Multiplying both sides by $2\dfrac{dy}{dx}$, we get

$$\frac{d}{dx}\left(\frac{dy}{dx}\right)^2 = 2\cot y \operatorname{cosec}^2 y \frac{dy}{dx} = -\frac{d}{dx}(\cot^2 y).$$

Integrating, we get $\quad \left(\dfrac{dy}{dx}\right)^2 = C - \cot^2 y$

or, $\quad \dfrac{dy}{dx} = \pm\sqrt{C - \cot^2 y} \quad$ or, $\quad \dfrac{dy}{\sqrt{C - \cot^2 y}} = \pm dx$

or, $\quad \dfrac{\sin y dy}{\sqrt{C\sin^2 y - \cos^2 y}} = \pm dx \quad$ or, $\quad \dfrac{d(\cos y)}{\sqrt{C - (1 + C)\cos^2 y}} = \pm dx.$

Integrating, we get $\quad \sin^{-1}\left(\dfrac{\sqrt{C+1}\cos y}{\sqrt{C}}\right) = \pm\sqrt{1 + C}x + D.$

(iii) *Equations of the form*: $f\left(\dfrac{d^ny}{dx^n}, \dfrac{d^{n-1}y}{dx^{n-1}}, \ldots, \dfrac{d^my}{dx^m}, x\right) = 0$, where $m < n$.

In such cases, the order of the differential equation can be decreased by setting the lowest differential coefficient, i.e., $\dfrac{d^my}{dx^m}$ as new dependent variable p. If we take $\dfrac{d^my}{dx^m} = p$, then the order of the differential equation will be reduced to $n - m$. If the reduced equation is solvable for p, then we have

$$\frac{d^my}{dx^m} = F(x),$$

and this can be solved by the method discussed in (i).

WORKED-OUT EXAMPLES

1. Solve: $(1 + x^2)\dfrac{d^2y}{dx^2} - 1 - \left(\dfrac{dy}{dx}\right)^2 = 0.$

Solution

Putting $\dfrac{dy}{dx} = p$, we rewrite the given equation as

$$\frac{dp}{1 + p^2} = \frac{dx}{1 + x^2}.$$

Integrating, we get $\tan^{-1} p = \tan^{-1} C + \tan^{-1} x$

or, $p = \frac{dy}{dx} = \frac{x+C}{1-xC}$

or, $dy = -\frac{1}{C}dx + \left(C + \frac{1}{C}\right)\frac{dx}{1-xC}.$

Integrating, we get $y = -\frac{x}{C} - \left(1 + \frac{1}{C^2}\right)\ln(1-xC) + D.$

2. Solve: $\frac{d^2y}{dx^2} = \sqrt{1 + \left(\frac{dy}{dx}\right)^2}.$

Solution

Putting $\frac{dy}{dx} = p$ in the given equation, we get

$$\frac{dp}{\sqrt{1+p^2}} = dx.$$

Integrating, we get $\ln\left(p + \sqrt{1+p^2}\right) = x + C$

or, $\sqrt{1+p^2} + p = e^{x+C},$

Then, $\sqrt{1+p^2} - p = e^{-(x+C)}.$

So, $p = \frac{dy}{dx} = \frac{1}{2}\left(e^{x+C} - e^{-(x+C)}\right) = \sinh(x+C).$

Integrating, we get $y = \cosh(x+C) + D.$

3. Solve: $(1+x^2)\frac{d^2y}{dx^2} + 2x\frac{dy}{dx} = 2.$

Solution

Putting $\frac{dy}{dx} = p$ in the given equation, we get

$$(1+x^2)\frac{dp}{dx} + 2xp = 2$$

or, $\frac{dp}{dx} + \frac{2x}{1+x^2}p = \frac{2}{1+x^2}.$

This is a linear differential equation, whose integrating factor is $1 + x^2$. Multiplying the given equation by this factor, we get

$$\frac{d}{dx}[p(1+x^2)] = 2.$$

Integrating, we get $\quad p(1+x^2) = 2x + C$

or, $\quad p = \dfrac{dy}{dx} = \dfrac{2x}{1+x^2} + \dfrac{C}{1+x^2}.$

Integrating we get $\quad y = \ln(1+x^2) + C\tan^{-1} x + D.$

(iv) *Equations of the form:* $f\left(\dfrac{d^n y}{dx^n}, \dfrac{d^{n-1}y}{dx^{n-1}}, \ldots, \dfrac{d^m y}{dx^m}, y\right) = 0$, where $m = 1, 2$.

In such cases, the order of the differential equation can be decreased by setting the lowest differential coefficient, i.e., $\dfrac{d^m y}{dx^m}$, as new dependent variable p. If we take

$$\frac{d^m y}{dx^m} = p,$$

then the order of the differential equation will be reduced to $n - m$. If the reduced equation is solvable for p, then we have

$$\frac{d^m y}{dx^m} = F(y).$$

If $m = 1$, then $\dfrac{dy}{dx} = F(y)$. Integrating, we get

$$\int \frac{dy}{F(y)} = x + C.$$

If $m = 2$, then this can be solved by the method discussed in (ii).

WORKED-OUT EXAMPLES

1. Solve: $y\dfrac{d^2y}{dx^2} - \left(\dfrac{dy}{dx}\right)^2 = y^2 \ln y.$

Solution

Putting $\dfrac{dy}{dx} = p$ in the given equation, we get

$$yp\frac{dp}{dy} - p^2 = y^2 \ln y$$

or, $$p\frac{dp}{dy} - \frac{p^2}{y} = y \ln y$$

or, $$\frac{dp^2}{dy} - \frac{2}{y}p^2 = 2y \ln y.$$

This is a linear differential equation, whose integrating factor is $\dfrac{1}{y^2}$. Multiplying the above equation by this factor, we get

$$\frac{d}{dy}\left(\frac{p^2}{y^2}\right) = \frac{2\ln y}{y}.$$

Integrating, we get $\quad \frac{p^2}{y^2} = \int \frac{2\ln y}{y} dy + C = (\ln y)^2 + C$

$$\text{or, } p = \frac{dy}{dx} = y\sqrt{(\ln y)^2 + C} \text{ or, } \frac{dy}{y\sqrt{(\ln y)^2 + C}} = dx.$$

Integrating, we get $\quad \ln\left[\ln y + \sqrt{(\ln y)^2 + C}\right] = x + D$

or, $\quad \ln y + \sqrt{(\ln y)^2 + C} = e^{x+D},$

then $\quad \ln y - \sqrt{(\ln y)^2 + C} = -Ce^{-(x+D)}.$

Adding, we get

$$\ln y = \frac{1}{2}\left[e^{x+D} - Ce^{-(x+D)}\right] = Ae^x - Be^{-x}, \text{ where } A = \frac{1}{2}e^D, B = \frac{1}{2}Ce^{-D}.$$

2. Solve: $y(1 - \ln y)\frac{d^2y}{dx^2} + (1 + \ln y)\left(\frac{dy}{dx}\right)^2 = 0.$

Solution

Putting $\frac{dy}{dx} = p$ in the given equation, we get

$$y(1 - \ln y)\frac{dp}{dx} + (1 + \ln y)p^2 = 0$$

$$\text{or, } \quad y(1 - \ln y)p\frac{dp}{dy} + (1 + \ln y)p^2 = 0$$

$$\text{or, } \quad \frac{dp}{p} + \frac{1 + \ln y}{y(1 - \ln y)}dy = 0$$

$$\text{or, } \quad \frac{dp}{p} = \frac{u+1}{u-1}du, \text{ where } u = \ln y.$$

Integrating we get $\quad \ln p = u + 2\ln(u - 1) + \ln C$

$$\text{or, } p = Ce^u(u-1)^2 \quad \text{or, } \frac{dy}{dx} = Cy(\ln y - 1)^2$$

$$\text{or, } \frac{d(\ln y)}{(\ln y - 1)^2} = Cdx.$$

Integrating, we get $\quad \frac{1}{1 - \ln y} = Cx + D.$

3. Solve: $\frac{d^2y}{dx^2} + 1 + \left(\frac{dy}{dx}\right)^2 = 0.$

Solution

Putting $\frac{dy}{dx} = p$ in the given equation we get

$$p\frac{dp}{dy} + 1 + p^2 = 0 \quad \text{or, } \quad \frac{pdp}{1 + p^2} + dy = 0.$$

Integrating, we get $\ln(1+p^2)+2y=C$

or, $1+p^2=A^2e^{-2y}$, where $A^2=e^C$,

or, $p=\frac{dy}{dx}=\sqrt{A^2e^{-2y}-1}$

or, $\frac{e^y dy}{\sqrt{A^2-e^{2y}}}=dx.$

Integrating, we get $\sin^{-1}\left(\frac{e^y}{A}\right)=x+D$

or, $y=\ln[A\sin(x+D)].$

(v) *Equation of the form*: $f\left(\frac{d^n y}{dx^n}, \frac{d^{n-1}y}{dx^{n-1}}, x\right)=0$, where $n\geq 1$.

The given equation can be reduced to $f\left(\frac{dq}{dx}, q, x\right)=0$, where $q=\frac{d^{n-1}y}{dx^{n-1}}$. This equation can be solved for q. If

$$q=\frac{d^{n-1}y}{dx^{n-1}}=F(x),$$

then method (i) can be applied for solving it.

WORKED-OUT EXAMPLES

1. Solve: $\frac{d^3y}{dx^3}+a\frac{d^2y}{dx^2}=0.$

Solution

Putting $\frac{d^2y}{dx^2}=p$ in the given equation, we get

$$\frac{dp}{dx}+ap=0.$$

Integrating, we get $\ln p+ax=C$

or, $p=\frac{d^2y}{dx^2}=Ae^{-ax}$, where $A=e^C$.

Integrating, successively, we get

$$\frac{dy}{dx}=-\frac{A}{a}e^{-ax}+B$$

and $$y=\frac{A}{a^2}e^{-ax}+Bx+C.$$

2. Solve: $x\dfrac{d^2y}{dx^2}+x\left(\dfrac{dy}{dx}\right)^2-\dfrac{dy}{dx}=0.$

Solution

Putting $\dfrac{dy}{dx}=p$ in the given equation, we get

$$x\frac{dp}{dx}+xp^2-p=0$$

or, $$\frac{1}{p^2}\frac{dp}{dx}-\frac{1}{xp}=-1$$

or, $$\frac{dq}{dx}+\frac{1}{x}q=1, \text{ where } q=\frac{1}{p}.$$

This is a linear differential equation whose integrating factor is x. Multiplying the above equation by this factor, we get

$$\frac{d}{dx}(xq)=x.$$

Integrating, we get $$xq=\frac{x}{p}=\frac{x^2}{2}+C$$

or, $$p=\frac{dy}{dx}=\frac{2x}{x^2+2C}.$$

Integrating, we get $$y=\ln(x^2+2C)+D.$$

(vi) *Equation of the form*: $f\left(\dfrac{d^ny}{dx^n},\dfrac{d^{n-2}y}{dx^{n-2}},x\right)=0$, where $n\geq 2$.

The given equation can be reduced to $f\left(\dfrac{d^2q}{dx^2},q,x\right)=0$, where $q=\dfrac{d^{n-2}y}{dx^{n-2}}$. This equation can be solved for q. If

$$q=\frac{d^{n-2}y}{dx^{n-2}}=F(x),$$

then method (i) can be applied for solving it.

WORKED-OUT EXAMPLES

1. Solve: $\dfrac{d^4y}{dx^4}=a^2\dfrac{d^2y}{dx^2}.$

Solution

Putting $\dfrac{d^2y}{dx^2}=p$ in the given equation, we get

$$\frac{d^2p}{dx^2}=a^2p.$$

Multiplying both sides by $2\frac{dp}{dx}$, we get

$$\frac{d}{dx}\left(\frac{dp}{dx}\right)^2 = \frac{d}{dx}(a^2p^2).$$

Integrating, we get $$\left(\frac{dp}{dx}\right)^2 = a^2p^2 + b^2$$

or, $$\frac{dp}{\sqrt{a^2p^2+b^2}} = dx.$$

Integrating, we get $$\ln\left(ap + \sqrt{a^2p^2+b^2}\right) = ax + C$$

or, $$ap + \sqrt{a^2p^2+b^2} = ke^{ax},$$

then $$\sqrt{a^2p^2+b^2} - ap = \frac{b^2}{k}e^{-ax}.$$

Subtracting, we get $$p = \frac{d^2y}{dx^2} = \frac{1}{2a}\left(ke^{ax} - \frac{b^2}{k}e^{-ax}\right).$$

Integrating successively, we get

$$\frac{dy}{dx} = \frac{1}{2a^2}\left(ke^{ax} + \frac{b^2}{k}e^{-ax}\right) + \alpha$$

and $$y = \frac{1}{2a^3}\left(ke^{ax} - \frac{b^2}{k}e^{-ax}\right) + \alpha x + \beta.$$

2. Solve: $x^2\frac{d^4y}{dx^4} + \frac{d^2y}{dx^2} = 0.$

Solution

Putting $\frac{d^2y}{dx^2} = p$ in the given above equation, we get

$$x^2\frac{d^2p}{dx^2} + p = 0.$$

Putting $z = \ln x$ in the above equation, we get

$$\frac{d^2p}{dz^2} - \frac{dp}{dz} + p = 0.$$

The auxiliary equation is $m^2 - m + 1 = 0$.

Roots of this equation are $m = \frac{1 \pm i\sqrt{3}}{2}$.

So, $$p = \frac{d^2y}{dx^2} = e^{\frac{z}{2}}\left[A\cos\left(\frac{\sqrt{3}z}{2}\right) + B\sin\left(\frac{\sqrt{3}z}{2}\right)\right] = C_1\sqrt{x}\cos\left(\frac{\sqrt{3}\ln x}{2} + C_2\right).$$

Integrating twice, we get $$y = \alpha + \beta x + \gamma x^{\frac{5}{2}}\cos\left(\frac{\sqrt{3}\ln x}{2} + \delta\right).$$

EXERCISE

- Solve the following equations:

1. $\dfrac{d^2y}{dx^2} = \ln x,$

2. $x^2\dfrac{d^2y}{dx^2} = \ln x,$

3. $\dfrac{d^3y}{dx^3} = xe^x,$

4. $\dfrac{d^4y}{dx^4} + \dfrac{1}{x^2} = 0,$

5. $\sqrt{y}\dfrac{d^2y}{dx^2} = 1,$

6. $y^3\dfrac{d^2y}{dx^2} = a,$

7. $x\dfrac{d^2y}{dx^2} + \dfrac{dy}{dx} = 0,$

8. $\dfrac{d^2y}{dx^2} + \dfrac{dy}{dx} + \left(\dfrac{dy}{dx}\right)^3 = 0,$

9. $x\dfrac{d^2y}{dx^2} + \dfrac{dy}{dx} = 4x,$

10. $x\dfrac{d^2y}{dx^2} + x\left(\dfrac{dy}{dx}\right)^2 - \dfrac{dy}{dx} = 0,$

11. $(1-x^2)\dfrac{d^2y}{dx^2} - x\dfrac{dy}{dx} = 2,$

12. $\dfrac{d^4y}{dx^4} - \cot x\dfrac{d^3y}{dx^3} = 0,$

13. $\dfrac{d^2y}{dx^2} + \dfrac{dy}{dx} = e^x,$

14. $y\dfrac{d^2y}{dx^2} + \left(\dfrac{dy}{dx}\right)^2 = y^2,$

15. $y\dfrac{d^2y}{dx^2} + \left(\dfrac{dy}{dx}\right)^2 = \dfrac{dy}{dx},$

16. $\dfrac{d^3y}{dx^3}\dfrac{d^2y}{dx^2} = 2,$

17. $(1-y^2)\dfrac{d^2y}{dx^2} = 2\dfrac{dy}{dx}\left(1 + y\dfrac{dy}{dx}\right),$

18. $y\dfrac{d^2y}{dx^2} + \left(\dfrac{dy}{dx}\right)^2 = 1,$

19. $y^2\dfrac{d^2y}{dx^2} = k,$

20. $\dfrac{d^2y}{dx^2} - a\left(\dfrac{dy}{dx}\right)^2 = 0,$

21. $\dfrac{d^3y}{dx^3} + \dfrac{d^2y}{dx^2} = 0,$

22. $\dfrac{d^3y}{dx^3} = 4\dfrac{dy}{dx},$

23. $\dfrac{d^4y}{dx^4} = \dfrac{d^2y}{dx^2}.$

ANSWERS

1. $y = \dfrac{x^2}{2}\ln x - \dfrac{5}{4}x^2 + Ax + B,$

2. $y = Ax + B - \ln x - \dfrac{1}{2}(\ln x)^2,$

3. $y = xe^x - 3e^x + \dfrac{A}{2}x^2 + Bx + C,$

4. $y = \dfrac{1}{2}x^2\ln x + Ax^3 + Bx^2 + Cx + D,$

5. $3x = 2(\sqrt{y} - 2A)\sqrt{\sqrt{y} + A} + B,$

6. $(Ax \pm B)^2 + a = Ay^2,$

7. $y = A\ln x + B,$

8. $x = A - \ln\sin(B - y),$

9. $y = x^2 + A\ln x + B,$

10. $y = \ln(x^2 + A) + B,$

11. $y = (\sin^{-1}x)^2 + A\sin^{-1}x + B,$

12. $y = A\cos x + \dfrac{1}{2}Bx^2 + Cx + D,$

13. $y = Ae^{-x} + B + \frac{1}{2}e^{x}$,

14. $\ln\left(y^2 + \sqrt{y^4 + A^2}\right) = \sqrt{2}(B \pm x)$,

15. $y = x + A\ln(A + y) + B$,

16. $15y = 8(x + A)^{\frac{5}{2}} + Bx + C$,

17. $8x = -2y^2 + 2Ay + (4 - A^2)\ln(2y + A) + B$,

18. $y^2 = x^2 + Ax + B$,

19. $Cy^2 = k + (Cx + D)^2$,

20. $y = A - \frac{1}{a}\ln(ax + B)$,

21. $y = Ae^{-x} + Bx + C$,

22. $y = A + Be^{2x} - Ce^{-2x}$,

23. $y = Ae^{x} + Be^{-x} + Cx + D$.

Chapter 12

EIGEN VALUE PROBLEMS

12.1 Introduction

A second or higher order differential equation together with the values of the unknown function and its derivatives given at two or more points constitute a boundary value problem. A second order non-homogeneous boundary value problem has the form:

$$\frac{d^2y}{dx^2} + P(x)\frac{dy}{dx} + Q(x)y = F(x)$$

with

$$Ay(a) + By'(a) = u,$$

$$Cy(b) + Dy'(b) = v,$$

where $y' = \dfrac{dy}{dx}$ and $P(x), Q(x), R(x)$ are functions of x only; A, B, C, D, u, v are real constants.

If $F(x) = 0$, then the above problem is called *homogeneous boundary value problem*. The most general type of homogeneous boundary value problem has the form:

$$\frac{d^2y}{dx^2} + P(x, \lambda)\frac{dy}{dx} + Q(x,\lambda)y = 0$$

with

$$Ay(a) + By'(a) = 0,$$

$$Cy(b) + Dy'(b) = 0,$$

where λ is an arbitrary constant.

Thus the solution of boundary value problems may sometimes be reduced to the solution of ordinary differential equation containing a parameter and subject to certain boundary conditions. Boundary value problems that admit of solutions only for special values of the constant occurring in the differential equation are known as *eigen value problems*. The special values of the constant are known as eigen values and the solutions obtained for special values of the constant are called *eigen functions*.

A simple example of this situation is the problem of finding the solution of the following differential equation in the interval [0, 1],

$$\frac{d^2x}{dt^2} + \lambda x = 0, \tag{1}$$

given that $x(0) = x(1) = 0$, where λ is a parameter (real or complex) and x is a scalar.

If $\lambda = 0$, then we get $x = A + Bt$, where A, B are arbitrary constants.

Using $x(0) = x(1) = 0$, we get $A = 0$, $A + B = 0$. These give $A = 0$, $B = 0$, i.e., $x = 0$. Thus $x = 0$ is a trivial solution of the equation.

If $\lambda < 0$, then we get $x = Ae^{\sqrt{-\lambda}t} + Be^{-\sqrt{-\lambda}t}$, where A, B are arbitrary constants.

Using $x(0) = x(1) = 0$, we get $Ae^{\sqrt{-\lambda}} + Be^{-\sqrt{-\lambda}}$, $A + B = 0$. These give $A = 0$, $B = 0$, i.e., $x = 0$.

Thus $x = 0$ is a trivial solution of the equation.

Hence, if $\lambda \leq 0$, then equation (1) has only trivial solution $x = 0$.

If $\lambda > 0$, then solving (1) we get

$$x = A\cos\sqrt{\lambda}t + B\sin\sqrt{\lambda}t.$$

Using the conditions $x(0) = x(1) = 0$, we get $A = 0$ and $B\sin\sqrt{\lambda} = 0$. Now if $B = 0$, then $x = 0$, which is a trivial solution of (1). If non-trivial solution exists, then we must have

$$\sin\sqrt{\lambda} = 0 \quad \lambda = n^2\pi^2,\ n = 1, 2, 3, \cdots.$$

These values are called eigen values and corresponding eigen functions are

$$x = B\sin n\pi t,\ n = 1, 2, 3, \cdots.$$

In general, $x = B_n \sin n\pi t,\ n = 1, 2, 3, \cdots$ is called eigen function, where B_n is a arbitrary constant.

It is to be noted that the sequence of eigen values $\{\lambda_n\}$ is an monotonically increasing sequence diverges to ∞ and the nth eigen function $\sin n\pi t$ has $(n + 1)$ zeros in the interval [0, 1].

12.2 Strum-Liouville Problem

A second order *Strum-Liouville problem* is a homogeneous boundary value problem of the form:

$$\frac{d}{dx}\left\{p(x)\frac{dy}{dx}\right\} + \{q(x) + \lambda r(x)\}y = 0 \tag{2}$$

with
$$Ay(a) + By'(a) = 0,$$
$$Cy(b) + Dy'(b) = 0,$$

where $p(x)$, $q(x)$, $r(x)$, $p'(x)$ are real valued continuous functions defined on $[a, b]$ such that $p(x)$ and $r(x)$ are positive on $[a, b]$ and λ is an arbitrary parameter. In the boundary conditions the constants A, B, C, D are such that at least one of A, C and at least one of B, D are non-zero.

12.3 Conversion to Strum-Liouville Problem

Consider the following second order differential equation:

$$p_0(x)\frac{d^2y}{dx^2}+p_1(x)\frac{dy}{dx}+p_2(x)y+\lambda p_3(x)y=0, \qquad (3)$$

where $p_0(x), p_3(x)$ assume positive values in the interval of interest. Multiplying the equation (3) by

$$I(x)=e^{\int\frac{p_1(x)}{p_0(x)}dx},$$

we get $$I(x)\,p_0(x)\frac{d^2y}{dx^2}+I(x)\,p_1(x)\frac{dy}{dx}+I(x)\,p_2(x)y+I(x)\lambda p_3(x)y=0$$

or, $$p_0(x)\frac{d}{dx}\left\{I(x)\frac{dy}{dx}\right\}+I(x)\,p_2(x)y+I(x)\lambda p_3(x)y=0$$

or, $$\frac{d}{dx}\left\{I(x)\frac{dy}{dx}\right\}+\left\{\frac{I(x)p_2(x)}{p_0(x)}+\lambda\frac{I(x)\,p_3(x)}{p_0(x)}\right\}y=0,$$

where $I(x), \lambda\dfrac{I(x)\,p_3(x)}{p_0(x)}$ are positive functions as $p_0(x), p_3(x)$ assume positive values in the interval of interest and $I'(x)=\dfrac{p_1(x)}{p_0(x)}e^{\int\frac{p_1(x)}{p_0(x)}dx}$ is a real valued continuous function defined in the interval of interest. Thus the second order homogeneous differential equation given by (3) has reduced to the Strum-Liouville form (2).

WORKED-OUT EXAMPLES

1. Find the eigen values λ_n and eigen functions $y_n(x)$ for the differential equation

$$\frac{d^2y}{dx^2}+\lambda y=0(\lambda>0)$$

satisfying the boundary conditions $y(0) = 0 = y(2\pi)$.

Solution

For $\lambda > 0$, the general solution of the given equation is

$$y=A\cos\sqrt{\lambda}x+B\sin\sqrt{\lambda}x.$$

Using the given conditions we get $A=0$ and $B\sin(2\pi\sqrt{\lambda})=0$.

If $B\neq 0, \sin(2\pi\sqrt{\lambda})=\sin n\pi \quad \lambda=\dfrac{n^2}{4}, n=1,2,3,\cdots$

So the eigen values are $\lambda_n=\dfrac{n^2}{4}$ and eigen functions are

$$y_n(x)=B_n\sin\left(\frac{nx}{2}\right), n=1,2,3,\cdots.$$

2. Find the eigen values λ_n and eigen functions $y_n(x)$ for the differential equation

$$\frac{d^2y}{dx^2} + \lambda y = 0 (\lambda > 0)$$

satisfying the boundary conditions $y'(0) = 0$ and $y'(l) = 0$.

Solution

The general solution of the given equation is

$$y = A\cos\sqrt{\lambda}x + B\sin\sqrt{\lambda}x,$$

which gives $$y' = \sqrt{\lambda}\left[B\cos\sqrt{\lambda}x - A\sin\sqrt{\lambda}x\right].$$

Using the boundary conditions $y'(0) = 0$ and $y'(l) = 0$, we get

$$B = 0 \text{ and } \sin\sqrt{\lambda}l = 0 (\text{if } A \neq 0) \quad \lambda = \frac{n^2\pi^2}{l^2}, \quad n = 1, 2, 3, \cdots.$$

So, the eigen values are $\lambda_n = \dfrac{n^2\pi^2}{l^2}$ and the eigen functions are

$$y_n(x) = A_n \cos\left(\frac{n\pi x}{l}\right), \quad n = 1, 2, 3, \cdots.$$

3. Find the eigen values λ_n and eigen functions $y_n(x)$ for the differential equation

$$x^2\frac{d^2y}{dx^2} + x\frac{dy}{dx} + \lambda y = 0, (\lambda > 0)$$

satisfying the boundary conditions $y(1) = 0 = y(e^{\pi})$.

Solution

To solve the equation we put $x = e^z$, i.e., $z = \log x$, then

$$\frac{dy}{dx} = \frac{dy}{dz}\frac{dz}{dx} = \frac{1}{x}\frac{dy}{dz}$$

$$\Rightarrow x\frac{dy}{dx} = \frac{dy}{dz}$$

$$\Rightarrow \frac{d}{dx}\left[x\frac{dy}{dx}\right] = \frac{d}{dz}\left[\frac{dy}{dz}\right]\frac{dz}{dx} = \frac{1}{x}\frac{d^2y}{dz^2}$$

$$x^2\frac{d^2y}{dx^2} + x\frac{dy}{dx} = \frac{d^2y}{dz^2}.$$

So the given equation becomes

$$\frac{d^2y}{dz^2} + \lambda y = 0,$$

where y satisfies the conditions $y(0) = 0 = y(\pi)$.

The general solution of the transformed equation is

$$y = A\cos\sqrt{\lambda}z + B\sin\sqrt{\lambda}z.$$

Using the transformed conditions, we get

$$A = 0 \text{ and } \sin\sqrt{\lambda}\pi = 0 \text{ (if } B \neq 0) \quad \lambda = n^2, n = 1, 2, 3, \cdots.$$

So the eigen values are $\lambda_n = n^2$ and the eigen functions are

$$y_n(x) = B_n \sin(n \log x), n = 1, 2, 3, \cdots.$$

4. Find the eigen values λ_n and eigen functions $y_n(x)$ for the differential equation

$$x^2 \frac{d^2y}{dx^2} + x\frac{dy}{dx} + \lambda y = 0, (\lambda > 0)$$

satisfying the boundary conditions $y'(1) = 0 = y'(e^{\frac{\pi}{2}})$.

Solution

To solve the equation we put $x = e^z$, i.e., $z = \log x$, then the given equation reduces to

$$\frac{d^2y}{dz^2} + \lambda y = 0 (\lambda > 0),$$

where y satisfies the conditions $y'(0) = 0 = y'(\pi/2)$.

The general solution of the transformed equation is

$$y = A\cos\sqrt{\lambda}z + B\sin\sqrt{\lambda}z, \text{ which gives}$$

$$y' = \sqrt{\lambda}\{B\cos\sqrt{\lambda}z - A\sin\sqrt{\lambda}z].$$

Using the transformed conditions, we get

$$B = 0 \text{ and } \sin\left(\frac{\sqrt{\lambda}\pi}{2}\right) = 0 \text{ (if } A \neq 0) \quad \lambda = 4n^2, \; n = 1, 2, 3, \cdots.$$

So the eigen values are $\lambda_n = 4n^2$ and the eigen functions are

$$y_n(x) = A_n \cos(2n \log x), \; n = 1, 2, 3, \cdots.$$

5. Find the eigen values λ_n and eigen functions $y_n(x)$ for the differential equation

$$\frac{d^2y}{dx^2} + \lambda y = 0 (\lambda > 0)$$

satisfying the boundary conditions $y(0) + y'(0) = 0$ and $y(l) + y'(l) = 0$.

Solution

The general solution of the given equation is

$$y = A\cos\sqrt{\lambda}x + B\sin\sqrt{\lambda}x, \text{ which gives}$$

$$y' = \sqrt{\lambda}\left[B\cos\sqrt{\lambda}x - A\sin\sqrt{\lambda}x\right].$$

Using the boundary conditions, we get

$$A + \sqrt{\lambda}B = 0 \text{ and } (1 + \lambda)\sin\sqrt{\lambda}l = 0, \text{ if } B \neq 0 \quad \lambda = \frac{n^2\pi^2}{l^2}, \; n = 1, 2, 3, \cdots.$$

So the eigen values are $\lambda_n = \frac{n^2\pi^2}{l^2}$ and eigen functions are given by

$$y_n(x) = B_n\left[\sin\left(\frac{n\pi x}{l}\right) - \frac{n\pi}{l}\cos\left(\frac{n\pi x}{l}\right)\right], \; n = 1, 2, 3, \cdots.$$

6. Find the eigen values λ_n and eigen functions $y_n(x)$ for the differential equation

$$\frac{d^2y}{dx^2} + \lambda y = 0 (\lambda > 0)$$

satisfying the boundary conditions $y(0) = y(\pi)$ and $y'(0) = y'(\pi)$.

Solution

The general solution of the given equation is

$$y = A\cos\sqrt{\lambda}x + B\sin\sqrt{\lambda}x, \text{ which gives}$$

$$y' = \sqrt{\lambda}\left[B\cos\sqrt{\lambda}x - A\sin\sqrt{\lambda}x\right].$$

Using the boundary conditions, we get

$$A(1-\cos\sqrt{\lambda}\pi) = B\sin\sqrt{\lambda}\pi \text{ and } B(1-\cos\sqrt{\lambda}\pi) = -A\sin\sqrt{\lambda}\pi.$$

From the above relations we get

$$1-\cos\sqrt{\lambda}\pi = 0 \quad \lambda = 4n^2.$$

Hence the eigen values are given by $\lambda_n = 4n^2 . n = 1, 2, 3, \cdots$ and the eigen functions are given by

$$y_n = C_n \cos(2nx + \alpha) \text{ where } \tan\alpha = -\frac{B}{A}.$$

7. Put, if possible, the following equations in Strum-Liouville form:

$$x\frac{d^2y}{dx^2} + 2x^2\frac{dy}{dx} + (x+\lambda)y = 0,$$

Solution

Comparing the given equation with the equation (3) we see that

$$p_0(x) = x, \; p_1(x) = 2x^2, \; p_2(x) = x \text{ and } p_3(x) = 1.$$

So $I(x) = e^{\int\frac{p_1(x)}{p_0(x)}dx} = e^{\int 2x dx} = e^{x^2}$. Multiplying the given equation by e^{x^2}, we get

$$\frac{d}{dx}\left(e^{x^2}\frac{dy}{dx}\right) + e^{x^2}y + \lambda\frac{e^{x^2}}{x}y = 0.$$

This is the Strum-Liouville form of the given equation.

EXERCISE

1. Find the eigen values λ_n and eigen functions $y_n(x)$ for the differential equation

$$\frac{d^2y}{dx^2} + \lambda y = 0 (\lambda > 0)$$

satisfying the boundary conditions:

(i) $y(0) = 0 = y(\pi)$, (ii) $y(0) = 0 = y(1)$,

(iii) $y(0) = 0, y'(\pi) = 0$, (iv) $y'(0) = 0 = y'(1)$,

(v) $y(0) = 0 = y\left(\frac{\pi}{2}\right)$, (vi) $y(0) = 0 = y(2\pi)$,

(vii) $y(0) = 0 = y(a), a > 0$, (viii) $y(\pm a) = 0, a > 0$,

(ix) $y(a) = 0 = y(b), a < b$,

[*Hints*: Complementary function is

$$y = A\cos\sqrt{\lambda}x + B\sin\sqrt{\lambda}x \quad \text{(i)}$$

Using the given conditions we get

$$A\cos\sqrt{\lambda}a + B\sin\sqrt{\lambda}a = 0 \quad \text{(ii)}$$

and

$$A\cos\sqrt{\lambda}b + B\sin\sqrt{\lambda}b = 0 \quad \text{(iii)}$$

From (ii) we get

$$\frac{\cos\sqrt{\lambda}a}{B} = \frac{\sin\sqrt{\lambda}a}{-A} = \frac{1}{\sqrt{A^2 + B^2}}$$

or,

$$A = -\sqrt{A^2 + B^2}\sin\sqrt{\lambda}a$$

and

$$B = \sqrt{A^2 + B^2}\cos\sqrt{\lambda}a$$

Putting the values of A, B in (i) and (iii), we get

$$y = \sqrt{A^2 + B^2}\sin\sqrt{\lambda}(x - a) \quad \text{(iv)}$$

and

$$\sin\sqrt{\lambda}(b - a) = 0 \text{ respectively,}$$

which gives

$$\lambda = \left(\frac{n\pi}{b - a}\right)^2.$$

Hence the eigen values are $\lambda_n = \left(\frac{n\pi}{b - a}\right)^2, n = 1, 2, 3, \cdots$ and the eigen functions are given by $y_n = C_n \sin\left(\frac{n\pi(x - a)}{b - a}\right), n = 1, 2, 3, \cdots$.]

(x) $y(0) = 0 = y'(a)$, (xi) $y'(0) = 0 = y'(a)$,

(xii) $y'(0) = 0 = y'(\pi)$, (xiii) $y(0) = 0, y'\left(\frac{\pi}{2}\right) = 0$,

(xiv) $y'(\pm\pi) = 0$, (xv) $y(0) + y'(0) = 0, y(1) + y'(1) = 0$.

2. Find the eigen values λ_n and eigen functions $y_n(x)$ for the differential equation

$$x^2\frac{d^2y}{dx^2}+\frac{dy}{dx}+\lambda y=0, (\lambda>0)$$

satisfying the boundary conditions:

(i) $y(1)=0=y(e^{\pi})$,

(ii) $y'(1)=0=y'(e^{2\pi})$.

3. Find the eigen value λ_n and eigen functions $y_n(x)$ for the differential equation

$$x^2\frac{d^2y}{dx^2}+3x\frac{dy}{dx}+(\lambda+1)y=0, (\lambda>0),$$

satisfying the boundary conditions $y(1)=0$, $y(e)=0$.

4. Determine the eigen values and corresponding eigen functions of the boundary value problem:

$$\frac{d^2y}{dx^2}+2\frac{dy}{dx}+(1+\lambda)y=0;\ y(0)=0 \text{ and } y'(a)=0.$$

5. Determine the eigen values and corresponding eigen functions of the boundary value problem:

$$\frac{d^2y}{dx^2}-\frac{dy}{dx}+\frac{1+\lambda}{4}y=0;\ y(0)=0=y(1).$$

6. Put, if possible, the following equations in Strum-Liouville form:

(i) $\frac{d^2y}{dx^2}+2x\frac{dy}{dx}+(x+\lambda)y=0$, (ii) $\frac{d^2y}{dx^2}+3x^2\frac{dy}{dx}+(x+\lambda)y=0$,

(iii) $\frac{d^2y}{dx^2}+\lambda y=0$, (iv) $4\frac{d^2y}{dx^2}-4\frac{dy}{dx}+(1+\lambda)y=0$.

ANSWERS

1.

(i) $\lambda_n=n^2;\ y_n(x)=a_n\sin(nx)$, (ii) $\lambda_n=n^2\pi^2;\ y_n(x)=a_n\sin(n\pi x)$,

(iii) $\lambda_n=\left(n-\frac{1}{2}\right)^2;\ y_n(x)=a_n\sin\left(n-\frac{1}{2}\right)x$, (iv) $\lambda_n=n^2\pi^2;\ y_n(x)=a_n\cos(n\pi x)$,

(v) $\lambda_n=4n^2;\ y_n(x)=a_n\sin(2nx)$, (vi) $\lambda_n=\frac{n^2}{4};\ y_n(x)=a_n\sin\left(\frac{nx}{2}\right)$,

(vii) $\lambda_n=\frac{n^2\pi^2}{a^2};\ y_n(x)=a_n\sin\left(\frac{n\pi x}{a}\right)$,

(viii) $\lambda_n=\frac{n^2\pi^2}{4a^2};\ y_n(x)=a_n\sin\left(\frac{n\pi(x+a)}{2a}\right)$,

(ix) $\lambda_n=\frac{n^2\pi^2}{(b-a)^2};\ y_n(x)=a_n\sin\left(\frac{n\pi(x-a)}{b-a}\right)$,

(x) $\lambda_n = \dfrac{(2n-1)^2\pi^2}{4a^2}$; $y_n(x) = a_n \sin\left(\dfrac{(2n-1)\pi x}{2a}\right)$,

(xi) $\lambda_n = \dfrac{n^2\pi^2}{a^2}$; $y_n(x) = a_n \cos\left(\dfrac{n\pi x}{a}\right)$,

(xii) $\lambda_n = n^2$; $y_n(x) = a_n \cos nx$,

(xiii) $\lambda_n = (2n-1)^2$; $y_n(x) = a_n \sin(2n-1)x$,

(xiv) $\lambda_n = \dfrac{n^2}{4}$; $y_n(x) = a_n \cos\left[\dfrac{n(\pi + x)}{2}\right]$,

(xv) $\lambda_n = n^2\pi^2$; $y_n(x) = a_n[\sin(n\pi x) - n\pi\cos(n\pi x)]$,

2. (i) $\lambda_n = n^2$; $y_n(x) = a_n \sin(n \ln x)$,

(ii) $\lambda_n = \dfrac{n^2}{4}$; $y_n(x) = a_n \cos\left(\dfrac{n}{2}\ln x\right)$

3. $\lambda_n = n^2\pi^2$; $y_n(x) = a_n \dfrac{1}{x}\sin(n\pi \ln x)$,

4. $\lambda_n = a_n^2$, where a_n is a root of the equation $\tan ax = x$ and $y_n = A_n e^{-x} \sin a_n x$.

5. $\lambda_n = 4n^2\pi^2$; $y_n(x) = a_n e^{x/2} \sin n\pi x$,

6. (i) $\dfrac{d}{dx}\left(e^{x^2}\dfrac{dy}{dx}\right) + xe^{x^2}y + \lambda e^{x^2}y = 0$,

(ii) $\dfrac{d}{dx}\left(e^{x^3}\dfrac{dy}{dx}\right) + xe^{x^3}y + \lambda e^{x^3}y = 0$,

(iii) $\dfrac{d}{dx}\left(e^{x^3}\dfrac{dy}{dx}\right) + xe^{x^3}y + \lambda e^{x^3}y = 0$,

(iv) $\dfrac{d}{dx}\left(4e^{-x}\dfrac{dy}{dx}\right) + e^{-x}y + \lambda e^{-x}y = 0$.

Chapter 13

GEOMETRICAL APPLICATIONS

13.1 Introduction

The Governing equations of many geometrical, physical problems are first ordered differential equations with some prescribed conditions on the dependent variable and/ or its first ordered derivatives. Solutions of such problems are actually the solutions of the differential equation associated with the problems. In this chapter we will consider some applications of differential equations to geometrical problems only. Before going into details, we give here a list of basic results for convenience.

(a) *Cartesian form*: If (x, y) be any point on the curve $f(x, y) = 0$, then

(i) length of the radius vector of the point (x, y) is $\sqrt{x^2 + y^2}$.

(ii) $\dfrac{dy}{dx}$ is the slope of the tangent to the curve at the point (x, y).

(iii) $-\dfrac{dx}{dy}$ is the slope of the normal to the curve at the point (x, y).

(iv) equation of tangent at (x, y) to the curve is $Y - y = \dfrac{dy}{dx}(X - x)$, where (X, Y) is any point on the tangent. Hence, $x - y\dfrac{dx}{dy}$, $y - x\dfrac{dy}{dx}$ are the x- and y- intercepts of the tangent.

(v) equation of normal at (x, y) to the curve is $Y - y = -\dfrac{dx}{dy}(X - x)$, where (X, Y) is any point on the normal. Hence, $x + y\dfrac{dy}{dx}$, $y + x\dfrac{dx}{dy}$ are the x- and y- intercepts of the normal.

(vi) $y\sqrt{1+\left(\dfrac{dx}{dy}\right)^2}$ and $x\sqrt{1+\left(\dfrac{dy}{dx}\right)^2}$ are the lengths of the tangent between the point of contact and the x-axis and y-axis respectively.

(vii) $y\sqrt{1+\left(\dfrac{dy}{dx}\right)^2}$ and $x\sqrt{1+\left(\dfrac{dx}{dy}\right)^2}$ are the lengths of the normal between the point of contact and the x-axis and y-axis respectively.

(viii) $y\dfrac{dx}{dy}$ and $y\dfrac{dy}{dx}$ are the lengths of sub-tangent and sub-normal respectively.

(ix) An elementary arc length is $dx\sqrt{1+\left(\dfrac{dy}{dx}\right)^2}$ or $dy\sqrt{1+\left(\dfrac{dx}{dy}\right)^2}$.

(x) An elementary area is ydx or xdy.

(xi) The curvature at the point (x, y) is $\dfrac{\dfrac{d^2y}{dx^2}}{\left(1+\left(\dfrac{dy}{dx}\right)^2\right)^{3/2}}$.

(b) *Polar form*: If (r, θ) be any point of the curve $r = f(\theta)$, then

(i) $\tan\phi = r\dfrac{d\theta}{dr}$ where ϕ is the angle between radius vector and the tangent drawn towards the initial line.

(ii) The tangent at any point makes an angle $\psi = \phi + \theta$ with the initial line.

(iii) Length of the polar tangent $= r\sqrt{1+\left(r\dfrac{d\theta}{dr}\right)^2}$.

(iv) Length of the polar normal $\sqrt{r^2+\left(\dfrac{dr}{d\theta}\right)^2}$

(v) Length of the polar sub-tangent $= r^2\dfrac{d\theta}{dr}$.

(vi) Length of the polar sub-normal $= \dfrac{dr}{d\theta}$.

(vii) An elementary arc length is $dr\sqrt{1+\left(r\dfrac{d\theta}{dr}\right)^2}$ or $d\theta\sqrt{r^2+\left(\dfrac{dr}{d\theta}\right)^2}$.

(viii) An elementary area is $\dfrac{1}{2}r^2d\theta$.

(ix) $r\sin\phi = r^2\dfrac{d\theta}{ds}$ is the length of the perpendicular from the pole to the tangent.

WORKED-OUT EXAMPLES

1. Find the equation of the curve for which the Cartesian sub-tangent is constant.

Solution

Length of the Cartesian sub-tangent to the curve $f(x, y) = 0$ at the point (x, y) is $y\dfrac{dx}{dy}$. So, given that

$$y\frac{dx}{dy} = k \text{ or, } dx = k\frac{dy}{y}.$$

Integrating we get

$$x + C = k\ln y \text{ or, } y = Ae^{x/k}, \text{ where } A \text{ is a constant.}$$

2. Find the equation of the curve for which the polar sub-tangent is constant.

Solution

Length of the polar sub-tangent to the curve $f(r, \theta) = 0$ at the point (r, θ) is $r^2 \dfrac{d\theta}{dr}$.
So given that

$$r^2 \frac{d\theta}{dr} = k \quad \text{or,} \quad \frac{dr}{r^2} = \frac{1}{k} d\theta.$$

Integrating, we get

$$-\frac{1}{r} = \frac{\theta}{k} + C, \text{ where } C \text{ is a constant.}$$

3. Find the equation of the curve for which the polar sub-tangent is equal to polar sub-normal.

Solution

Length of the polar sub-tangent and polar sub-normal to the curve $f(r, \theta) = 0$ at the point (r, θ) is $r^2 \dfrac{d\theta}{dr}$ and $\dfrac{dr}{d\theta}$ repectivey. So given that

$$r^2 \frac{d\theta}{dr} = \frac{dr}{d\theta} \quad \text{or,} \quad \frac{dr}{d\theta} = \pm r \quad \text{or,} \quad \frac{dr}{r} = \pm d\theta.$$

Integrating we get

$$\ln r = C \pm \theta \text{ or, } r = Ae^{\pm\theta}, \text{ where } A \text{ is a constant.}$$

4. Prove that the curve with the property that all its normal pass through a fixed point is a circle.

Solution

Equation of the normal to the curve $f(x, y) = 0$ at the point (x, y) is

$$Y - y = -\frac{dx}{dy}(X - x), \text{ where } (X, Y) \text{ is any point on the normal.}$$

If it passes through the fixed point (α, β), then

$$\beta - y = -\frac{dx}{dy}(\alpha - x) \text{ or, } (x-\alpha)dx + (y-\beta)dy = 0.$$

Integrating, we get $\quad (x-\alpha)^2 + (y-\beta)^2 = a^2$, which is a circle whose centre is the fixed point (α, β).

5. Show that equation of the curve, whose slope at any point (x, y) is equal to $xy(x^2y^2 - 1)$ and which passes through the point $(0, 1)$ is $y^2(x^2 + 1) = 1$.

Solution

Slope of the tangent to the curve $f(x, y) = 0$ at the point (x, y) is $\dfrac{dy}{dx}$. So it is given that

$$\frac{dy}{dx} = xy(x^2y^2 - 1) \text{ or, } \frac{dy}{dx} + xy = x^3y^3 \text{ or, } \frac{1}{y^3}\frac{dy}{dx} + \frac{x}{y^2} = x^3.$$

Putting $z = \dfrac{1}{y^2}$, i.e., $\dfrac{dz}{dx} = -\dfrac{2}{y^3}\dfrac{dy}{dx}$ in the given equation, we get

$$\frac{dz}{dx} - 2xz = -2x^3.$$

This is linear differential equation whose integrating factor is

$$e^{\int -2x dx} = e^{-x^2}.$$

Multiplying the above equation by e^{-x^2}, we get

$$e^{-x^2}\left(\frac{dz}{dx} - 2xz\right) = -2x^3 e^{-x^2} \text{ or, } \frac{d}{dx}\left(e^{-x^2} z\right) = -2x^3 e^{-x^2}.$$

Integrating, we get

$$e^{-x^2} z = -\int 2x^3 e^{-x^2} dx \quad = -\int u e^{-u} du = (u+1)e^{-u} + C$$

$$= (1 + x^2) e^{-x^2} + C$$

$$\text{or, } \frac{1}{y^2} = 1 + x^2 + Ce^{x^2}.$$

If it passes through (0, 1), then $C = 0$. Hence the equation of the curve is

$$y^2(x^2 + 1) = 1.$$

6. Show that the curve for which the tangent at every point makes a constant angle with the radius vector is an equiangular spiral.

Solution

If the tangent to the curve $f(r, \theta) - 0$ at the point (r, θ) makes a constant angle α, then

$$r\frac{d\theta}{dr} = \tan\alpha = \text{constant} \text{ or, } \frac{dr}{r} = \cot\alpha\, d\theta.$$

Integrating, we get

$$\ln r - \ln C = \theta \cot\alpha \text{ or, } r = Ce^{\theta\cot\alpha}, \text{ which is an equiangular spiral.}$$

7. Show that the curve in which the angle between tangent and the radius vector at every point is one-half of the vectorial angle, is a cardioid.

Solution

If the tangent to the curve $f(r, \theta) = 0$ at the point (r, θ) makes a constant angle ϕ, then

$$r\frac{d\theta}{dr} = \tan\phi.$$

Given that $\phi = \frac{\theta}{2}$. So

$$r\frac{d\theta}{dr} = \tan\frac{\theta}{2} \text{ or, } \frac{dr}{r} = \cot\frac{\theta}{2} d\theta.$$

Integrating, we get

$$\ln r - \ln C = 2\ln\sin\frac{\theta}{2} \text{ or, } \frac{r}{C} = \sin^2\frac{\theta}{2} \text{ or, } r = \alpha(1 - \cos\theta), \text{ where } \alpha = \frac{C}{2}.$$

This is a cardioid.

8. Show that the curve in which the portion of the tangent included between the coordinate axes is bisected by the point of contact is a rectangular hyperbola.

Solution

Let the equation of the curve be $f(x, y) = 0$, then equation of the tangent at the point (x, y) to the curve is

$$Y - y = \frac{dy}{dx}(X - x).$$

It intersects the axes at $\left(x - y\frac{dx}{dy}, 0\right)$ and $\left(0, y - x\frac{dy}{dx}\right)$ respectively. Given that

$$x - y\frac{dx}{dy} = 2x \text{ and } y - x\frac{dy}{dx} = 2y.$$

From both the equations, we get

$$xdy + ydx = 0.$$

Integrating, we get $\quad xy = c^2$, which is a rectangular hyperbola.

EXERCISE

1. Find the equation of the curve for which the Cartesian sub-normal is constant.
2. Find the equation of the curve for which the polar sub-normal is constant.
3. Find the equation of the curve for which the Cartesian sub-tangent is proportional to the square of the abscissa and passes through $(1, e)$.

   [*Hints*: Given that $y\frac{dx}{dy} = kx^2$.]
4. Show that the curve in which the slope of the tangent at any point equal to the ratio of abscissa to the ordinate of the point is a rectangular hyperbola.
5. Find the equation of the curve whose polar sub-normal is twice the sine of the vectorial angle.

   [*Hints*: Given that $\frac{dr}{d\theta} = 2\sin\theta$]
6. Find the equation of the curve for which the Cartesian sub-tangent is equal to Cartesian sub-normal.

   [*Hints*: Given that $y\frac{dx}{dy} = y\frac{dy}{dx}$, i.e., $\frac{dy}{dx} = \pm 1$.]
7. Find the equation of the curve for which the y - intercepts of the tangent at any point is proportional to the square of the ordinate of the point.

   [*Hints*: Given that $y - x\frac{dy}{dx} = ky^2$, i.e., $\frac{1}{y^2}\frac{dy}{dx} - \frac{1}{xy} = -\frac{k}{x}$.

   Putting $z = \frac{1}{y}$, i.e., $\frac{1}{y^2}\frac{dy}{dx} = -\frac{dz}{dx}$ in the equation, we get

   $$\frac{dz}{dx} + \frac{z}{x} = \frac{k}{x}.$$

This is a linear differential equation whose integrating factor is $e^{\int \frac{dx}{x}} = e^{\ln x} = x.$ Multiplying the above equation by x, we get

$$\frac{d}{dx}(xz) = k$$

Integrating we get

$$xz = kx + C, \text{ or, } \frac{x}{y} = kx + C.]$$

8. Find the equation of the curve having the slope of the tangent to any point proportional to the abscissa of the point of tangency.

9. Find the equation of the curve through the point (0, –2) such that the slope of the tangent at any point of it is equal to the ordinate of this point increased by 3.

[*Hints*: Given that $\frac{dy}{dx} = y + 3$]

10. Find the equation of the curve through the point (1, 1) so that slope of the tangent to the curve at any point is proportional to the square of the ordinate of this point.

11. Find the equation of the curve which passes through the point (2, 1) and has its slope $-\frac{x+y}{x}$ at the point (x, y).

[*Hints*: At any point (x, y) to the curve $y = f(x)$, slope of the tangent is given by $\frac{dy}{dx}$. So, given that $\frac{dy}{dx} = -\frac{x+y}{x}$, i.e., $\frac{dy}{dx} + \frac{y}{x} = -1$. This is a linear differential equation whose integrating factor is x, etc.]

12. Find the equation of the curve for which the slope of the tangent at any point is n times the slope of the straight line joining the point to the origin.

[*Hints*: Given that $\frac{dy}{dx} = n\frac{y}{x}$.]

13. Find a curve such that at each point of it length of the sub-tangent is equal to doubled abscissa.

14. Find a curve for which the radius vector equal to the length of the tangent between the point of tangency and the x-axis.

[*Hints*: Given that $y\sqrt{1+\left(\frac{dx}{dy}\right)^2} = \sqrt{x^2+y^2}$, i.e., $y\frac{dx}{dy} = x$, or, $\frac{dx}{x} = \frac{dy}{y}$.]

15. Find the curve whose sub-normal is the arithmetic mean between the abscissa and the ordinate.

[*Hints*: Given that $y\frac{dy}{dx} = \frac{x+y}{2}$. Putting $y = vx$, i.e., $\frac{dy}{dx} = v + x\frac{dv}{dx}$, we get

$$v + x\frac{dv}{dx} = \frac{1+v}{2v} \quad \text{or,} \quad x\frac{dv}{dx} = \frac{1-v-2v^2}{2v}$$

$$\text{or,} \quad \frac{2v}{1+v-2v^2}dv = \frac{dx}{x} \quad \text{or,} \quad \frac{2}{3}\left[\frac{dv}{v-1} + \frac{dv}{2v+1}\right] + \frac{dx}{x} = 0$$

Integrating, we get $\quad (x-y)^2(2y+x) = C.$]

16. Find the equation of the curve in which the ratio of the y-intercepts of the tangent to the radius vector is equal to a constant.

[*Hints*: Given that $y - x\dfrac{dy}{dx} = k\sqrt{x^2 + y^2}$. Simplifying, we get

$$\frac{d\left(\dfrac{y}{x}\right)}{\sqrt{1+\left(\dfrac{y}{x}\right)^2}} + k\frac{dx}{x} = 0.$$

Integrating, we get $\quad \ln\left|y + \sqrt{x^2+y^2}\right| = (1-k)\ln x + \ln C$

or, $\quad y + \sqrt{x^2+y^2} = Cx^{1-k}.$

So, $\quad -y + \sqrt{x^2+y^2} = \dfrac{x^2}{y+\sqrt{x^2+y^2}} = \dfrac{x^{1+k}}{C}.$

Subtracting, we get $\quad 2y = Cx^{1-k} - \dfrac{x^{1+k}}{C}$ or, $2\dfrac{y}{x} = \dfrac{C}{x^k} - \dfrac{x^k}{C}$.]

17. Find the equation of the curve in which the length of the y-intercepts of the tangent line is equal to $a\sec\theta$, where θ is the angle between the radius vector and X-axis.

[*Hints*: Given that $y - x\dfrac{dy}{dx} = a\sec\theta$, where $\tan\theta = \dfrac{y}{x}$. So

$$y - x\frac{dy}{dx} = \frac{a\sqrt{x^2+y^2}}{x}.$$

Simplifying, we get

$$\frac{d\left(\dfrac{y}{x}\right)}{\sqrt{1+\left(\dfrac{y}{x}\right)^2}} + a\frac{dx}{x^2} = 0.$$

Integrating, we get

$$\ln\left|y+\sqrt{x^2+y^2}\right| - \frac{a}{x} - \ln x = \ln C$$

or, $\quad y + \sqrt{x^2+y^2} = Cxe^{\frac{a}{x}}$

So, $\quad -y + \sqrt{x^2+y^2} = \dfrac{x^2}{y+\sqrt{x^2+y^2}} = \dfrac{xe^{-\frac{a}{x}}}{C}.$

Subtracting, we get $\quad y = \dfrac{x}{2}\left[Ce^{\frac{a}{x}} - \dfrac{1}{C}e^{-\frac{a}{x}}\right].$

18. Find the equation of the curve in which the ratio of the x-intercepts of the normal to the radius vector is equal to a constant.

[*Hints*: Given that $x + y\dfrac{dy}{dx} = k\sqrt{x^2 + y^2}$.]

19. Determine the curve for which y-intercept of the normal drawn to some point of the curve is equal to distance of the point from origin.

[*Hints*: Given that $y + x\dfrac{dx}{dy} = \sqrt{x^2 + y^2}$.]

20. Show that equation of the curve, whose slope at any point (x, y) is equal to $y + 2x$ and which passes through the origin is, $y = 2(e^x - x - 1)$.

[*Hints*: Given that $\dfrac{dy}{dx} = y + 2x$ i.e., $\dfrac{dy}{dx} - y = 2x$. This is a linear differential equation whose I.F. is $e^{-\int dx} = e^{-x}$.]

21. Find the curve that has the property that the product of the square of the distance of any point of it from the origin and the x- intercepts of the normal at this point is equal to the cube of the abscissa of this point.

[*Hints*: Given that, $\left(x + y\dfrac{dy}{dx}\right)(x^2 + y^2) = x^3$, i.e., $\dfrac{dy}{dx} = -\dfrac{xy}{x^2 + y^2}$. Now, put $y = vx$]

22. Show that all the curves for which the length of normal is equal to the radius vector are either circles or rectangular hyperbolas.

[*Hints*: Given that $y\sqrt{1 + \left(\dfrac{dy}{dx}\right)^2} = \sqrt{x^2 + y^2}$, i.e., $\dfrac{dy}{dx} = \pm\dfrac{x}{y}$, i.e., $xdx \pm ydy = 0$. Integrating, we get

$$x^2 \pm y^2 = a^2.]$$

23. Show that the curve for which the radius of curvature at every point is constant is a circle.

[*Hints*: Given that $\dfrac{ds}{d\psi} = a$ (constant). Now

$$\frac{dx}{d\psi} = \frac{dx}{ds}\frac{ds}{d\psi} = a\cos\psi \text{ and } \frac{dy}{d\psi} = \frac{dy}{ds}\frac{ds}{d\psi} = a\sin\psi.$$

Integrating, we get

$$x = a\sin\psi + \alpha,$$

and $$y = -a\cos\psi + \beta.$$

Eliminating ψ, we get $\quad (x - \alpha)^2 + (y - \beta)^2 = a^2$.]

24. Find the curve for which the sum of the reciprocals of the radius vector and polar sub-tangent is constant.

25. Show that the curves whose radius of curvature at any point is equal to the normal at that point are either circles or catenaries.

[*Hints*: Given that $\dfrac{\left\{1+\left(\dfrac{dy}{dx}\right)^2\right\}^{3/2}}{\dfrac{d^2y}{dx^2}} = \mp y\sqrt{1+\left(\dfrac{dy}{dx}\right)^2}$, according as the direction of radius of curvature in the direction of normal or in the opposite direction of normal.

$$\text{i.e., } \mp y\frac{d^2y}{dx^2} - \left(\frac{dy}{dx}\right)^2 - 1 = 0,$$

Taking positive sign, we get $\dfrac{pdp}{1+p^2} = \dfrac{dy}{y}$, where $p = \dfrac{dy}{dx}$.

Integrating, we get $\quad 1+p^2 = a^2y^2$, i.e., $\dfrac{dy}{\sqrt{a^2y^2-1}} = \pm dx$.

Again integrating, we get $y = \dfrac{1}{2a}\left[e^{\pm ax+b} + e^{\mp ax-b}\right]$, which are catenaries.

Again taking negative sign, we get

$$y\frac{d^2y}{dx^2} + \left(\frac{dy}{dx}\right)^2 + 1 = 0,$$

$$\text{or, } \frac{d}{dx}\left(y\frac{dy}{dx}\right) + 1 = 0, \quad \text{or, } y\frac{dy}{dx} + x - a = 0,$$

$$\text{or, } (x-a)dx + ydy = 0,$$

Integrating, we get $\quad (x-a)^2 + y^2 = b^2$, which are circles.]

26. Find the curve in which the sub-tangent is always bisected at the origin.

27. Show that the curve in which the angle between tangent and the radius vector at every point is one-third of the inclination of the tangent to the initial line, is a cardioid.

28. In polar coordinates, find the equation of the curve at each point of which the tangent of the angle between the radius vector and tangent line is equal to the reciprocal of the radius vector with sign reversed.

29. In polar coordinates, find the equation of the curve at each point of which the tangent of the angle between the radius vector and tangent line is equal to the square of the radius vector.

30. Find the curve for which each of its tangent lines forms a triangle with the coordinates axes of constant area a^2.

31. Find the curve for which the sum of the intercepts of the tangent line on the coordinate axes is constant.

[*Hints*: Intercepts of the tangent to the curve $f(x, y) = 0$ at the point (x, y) are $x - y\dfrac{dx}{dy}$ and $y - x\dfrac{dy}{dx}$ respectively. Given that

$$x - y\frac{dx}{dy} + y - x\frac{dy}{dx} = a \text{ (constant)}.$$

Simplifying, we get

$$y = px + \frac{ap}{p-1}, \text{ where } p = \frac{dy}{dx}.$$

The above equation is the Clairaut's form, whose general solution is

$$y = Cx + \frac{Ca}{C-1}, \ C \text{ being an arbitrary constant.}$$

Hence the required curve is the envelope of the above family of straight lines, whose equation is obtained by equating the discriminant of the quadratic equation in C to zero, i.e., equation of the required curve is

$$(x+y-a)^2 = 4xy \text{ or, } \sqrt{x} \pm \sqrt{y} = \sqrt{a}.]$$

32. Find the curve for which the portion of the tangent line intercepted by the coordinate axes is of constant length.

33. Find the curve for which the product of the intercepts of the tangent line on the coordinate axes is constant.

34. Find the polar equation of the family of curves for which the sum of the radius vector and the polar sub-normal varies as the k th power of the radius vector.

35. Find the equation of the curve in which the perpendicular from origin on any tangent is equal to the abscissa of the point of contact.

ANSWERS

1. $y^2 = 2ax + C;$

2. $r = a\theta + C,$

8. $y = ax^2 + C,$

10. $yk(x-1) - y + 1 = 0,$

11. $x^2 + 2xy = 8,$

13. $y = C\sqrt{x}$

21. $y^2(2x^2 + y^2) = C,$

24. $\frac{1}{r} = a + Ce^{\theta},$

26. $y^2 = xC,$

28. $r(\theta + C) = 1,$

29. $r^2 = 2(\theta + C),$

30. $2xy = a^2,$

31. $\sqrt{x} \pm \sqrt{y} = \sqrt{a},$

32. $x^{2/3} + y^{2/3} = a^{2/3},$

33. $xy = C^2,$

34. $\frac{1}{r^{k-1}} = Ce^{(k-1)\theta} + A,$

35. $x^2 + y^2 = Cx.$

13.2 Trajectories

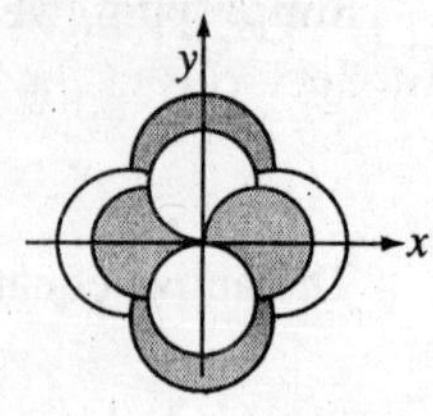

If a curve intersects every member of a family of curves according to some definite law, then it is called *trajectory* of the family of curves. Angle between two curves is measured by the angle between the tangents to the curves drawn at the point of intersection. A trajectory of a family of curves is said to be an α-trajectory or *isogonal trajectory*, if it intersects each member of a family of the curves at an angle α.

In particular, if a curve intersects each member of a family of the curves at an angle 90°, then it is said to be an *orthogonal trajectory* of the family of curves. If two families of curves are such that each member of a family cuts all the members of other family at right angle, then each family is called *orthogonal trajectories* of the other family. In the above figure, we see that the circles through the origin with centres on x-axis are the orthogonal trajectories of the circles through the origin with centres on y-axis and conversely.

13.2.1 Isogonal trajectories

In Cartesian coordinate, equation of a family of curves is

$$f(x, y, c) = 0, \tag{1}$$

where c is a parameter. Differentiating (1) with respect to x, we get

$$\frac{\partial f}{\partial x} + \frac{\partial f}{\partial y}\frac{dy}{dx} = 0. \tag{2}$$

Eliminating c from (2) and (3), we get an equation of the form

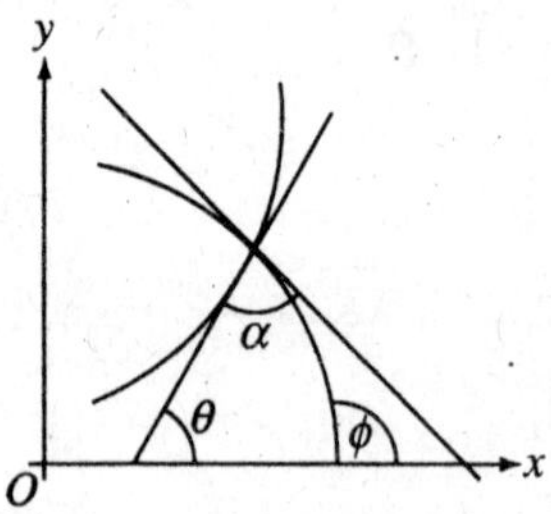

$$F\left(x, y, \frac{dy}{dx}\right) = 0. \tag{3}$$

Let the trajectories cut the curve of a family of curves at an angle α. If the tangents to a member of the family of curves and of its trajectories at the point of intersection make angle θ, ϕ with positive direction of x-axis, then

$$\tan\alpha = \tan(\phi - \theta) = \frac{\tan\phi - \tan\theta}{1 + \tan\phi\tan\theta} = \frac{\dfrac{dy_T}{dx} - \dfrac{dy_C}{dx}}{1 + \dfrac{dy_T}{dx}\dfrac{dy_C}{dx}},$$

where $\dfrac{dy_C}{dx}, \dfrac{dy_T}{dx}$ represent the slope of the tangents to the curve and its tarjectory respectively.

$$\text{or,} \qquad \frac{dy_C}{dx} = \frac{\dfrac{dy_T}{dx} - \tan\alpha}{1 + \dfrac{dy_T}{dx}\tan\alpha}.$$

Now dropping the subscripts C, T, from (3) we get the differential equation of the α-trajectories of the family of curves given by equation (1) is

$$F\left(x, y, \frac{\frac{dy}{dx} - \tan\alpha}{1 + \frac{dy}{dx}\tan\alpha}\right) = 0. \tag{4}$$

Thus we see that differential equation of α-trajectories is obtained from the differential equation of the family of curves on replacing $\frac{dy}{dx}$ by $\frac{\frac{dy}{dx} - \tan\alpha}{1 + \frac{dy}{dx}\tan\alpha}$.

Equation (4) can be integrated to get the equation of α-trajectories of the family of curves given by (1). In polar coordinate, equation of a family of curves is

$$f(r, \theta, c) = 0, \tag{5}$$

where c is a parameter. Differentiating (5) with respect to θ and eliminating c from it with the aid of (5), we get

$$F\left(r, \theta, \frac{dr}{d\theta}\right) = 0. \tag{6}$$

Now proceeding as before we see that differential equation of α-trajectories is obtained from the differential equation of the family of curves on replacing

$r\frac{d\theta}{dr}$ by $\frac{r\frac{d\theta}{dr} + \tan\alpha}{1 - r\frac{d\theta}{dr}\tan\alpha}$, i.e., $\frac{dr}{d\theta}$ by $\frac{r - r^2\frac{d\theta}{dr}\tan\alpha}{r\frac{d\theta}{dr} + \tan\alpha}$.

Thus the differential equation of the α-trajectories of the family of curves given by equation (5) is

$$F\left(r, \theta, \frac{r - r^2\frac{d\theta}{dr}\tan\alpha}{r\frac{d\theta}{dr} + \tan\alpha}\right) = 0. \tag{7}$$

13.2.2 Orthogonal trajectories

To get equation of the orthogonal trajectories of the family of curves given by (1) we put $\alpha = 90°$ in the differential equation given by (4). When $\alpha = 90°$, the expression

$\frac{\frac{dy}{dx} - \tan\alpha}{1 + \frac{dy}{dx}\tan\alpha}$ simplifies to $\frac{\frac{dy}{dx}\cot 90° - 1}{\cot 90° + \frac{dy}{dx}} = -\frac{dx}{dy}$.

Thus, in Cartesian coordinate, the differential equation of the orthogonal trajectories of the family of curves given by equation (1) is

$$F\left(x, y, -\frac{dx}{dy}\right) = 0. \tag{8}$$

Thus we see that differential equation of orthogonal trajectories is obtained from the differential equation of the family of curves on replacing $\frac{dy}{dx}$ by $-\frac{dx}{dy}$.

Similarly, in polar coordinate when $\alpha = 90°$ the expression $\dfrac{r - r^2\frac{d\theta}{dr}\tan\alpha}{r\frac{d\theta}{dr} + \tan\alpha}$ reduces to

$$\frac{r\cot 90° - r^2\frac{d\theta}{dr}}{r\frac{d\theta}{dr}\cot 90° + 1} = -r^2\frac{d\theta}{dr}.$$

Thus the differential equation of the orthogonal trajectories of the family of curves given by equation (5) is

$$F\left(r, \theta, -r^2\frac{d\theta}{dr}\right) = 0. \tag{9}$$

Thus, in polar coordinate, we see that differential equation of orthogonal trajectories is obtained from the differential equation of the family of curves on replacing $\frac{dr}{d\theta}$ by $-r^2\frac{d\theta}{dr}$.

WORKED-OUT EXAMPLES

1. Find the orthogonal trajectories of the family of concurrent straight lines $y = mx$.

Solution

Given family of straight lines is $y = mx$. Differentiating with respect to x, we get

$$\frac{dy}{dx} = m.$$

Eliminating m, we get $\frac{dy}{dx} = \frac{y}{x}$. This is the differential equation whose primitive is the given family of straight lines.

Changing $\frac{dy}{dx}$ by $-\frac{dx}{dy}$ in the above equation, we get

$$-\frac{dx}{dy} = \frac{y}{x} \quad \text{or,} \quad xdx + ydy = 0.$$

Integrating, we get $\quad x^2 + y^2 = a^2$, where a^2 is an arbitrary constant.

This is the orthogonal trajectories of the given family of concurrent straight lines.

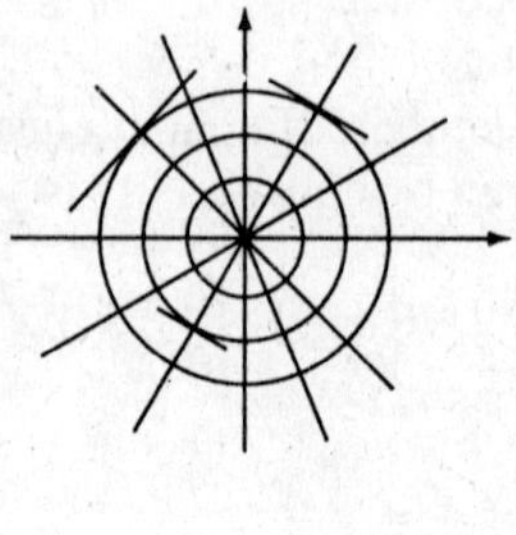

2. Find the orthogonal trajectory of the rectangular hyperbola $xy = c^2$.

Solution

Given family of rectangular hyperbola is $xy = c^2$. Differentiating with respect to x, we get

$$x\frac{dy}{dx} + y = 0.$$

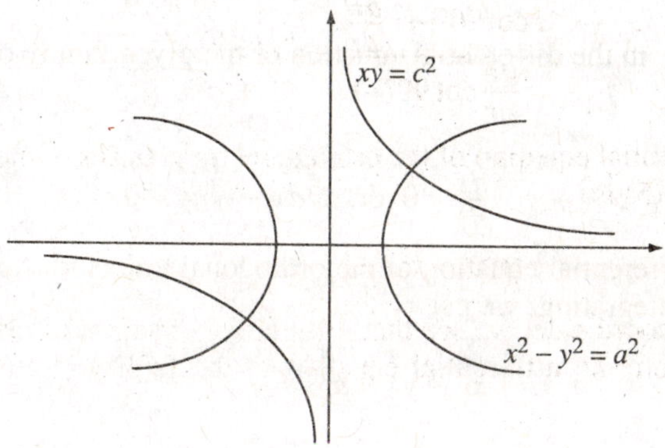

Now changing $\frac{dy}{dx}$ by $-\frac{dx}{dy}$ in the above equation, we get

$$-x\frac{dx}{dy} + y = 0 \text{ or, } xdx - ydy = 0.$$

Integrating, we get $\qquad x^2 - y^2 = a^2$, where a^2 is an arbitrary constant.

This is the orthogonal trajectories of the given family of rectangular hyperbolas.

3. Find the orthogonal trajectories of the family of curves given by $x^{\frac{3}{2}} + y^{\frac{3}{2}} = a^{\frac{3}{2}}$.

Solution

Differentiating $x^{\frac{3}{2}} + y^{\frac{3}{2}} = a^{\frac{3}{2}}$ with respect to x, we get

$$\sqrt{x} + \sqrt{y}\frac{dy}{dx} = 0.$$

Replacing $\frac{dy}{dx}$ in the differential equation of the given family of curves by $-\frac{dx}{dy}$, we get

$$\sqrt{x} - \sqrt{y}\frac{dx}{dy} = 0 \text{ or, } \frac{dy}{\sqrt{y}} - \frac{dx}{\sqrt{x}} = 0.$$

This is the differential equation of the orthogonal trajectories of the given family of curves. Now integrating we get

$$\sqrt{y} - \sqrt{x} = \sqrt{a}.$$

4. Find the orthogonal trajectories of the family of curves given by $x^{\frac{2}{3}} + y^{\frac{2}{3}} = a^{\frac{2}{3}}$.

Solution

Differentiating $x^{\frac{2}{3}} + y^{\frac{2}{3}} = a^{\frac{2}{3}}$ with respect to x, we get

$$x^{-\frac{1}{3}} + y^{-\frac{1}{3}} \frac{dy}{dx} = 0.$$

Replacing $\frac{dy}{dx}$ in the differential equation of the given family of curves by $-\frac{dx}{dy}$, we get

$$x^{-\frac{1}{3}} - y^{-\frac{1}{3}} \frac{dx}{dy} = 0 \text{ or, } x^{\frac{1}{3}} dx - y^{\frac{1}{3}} dy = 0.$$

This is the differential equation of the orthogonal trajectories of the given family of curves. Now integrating, we get

$$x^{\frac{4}{3}} - y^{\frac{4}{3}} = a^{\frac{4}{3}}.$$

5. Find the orthogonal trajectories of the family of curves $x^2 + \frac{y^2}{\lambda + 1} = 1$.

Solution

Differentiating the given equation with respect to x, we get

$$x + \frac{y\frac{dy}{dx}}{\lambda + 1} = 0 \text{ or, } \lambda + 1 = -\frac{y}{x}\frac{dy}{dx}.$$

This is the differential equation of the given family of curves. Eliminating λ from the given equation with the aid of above result, we get

$$x^2 - xy\frac{dx}{dy} = 1.$$

Replacing $\frac{dy}{dx}$ by $-\frac{dx}{dy}$ in the above equation, we get

$$x^2 + xy\frac{dy}{dx} = 1 \text{ or, } x^2 dx + xydy = dx \quad \text{or, } xdx + ydy = \frac{dx}{x}.$$

Integratin we get $\qquad x^2 + y^2 - 2\ln x + C = 0.$

6. Show that the orthogonal trajectories of the system of co-axial circles

$$x^2 + y^2 + 2\lambda x + C = 0$$

from another system of coaxial circles

$$x^2 + y^2 + 2\mu y - C = 0$$

where λ, μ are parameters and C is a given constant.

Solution

Differentiating $x^2+y^2+2\lambda x+C=0$ with respect to x, we get

$$x+y\frac{dy}{dx}+\lambda=0.$$

Eliminating λ from the given equation with aid of this result, we get

$$x^2+y^2-2\left(x+y\frac{dy}{dx}\right)x+C=0.$$

Now changing $\frac{dy}{dx}$ by $-\frac{dx}{dy}$ in the above equation we get

$$x^2+y^2-2\left(x-y\frac{dx}{dy}\right)x+C=0$$

$$\text{or, } (y^2-x^2)dy+2xydx+Cdy=0$$

$$\text{or, } \frac{ydx^2-x^2dy}{y^2}+dy+C\frac{dy}{y^2}=0.$$

Integrating, we get

$$\frac{x^2}{y}+y-\frac{C}{y}=-2\mu \quad \text{or, } x^2+y^2+2\mu y-C=0.$$

7. Prove that the system of confocal conics $\frac{x^2}{a^2+\lambda}+\frac{y^2}{b^2+\lambda}=1$ is self-orthogonal.

Solution

Differentiating $\frac{x^2}{a^2+\lambda}+\frac{y^2}{b^2+\lambda}=1$ with respect to x, we get

$$\frac{x}{a^2+\lambda}+\frac{y}{b^2+\lambda}\frac{dy}{dx}=0 \quad \text{or, } x(b^2+\lambda)+y(a^2+\lambda)\frac{dy}{dx}=0$$

$$\text{or, } \lambda\left(x+y\frac{dy}{dx}\right)=-\left(xb^2+ya^2\frac{dy}{dx}\right) \quad \text{or, } \lambda=-\frac{\left(xb^2+ya^2\frac{dy}{dx}\right)}{\left(x+y\frac{dy}{dx}\right)}.$$

So,
$$a^2+\lambda=a^2-\frac{\left(xb^2+ya^2\frac{dy}{dx}\right)}{\left(x+y\frac{dy}{dx}\right)}=\frac{x(a^2-b^2)}{\left(x+y\frac{dy}{dx}\right)}$$

and
$$b^2+\lambda=b^2-\frac{\left(xb^2+ya^2\frac{dy}{dx}\right)}{\left(x+y\frac{dy}{dx}\right)}=\frac{(b^2-a^2)y\frac{dy}{dx}}{\left(x+y\frac{dy}{dx}\right)}.$$

Using these in the given equation of conics, we get

$$\frac{x^2}{\dfrac{x(a^2-b^2)}{\left(x+y\dfrac{dy}{dx}\right)}}+\frac{y^2}{\dfrac{(b^2-a^2)y\dfrac{dy}{dx}}{\left(x+y\dfrac{dy}{dx}\right)}}=1 \quad \text{or,} \quad \left(x-y\frac{dx}{dy}\right)\left(x+y\frac{dy}{dx}\right)=a^2-b^2.$$

It is clear that the above equation remain unchanged on replacing $\dfrac{dy}{dx}$ by $-\dfrac{dx}{dy}$. This indicates that orthogonal trajectories of the given family of conics is the same family of conics, i.e., the given system of confocal conics is self-orthogonal.

8. Find the orthogonal trajectories of the family of curves $r(1 + \cos\theta) = 2a$.

Solution

Differentiating $r(1 + \cos\theta) = 2a$, i.e., $r = a\sec^2\dfrac{\theta}{2}$ with respect to θ, we get

$$\frac{dr}{d\theta}=a\sec^2\frac{\theta}{2}\tan\frac{\theta}{2}=r\tan\frac{\theta}{2}.$$

This is the differential equation of the given family of curves. To get the differential equation of the orthogonal trajectories of the given family of curves, we replace $\dfrac{dr}{d\theta}$ by $-r^2\dfrac{d\theta}{dr}$ in the above equation and get

$$-r^2\frac{d\theta}{dr}=r\tan\frac{\theta}{2} \quad \text{or,} \quad \frac{dr}{r}+\cot\frac{\theta}{2}d\theta=0.$$

Integrating we get

$$\ln r+2\ln\sin\frac{\theta}{2}=\ln\frac{C}{2} \quad \text{or,} \quad r\sin^2\frac{\theta}{2}=\frac{C}{2} \quad \text{or,} \quad r(1-\cos\theta)=C.$$

9. Find the orthogonal trajectories of the family of curves $r^n = a^n \sin n\theta$.

Solution

From $r^n = a^n \sin n\theta$, we get $n\ln r = n\ln a + \ln\sin n\theta$.

Differentiating both sides with respect to θ, we get

$$\frac{1}{r}\frac{dr}{d\theta}=\cot n\theta \quad \text{or,} \quad \frac{dr}{d\theta}=r\cot n\theta.$$

This is the differential equation of the given family of curves. Changing $\dfrac{dr}{d\theta}$ by $-r^2\dfrac{d\theta}{dr}$ in the above equation we get

$$-r^2\frac{d\theta}{dr}=r\cot n\theta \quad \text{or,} \quad \frac{dr}{r}+\tan n\theta\, d\theta=0.$$

Integrating we get

$$\ln r-\frac{1}{n}\ln\cos n\theta=\ln C \quad \text{or,} \quad n\ln r-\ln\cos n\theta=n\ln C \quad \text{or,} \quad r^n=C^n\cos n\theta.$$

10. Find the 45°-trajectories of the family of concentric circles $x^2 + y^2 = a^2$.

Solution

Differentiating $x^2 + y^2 = a^2$ with respect to x, we get

$$\frac{dy}{dx} = -\frac{x}{y}.$$

Now changing $\frac{dy}{dx}$ in the above equation by $\dfrac{\frac{dy}{dx}-1}{1+\frac{dy}{dx}}$, we get

$$\frac{\frac{dy}{dx}-1}{1+\frac{dy}{dx}} = -\frac{x}{y} \text{ or, } y\frac{dy}{dx} - y = -x - x\frac{dy}{dx}$$

$$\text{or, } (y+x)\frac{dy}{dx} = y - x \text{ or, } \frac{dy}{dx} = \frac{y-x}{y+x}.$$

Putting $y = vx$, i.e., $\frac{dy}{dx} = v + x\frac{dv}{dx}$ in the above equation, we get

$$v + x\frac{dv}{dx} = \frac{v-1}{v+1} \text{ or, } \frac{(v+1)\,dv}{1+v^2} + \frac{dx}{x} = 0.$$

Integrating, we get $$\frac{1}{2}\ln(v^2+1) + \tan^{-1} v + \ln x = \frac{1}{2}\ln C$$

$$\text{or, } \frac{1}{2}\ln(y^2+x^2) + \tan^{-1}\frac{y}{x} = \frac{1}{2}\ln C$$

$$\text{or, } y^2 + x^2 = Ce^{-2\tan^{-1}\frac{y}{x}}.$$

11. Find the α-trajectories of the family of straight lines $y = mx$.

Solution

Differentiating $y = mx$ with respect to x, we get

$$\frac{dy}{dx} = m.$$

Eliminating m from this equation with the aid of given equation, we get

$$\frac{dy}{dx} = \frac{y}{x}.$$

Now changing $\frac{dy}{dx}$ in the above equation by $\dfrac{\frac{dy}{dx}-\tan\alpha}{1+\frac{dy}{dx}\tan\alpha}$, we get

$$\frac{\frac{dy}{dx}-\tan\alpha}{1+\frac{dy}{dx}\tan\alpha}=\frac{y}{x} \text{ or, } x\frac{dy}{dx}-xk=y+yk\frac{dy}{dx}, \text{ where } k=\tan\alpha.$$

$$\text{or, } (x-yk)\frac{dy}{dx}=xk+y \quad \text{or, } \frac{dy}{dx}=\frac{xk+y}{x-yk}.$$

Putting $y=vx$, i.e., $\frac{dy}{dx}=v+x\frac{dv}{dx}$ in the above equation, we get

$$v+x\frac{dv}{dx}=\frac{k+v}{1-kv} \text{ or, } x\frac{dv}{dx}=\frac{k(1+v^2)}{1-kv} \text{ or, } \frac{1}{1+v^2}dv-\frac{kv}{1+v^2}dv=k\frac{dx}{x}.$$

Integrating we get $\qquad \tan^{-1}v-\frac{k}{2}\ln(1+v^2)=k\ln x+\ln C$

$$\text{or, } \quad \tan^{-1}\frac{y}{x}-\frac{k}{2}\ln(x^2+y^2)=\ln C$$

$$\text{or, } \quad C(x^2+y^2)^{\frac{1}{2}\tan\alpha}=e^{\tan^{-1}\frac{y}{x}}.$$

12. Find the α-trajectories of the family of curves $r^2=a^2\cos 2\theta$.

Solution

From the equation of the curve, we get

$$2\ln r=2\ln a+\ln\cos 2\theta.$$

Differentiating with respect to θ we get

$$\frac{1}{r}\frac{dr}{d\theta}=-\tan 2\theta.$$

Changing $\frac{dr}{d\theta}$ by $\dfrac{r-r^2\frac{d\theta}{dr}\tan\alpha}{r\frac{d\theta}{dr}+\tan\alpha}$ in the above equation, we get

$$\frac{1-r\frac{d\theta}{dr}\tan\alpha}{r\frac{d\theta}{dr}+\tan\alpha}=-\tan 2\theta \quad \text{or, } 1-r\frac{d\theta}{dr}\tan\alpha=-\tan 2\theta\left(r\frac{d\theta}{dr}+\tan\alpha\right)$$

$$\text{or, } r\frac{d\theta}{dr}(\tan\alpha-\tan 2\theta)=1+\tan 2\theta\tan\alpha \quad \text{or, } \tan(2\theta-\alpha)d\theta+\frac{dr}{r}=0.$$

Integrating, we get

$$\frac{1}{2}\ln\sec(2\theta-\alpha)+\ln r=\ln c \text{ or, } r^2\sec(2\theta-\alpha)=c^2 \text{ or, } r^2=c^2\cos(2\theta-\alpha).$$

EXERCISE

1. Find the orthogonal trajectories of the following family of curves:

(i) $x^2 + 2y = k$, (ii) $xy = c^2$,

(iii) $y^2 = 4ax$, (iv) $y^2 = 4a(x+a)$,

(v) $x^2 + y^2 - 2ax = 0$, (vi) $x^2 + y^2 - 2ay = 0$

(vii) $y = ax^2$, (viii) $x^2 + 2y^2 = a^2$,

(ix) $x^2 - y^2 = a^2$, (x) $ay^2 = x^3$,

(xi) $y = x - 1 + Ce^{-x}$.

2. Find the orthogonal trajectories of the following family of curves:

(i) $r = a\cos\theta$, (ii) $r = a(1+\sin\theta)$,

(iii) $r = a(1+\cos\theta)$, (iv) $r = a(\sec\theta + \tan\theta)$,

(v) $r^n \sin n\theta = a^n$, (vi) $r^n = a^n \sin n\theta$,

(vii) $r^n = a^n \cos n\theta$, (viii) $r = a\theta$,

(ix) $r = 4a\sec\theta\tan\theta$, (x) $r(1+\cos\theta) = 2a$,

(xi) $r = c\sin^2\theta$.

3. Find the orthogonal trajectories of $\dfrac{x^2}{A^2} + \dfrac{y^2}{B^2+\mu} = 1$.

4. Find the orthogonal trajectories of $\dfrac{x^2}{A^2} + \dfrac{y^2}{A^2+\mu} = 1$.

5. Find the orthogonal trajectories of $\sqrt{x} + \sqrt{y} = \sqrt{a}$.

6. Find the orthogonal trajectories of $r(1-\cos\theta) = 2a$.

7. Show that the family of parabolas $y^2 = 4a(x+a)$ is self orthogonal.

[*Hints*: Differentiating $y^2 = 4a(x+a)$ with respect to x, we obtain

$$y\frac{dy}{dx} = 2a.$$

Eliminating a from the given equation we see that the differential equation of the given family of parabolas is

$$y^2 = 2y\frac{dy}{dx}\left(x + \frac{1}{2}y\frac{dy}{dx}\right)$$

or, $$y = 2x\frac{dy}{dx} + y\left(\frac{dy}{dx}\right)^2,$$

or, $$y\left[1 - \left(\frac{dy}{dx}\right)^2\right] = 2x\frac{dy}{dx}. \tag{1}$$

Now changing $\frac{dy}{dx}$ by $-\frac{dx}{dy}$, we get

$$y\left[1-\left(\frac{dx}{dy}\right)^2\right]=-2x\frac{dx}{dy}$$

or, $$-y\left(\frac{dx}{dy}\right)^2\left[1-\left(\frac{dy}{dx}\right)^2\right]=-2x\frac{dx}{dy}$$

or, $$y\left[1-\left(\frac{dy}{dx}\right)^2\right]=2x\frac{dy}{dx}.$$

Thus the differential equation of the orthogonal trajectories of the given family of parabolas is same as the differential equation of the given family of parabolas as shown by (1). Hence, the family of parabolas $y^2 = 4a(x + a)$ is self-orthogonal]

8. Show that the system of confocal conics $\frac{x^2}{A}+\frac{y^2}{A-\mu}=1$, where A is an arbitrary constant, is self-orthogonal.

[*Hints*: Differentiating the given family of conics w.r.t. x, we get

$$\frac{x}{A}+\frac{y}{A-\mu}\frac{dy}{dx}=0.$$

This gives $$A=\frac{\mu x}{x+y\frac{dy}{dx}}.$$

So, $$A-\mu=\frac{-\mu y\frac{dy}{dx}}{x+y\frac{dy}{dx}}.$$

Using these in the given equation, we get

$$\left(x-y\frac{dx}{dy}\right)\left(x+y\frac{dy}{dx}\right)=\mu.$$

It is clear that the above equation remains unchanged on replacing $\frac{dy}{dx}$ by $-\frac{dx}{dy}$. This indicates that orthogonal trajectories of the given family of conics is the same family of conics, i.e., the given system of confocal conics is self-orthogonal.]

9. Find the orthogonal trajectories of the family of curves $x^2-y^2+2kxy=1$, k being the parameter.

10. Find the orthogonal trajectories of the curves $r^2\cos\theta=a$.

11. Find the orthogonal trajectories of the curves $r=a+\sin 5\theta$, a being the parameter.

12. Find the orthogonal trajectories of the curves $r=a(\sin\theta+\cos\theta)$, a being the parameter.

13. Find the α-trajectories of the family of parabolas $y^2 = 4ax$.

14. Find the 45°-trajectories of the family of parabolas $y^2 = 4ax$.

15. Find the α-trajectories of the family of circles $x^2 + y^2 = a^2$.

16. Find the 45°-trajectories of the family of hyperbolas $y^2 + 2xy - x^2 = a^2$.

17. Find the α-trajectories of the family of the curves $r^2 = a^2\cos 2\theta$.

18. Find the 45°-trajectories of the family of circles $x^2 + y^2 = a^2$.

19. Find the 45°-trajectories of the family of hyperbolas $xy = a^2$.

ANSWERS

1. (i) $y - 2x = C$, (ii) $x^2 - y^2 = K$,
(iii) $2x^2 + y^2 = C$, (iv) Self-orthogonal,
(v) $x^2 + y^2 - 2Cy = 0$, (vi) $x^2 + y^2 - 2ax = 0$,
(vii) $x^2 + 2y^2 = C$, (viii) $y = Cx^2$,
(ix) $xy = C^2$, (x) $2x^2 + 3y^2 = C^2$,
(xi) $x = y - 1 + Ce^{-y}$.

2. (i) $r = C\sin\theta$, (ii) $r = b(1 - \sin\theta)$,
(iii) $r = b(1 - \cos\theta)$, (iv) $r = be^{-\sin\theta}$,
(v) $r^n \cos n\theta = b^n$, (vi) $r^n = b^n \cos n\theta$,
(vii) $r^n = b^n \sin n\theta$, (viii) $r^2 = Ce^{-\theta^2}$,
(ix) $r^2(1 + \sin^2\theta) = C^2$, (x) $r(1 - \cos\theta) = 2C$,
(xi) $r^2 = a\cos\theta$.

3. $x^2 + y^2 - 2a^2 \ln x = C$, **4.** $x^2 + y^2 - 2a^2 \ln x = C$,

5. $x^{\frac{3}{2}} - y^{\frac{3}{2}} = a^{\frac{3}{2}}$, **6.** $r(1 - \cos\theta) = 2b$,

9. $(x^2 + y^2)^2 + 2(y^2 - x^2) = C$,

10. $r\sin^2\theta = b$, **11.** $e^{\frac{25}{r}} = C(\sec 5\theta + \tan 5\theta)$,

12. $r = b(\sin\theta - \cos\theta)$,

13. $$\frac{6\sqrt{\cot\alpha}}{\sqrt{8\tan^2\alpha - 1}}\tan^{-1}\left(\frac{2y - x\cot\alpha}{x\sqrt{8\tan^2\alpha - 1}}\right) + C = \ln(y^2\tan\alpha - xy + 2x^2\tan\alpha)$$

14. $$\frac{6}{\sqrt{7}}\tan^{-1}\left(\frac{2y - x}{x\sqrt{7}}\right) + C = \ln(y^2 - xy + 2x^2)$$

15. $x^2 + y^2 = Ae^{-2\tan\alpha\tan^{-1}\frac{y}{x}}$, **16.** $x^2 - y^2 = C^2$,

17. $r^2 = b^2\cos(2\theta + \alpha)$, where $b^2c^2 = -\sec\alpha$

18. $x^2 + y^2 = Ae^{-2\tan^{-1}\frac{y}{x}}$, **19.** $y^2 - 2xy - x^2 = C$.

Chapter 14

PARTIAL DIFFERENTIAL EQUATION

14.1 Introduction

So far we have considered the differential equations with one independent variable. But equations in many mathematical and physical problems like Heat equation, Wave equation, etc. arise with more than one independent variable, and partial derivatives occur only when there are two or more independent variables. Partial differential equations are those, which contain one or more partial derivatives. So in partial differential equation, there are more than one independent variables. Considering z as dependent variable and x, y as independent variables, the following are the examples of partial differential equation.

$$\frac{\partial z}{\partial x}+\frac{\partial z}{\partial y}=2z; \quad \frac{\partial^2 z}{\partial x^2}+2\frac{\partial^2 z}{\partial x \partial y}+\frac{\partial^2 z}{\partial y^2}=0.$$

We shall denote $\frac{\partial z}{\partial x}, \frac{\partial z}{\partial y}, \frac{\partial^2 z}{\partial x^2}, \frac{\partial^2 z}{\partial x \partial y}, \frac{\partial^2 z}{\partial y^2}$ by p, q, r, s and t respectively.

14.2 Derivation of Partial Differential Equation

14.2.1 By elimination of arbitrary constants

Let us have a function

$$f(x, y, z, a, b) = 0, \tag{1}$$

where a, b are arbitrary constants.

Differentiating partially w.r.t. x and y, we have

$$\frac{\partial f}{\partial x}+p\frac{\partial f}{\partial z}=0, \tag{2}$$

$$\frac{\partial f}{\partial y}+q\frac{\partial f}{\partial z}=0, \tag{3}$$

Eliminating a, b from (1)–(3), we get the differential equation of the form

$$F(x, y, z, p, q) = 0.$$

This is a partial differential equation of first order.

WORKED-OUT EXAMPLES

1. Eliminate the arbitrary constants a, b from $z = (x + a)(y + b)$.

Solution

Differentiating the given equation w.r.t. x and y, we get

$$p = y + b \text{ and } q = x + a.$$

So, $z = pq$, which is the required differential equation.

2. Eliminate the arbitrary constants a, b from $z = ax + a^2y^2 + b$.

Solution

Differentiating the given equation w.r.t x and y we get

$$p = a \text{ and } q = 2a^2y = 2p^2y.$$

So $q = 2p^2y$, which is the required differential equation.

3. Eliminate the arbitrary constants a, b from $z = \frac{x^2}{a} + \frac{y^2}{b}$.

Solution

Differentiating the given equation w.r.t. x and y, we get

$$p = \frac{2x}{a} \text{ and } q = \frac{2y}{b}, \text{ or, } a = \frac{2x}{p} \text{ and } b = \frac{2y}{q}.$$

Eliminating a, b from the given equation with the aid of above results, we obtain

$$z = \frac{px^2}{2x} + \frac{qy^2}{2y} \quad \text{or, } 2z = px + qy.$$

4. Eliminate the arbitrary constants a, b, c from $z = a(x + y) + b(x - y) + abt + c$.

Solution

Differentiating the given equation w.r.t x, y and t, we get

$$p = \frac{\partial z}{\partial x} = a + b,$$

$$q = \frac{\partial z}{\partial y} = a - b,$$

and $$\frac{\partial z}{\partial t} = ab.$$

Now using the result $(a + b)^2 - (a - b)^2 = 4ab$, we get

$$p^2 - q^2 = 4\frac{\partial z}{\partial t}.$$

5. Eliminate the arbitrary constants from $z = ae^{kt}\sin kx$, a, k are arbitrary constants.

Solution

Differentiating twice with respect to x and t, we get

$$\frac{\partial z}{\partial t} = ake^{kt}\sin kx,\ \frac{\partial^2 z}{\partial t^2} = ak^2e^{kt}\sin kx \text{ and } \frac{\partial z}{\partial x} = ake^{kt}\cos kx,\ \frac{\partial^2 z}{\partial x^2} = -ak^2e^{kt}\sin kx.$$

Adding, we get $$\frac{\partial^2 z}{\partial x^2} + \frac{\partial^2 z}{\partial t^2} = ak^2e^{kt}\sin kx - ak^2e^{kt}\sin kx = 0.$$

6. Eliminate the arbitrary constants from $z = ae^{-k^2t}\sin kx$, a, k are arbitrary constants.

Solution

Differentiating with respect to x and t, we get

$$\frac{\partial z}{\partial t} = -ak^2e^{-k^2t}\sin kx,\ \frac{\partial z}{\partial x} = ake^{-k^2t}\cos kx.$$

Again differentiating $\frac{\partial z}{\partial x} = ake^{-k^2t}\cos kx$ with respect to x, we get

$$\frac{\partial^2 z}{\partial x^2} = -ak^2e^{-k^2t}\sin kx = \frac{\partial z}{\partial t}.$$

14.2.2 By elimination of arbitrary functions

Let u, v be two functions of x, y, z and are connected by the relation

$$f(u, v) = 0. \tag{4}$$

Differentiating (4) partially w.r.t. x and y we get

$$\frac{\partial f}{\partial u}\left(\frac{\partial u}{\partial x} + p\frac{\partial u}{\partial z}\right) + \frac{\partial f}{\partial v}\left(\frac{\partial v}{\partial x} + p\frac{\partial v}{\partial z}\right) = 0 \tag{5}$$

$$\frac{\partial f}{\partial u}\left(\frac{\partial u}{\partial y} + q\frac{\partial u}{\partial z}\right) + \frac{\partial f}{\partial v}\left(\frac{\partial v}{\partial y} + q\frac{\partial v}{\partial z}\right) = 0. \tag{6}$$

Eliminating $\frac{\partial f}{\partial u}$ and $\frac{\partial f}{\partial v}$ from (5) and (6), we get

$$\left(\frac{\partial u}{\partial x} + p\frac{\partial u}{\partial z}\right)\left(\frac{\partial v}{\partial y} + q\frac{\partial v}{\partial z}\right) = \left(\frac{\partial u}{\partial y} + q\frac{\partial u}{\partial z}\right)\left(\frac{\partial v}{\partial x} + p\frac{\partial v}{\partial z}\right)$$

which can be written as $Pp + Qq = R$, (7)

where $$P = \frac{\partial u}{\partial y}\frac{\partial v}{\partial z} - \frac{\partial u}{\partial z}\frac{\partial v}{\partial y} = \frac{\partial(u, v)}{\partial(y, z)},$$

$$Q = \frac{\partial u}{\partial z}\frac{\partial v}{\partial x} - \frac{\partial u}{\partial x}\frac{\partial v}{\partial z} = \frac{\partial(u, v)}{\partial(z, x)},$$

and $$R = \frac{\partial u}{\partial x}\frac{\partial v}{\partial y} - \frac{\partial u}{\partial y}\frac{\partial v}{\partial x} = \frac{\partial(u, v)}{\partial(x, y)}.$$

Equation (7) is the required differential equation.

WORKED-OUT EXAMPLES

1. Eliminate f from $z = f\left(\frac{y}{x}\right)$.

Solution

Here $p = -\frac{y}{x^2} f'\left(\frac{y}{x}\right)$ and $q = \frac{1}{x} f'\left(\frac{y}{x}\right)$.

So, $xp + yq = -\frac{y}{x} f'\left(\frac{y}{x}\right) + \frac{y}{x} f'\left(\frac{y}{x}\right) = 0$, which is the required differential equation.

2. Eliminate f from $z = f(x^2 + y^2)$.

Solution

Differentiating partially w.r.t x and y, we get

$$p = 2xf'(x^2 + y^2) \text{ and } q = 2yf'(x^2 + y^2).$$

Dividing the above results, we get

$$\frac{p}{q} = \frac{x}{y} \text{ or, } py - qx = 0.$$

3. Eliminate f from $z = xy + f(x^2 + y^2)$.

Solution

Differentiating partially w.r.t x and y we get

$$p = y + 2xf'(x^2 + y^2) \text{ and } q = x + 2yf'(x^2 + y^2)$$

or, $\quad p - y = 2xf'(x^2 + y^2)$ and $q - x = 2yf'(x^2 + y^2)$.

Dividing the above results we get

$$\frac{p - y}{q - x} = \frac{x}{y} \text{ or, } py - qx + x^2 - y^2 = 0.$$

4. Eliminate f from $f(x + y + z, x^2 + y^2 - z^2) = 0$.

Solution

Let $u = x + y + z$ and $v = x^2 + y^2 - z^2$, so $f(u, v) = 0$.

Differentiating partially w.r.t x and y, we get

$$(1 + p)\frac{\partial f}{\partial u} + 2(x - zp)\frac{\partial f}{\partial v} = 0, \text{ and } (1 + q)\frac{\partial f}{\partial u} + 2(y - zp)\frac{\partial f}{\partial v} = 0.$$

Eliminating $\frac{\partial f}{\partial u}$ and $\frac{\partial f}{\partial v}$ from the above relations, we get

$$(y - zq)(1 + p) = (x - pz)(1 + q)$$

or, $\quad p(y + z) - q(x + z) = x - y$.

This is the required differential equation.

5. Eliminate f from $xyz = f(x^2 + y^2 + z^2)$.

Solution

Differentiating partially w.r.t. x and y, we get

$$pxy + yz = f'(x^2 + y^2 + z^2)(2x + 2zp)$$

and $$qxy + xz = f'(x^2 + y^2 + z^2)(2y + 2zq).$$

Dividing, we get $$\frac{pxy + yz}{qxy + xz} = \frac{x + zp}{y + zq}$$

or, $$(pxy + yz)(y + zq) = (qxy + xz)(x + zp)$$

or, $$x(y^2 - z^2)p + y(z^2 - x^2)q = z(x^2 - y^2).$$

This is the required differential equation.

6. Eliminate f from $x + y + z = f(x^2 + y^2 + z^2)$.

Solution

Differentiating partially w.r.t x and y, we get

$$p + 1 = f'(x^2 + y^2 + z^2)(2x + 2zp),$$

and $$q + 1 = f'(x^2 + y^2 + z^2)(2y + 2zq).$$

Dividing we get $$\frac{p+1}{q+1} = \frac{x + zp}{y + zq}$$

or, $$(p+1)(y + zq) = (q+1)(x + zp)$$

or, $$(y - z)p + (z - x)q = x - y.$$

This is the required differential equation.

7. Find the differential equation of all spheres of radius k with centre on the plane $x = y$.

Solution

Equation of the family of spheres having centre on the plane $x = y$ and of radius k is

$$(x - a)^2 + (y - a)^2 + (z - b)^2 = k^2,$$

where a, b are parameters.

Differentiating the above equation with respect to x, y, we get

$$x - a + (z - b)p = 0 \text{ and } y - a + (z - b)q = 0.$$

Subtracting, we get $$x - y + (z - b)(p - q) = 0.$$

Using the above results in the equation of the spheres, we get

$$(z - b)^2[p^2 + q^2 + 1] = k^2$$

or, $$(x - y)^2[p^2 + q^2 + 1] = k^2(p - q)^2.$$

This is the required differential equation.

8. Find the differential equation of all planes having equal x- and y- intercepts.

Solution

Equation of all planes having equal x- and y-intercepts are

$$\frac{x}{a}+\frac{y}{a}+\frac{z}{b}=1,$$

where a, b are parameters.

Differentiating partially with respect to x, y respectively, we get

$$p+\frac{b}{a}=0 \text{ and } q+\frac{b}{a}=0.$$

Subtracting, we get $\quad p-q=0.$

9. Find the differential equation of all surfaces of revolution having z-axis as the axis of revolution.

Solution

Equation of the revolution of surfaces having z-axis as the axis of revolution is

$$z=f\left(\sqrt{x^2+y^2}\right), \text{ where } f \text{ is an arbitrary function.}$$

Differentiating partially with respect to x, y respectively, we get

$$p=\frac{\partial z}{\partial x}=f'\left(\sqrt{x^2+y^2}\right)\frac{x}{\sqrt{x^2+y^2}},$$

$$q=\frac{\partial z}{\partial y}=f'\left(\sqrt{x^2+y^2}\right)\frac{y}{\sqrt{x^2+y^2}}.$$

Dividing, we get $\quad \frac{p}{q}=\frac{x}{y}$, i.e., $py=qx$.

10. Find the differential equation of the cones having vertices at the origin.

Solution

Equation of the cones having vertices at the origin is a second degree homogeneous equation in x, y, z. Let it be

$$ax^2+by^2+cz^2+2fyz+2gxz+2hxy=0,$$

where a, b, c, f, g, h are parameters.

Differentiating partially with respect to x, y respectively, we get

$$ax+czp+fyp+g(z+xp)+hy=0$$

or, $$ax+hy+gz+p(xg+yf+zc)=0 \quad (1)$$

and $$by+czq+f(z+yq)+gxq+hx=0$$

or, $$hx+by+fz+q(gx+fy+cz)=0. \quad (2)$$

Multiplying (1) by x and (2) by y and then adding the results, we get

$$ax^2 + 2hxy + by^2 + z(gx + fy) + (px + qy)(xg + yf + zc) = 0$$

or, $\quad -(cz + fy + gx)z + (px + qy)(xg + yf + zc) = 0$

or, $\quad (px + qy - z)(xg + yf + zc) = 0$

or, $\quad px + qy = z.$

This is the required differential equation.

EXERCISE

- Eliminate a, b from the following relations:

1. $z = ax + by + ab,$

2. $z = a(x + y) + b,$

3. $az + b = a^2x + y,$

4. $z = (x^2 + a)(y^2 + b),$

5. $z = axy + b,$

6. $z = (x - a)^2 + (y - b)^2,$

7. $z = ax^3 + by^3,$

8. $z = xy + y\sqrt{x^2 - a^2} + b,$

9. $z = ax^2 + by^2 + ab,$

10. $z = ax + by + a^2 + b^2,$

11. $z = axe^y + \frac{1}{2}a^2e^{2y} + b,$

12. $2z = (ax + y)^2 + b,$

13. $\ln(az - 1) = x + ay + b,$

14. $(x - a)^2 + (y - b)^2 + z^2 = 1.$

- Form partial differential equation by eliminating the arbitrary functions from the following relations:

15. $z = f(x^2 - y^2),$

16. $z = f(x + ay) + F(x - ay),$

17. $z = e^y f(x + y),$

18. $z = e^{my}\phi(x - y),$

19. $z = e^{mx+ny}\phi(mx - ny),$

20. $xyz = \phi(x + y + z),$

21. $\phi\left(\frac{z}{x^2}, x - y\right) = 0,$

22. $\phi\left(\frac{z}{x + y}, x^2 - y^2\right) = 0,$

23. $x = f(z) + g(y),$

24. $z = x\phi\left(\frac{y}{x}\right),$

25. $z = x^2\phi(x - y),$

26. $z = (x + y)\phi(x^2 - y^2),$

27. $lx + my + nz = \phi(x^2 + y^2 + z^2),$

28. $z = xf(y) + yg(x),$

29. $z = x + y + f(xy),$

30. $z = f\left(\frac{xy}{z}\right),$

31. $z = f(x) + e^y g(x).$

32. Find the differential equation of all right circular cones having z-axis as the axis of the cones.

[*Hints*: Equation of all right circular cones having z-axis as the axis of the cones is

$$x^2 + y^2 = (z - c)^2 \tan^2 \theta,$$

where $(0, 0, c)$ is the vertex and θ is the semi-vertical angle of the cone]

33. Find the differential equation of all spheres of radius r, having centre in the xy-plane.

[*Hints*: Equation of all spheres of radius r, having centre at the point $(a, b, 0)$, i.e., in the xy-plane is $(x-a)^2+(y-b)^2+z^2=r^2$.]

34. Find the partial differential equation of all planes which are at a constant distance k from the origin.

[*Hints*: Equation of the planes, which are at a constant distance k from the origin is $lx+my+nz=k$, where $l^2+m^2+n^2=1$.]

35. Find the partial differential equation of all spheres whose centres lie on z-axis.

[*Hints*: Equation of all spheres of radius r, having centre at the point $(0, 0, c)$, i.e., in the z-axis is $x^2+y^2+(z-c)^2=r^2$.]

36. Find the partial differential equation for the surfaces whose tangent planes form a tetrahedron of constant volume with coordinate plane.

[*Hints*: At any point (x, y, z) on the surface $z = f(x, y)$ the direction ratios of the normal to the plane are $(p, q, -1)$. We assume the equation of the plane as $px + qy - z = k$. Given that $\frac{1}{6}\left(\frac{k}{p}\right)\left(\frac{k}{q}\right)(-k) = \text{constant}$.]

37. Verify that the equation $p^2 - q^2 = 0$ is reduced to the form $\frac{\partial^2 z}{\partial\alpha\,\partial\beta} = 0$ by the transformations $\alpha = \frac{1}{2}(x+y), \beta = \frac{1}{2}(x-y)$.

ANSWERS

1. $z = px + yq + pq$,
2. $p = q$,
3. $pq = 1$,
4. $pq = 4xyz$,
5. $xp = yq$,
6. $4z = p^2 + q^2$,
7. $3z = xp + yq$,
8. $px + qy = pq$,
9. $pq + 2px^2y + 2qxy^2 = 4xyz$,
10. $z = px + qy + p^2 + q^2$,
11. $q = xp + p^2$
12. $px + qy = q^2$,
13. $(1 + q)p = zq$,
14. $z^2(1 + p^2 + q^2) = 1$,
15. $py + xq = 0$,
16. $t = a^2r$,
17. $z = q - p$,
18. $p + q = mz$,
19. $np + mq = 2mnz$,
20. $x(y - z)p + y(z - x)q = z(x - y)$,
21. $2z = x(p + q)$,
22. $z = yp + xq$,
23. $ps = qr$,
24. $z = px + qy$,
25. $2z = px + qx$,
26. $z = qx + py$,
27. $(ny - mz)p + (lz - nx)q = mx - ly$,
28. $xys = xp + yq - z$,
29. $px - qy = x - y$,
30. $px - qy = 0$,
31. $t - q = 0$,
32. $xq = yp$,
33. $z^2(1 + p^2 + q^2) = r^2$,
34. $z = px + qy + k\sqrt{p^2+q^2+1}$,
35. $xq = yp$,
36. $pq = C$.

14.3 Linear Partial Differential Equation of Order One

A differential equation involving partial derivatives p and q only and no other higher order derivatives is called of order one. If degree of p and q is one then it is called linear partial differential equation of order one. We have already seen that if two function u, v, of x, y, z are connected by $f(u, v) = 0$, then eliminating f we get a partial differential equation

$$Pp + Qq = R.$$

The above equation is a linear partial differential equation of order one and is called Lagrange's linear equation.

A *solution* or *integral* of a differential equation is a relation between the variables which when substituted in the equation, satisfies it. If the number of constants occurring in a solution is equal to the number of independent variables, then it is called *complete integral* or *complete solution* or *general solution*. By giving particular values to the constants in a complete solution we get a *particular solution*. The equation of the envelope of the surfaces represented by the general solution of a differential equation is called *singular integral*. If we express a differential equation in the form

$$f(x, y, z, p, q) = 0.$$

Then its complete integral is given by

$$g(x, y, z, a, b) = 0$$

and singular solution is envelope of the curves represented by the above equation and hence singular solution will be obtained by eliminating of a, b from

$$g(x, y, z, a, b) = 0, \quad \frac{\partial g}{\partial a} = 0 \text{ and } \frac{\partial g}{\partial b} = 0.$$

14.4 Lagrange's Method

We have already seen that eliminating arbitrary function f from

$$f(u, v) = 0 \tag{8}$$

connecting the functions u, v we get the partial differential equation

$$Pp + Qq = R. \tag{9}$$

Thus (8) is the general solution of the equation (9) and we have to find the values of u, v. Let $u = a$ and $v = b$ be two equations where a, b are two arbitrary constants. Differentiating, we get

$$\frac{\partial u}{\partial x}dx + \frac{\partial u}{\partial y}dy + \frac{\partial u}{\partial z}dz = 0,$$

$$\frac{\partial v}{\partial x}dx + \frac{\partial v}{\partial y}dy + \frac{\partial v}{\partial z}dz = 0.$$

Solving we get $$\frac{dx}{p} = \frac{dy}{Q} = \frac{dz}{R}, \tag{10}$$

where $P = \frac{\partial u}{\partial y}\frac{\partial v}{\partial z} - \frac{\partial u}{\partial z}\frac{\partial v}{\partial y}$ $Q = \frac{\partial u}{\partial z}\frac{\partial v}{\partial x} - \frac{\partial u}{\partial x}\frac{\partial v}{\partial z}$, $R = \frac{\partial u}{\partial x}\frac{\partial v}{\partial y} - \frac{\partial u}{\partial y}\frac{\partial v}{\partial x}$.

Solution of the differential equation given by (10) are $u = a$ and $v = b$. Thus the solution of the equation (9) is found as $f(u, v) = 0$. Equations (10) are called Lagrange's auxiliary equation.

WORKED-OUT EXAMPLES

1. Find the general solution of $px + qy = 2z$.

Solution

Lagrange's auxiliary equations are $\frac{dx}{x} = \frac{dy}{y} = \frac{dz}{2z}$

From $\frac{dx}{x} = \frac{dz}{2z}$, we get $2\log x = \log z + \log A \quad A = \frac{x^2}{z}$.

Again, from $\frac{dy}{y} = \frac{dz}{2z}$, we get $2\log y = \log z + \log B \quad B = \frac{y^2}{z}$.

So the general solutions is $\phi\left(\frac{x^2}{z}, \frac{y^2}{z}\right) = 0$.

2. Solve : $yzp + xzq = xy$.

Solution

The auxiliary equation, are

$$\frac{dx}{yz} = \frac{dy}{xz} = \frac{dz}{xy}.$$

From $\frac{dx}{yz} = \frac{dy}{xz}$, we get $x^2 - y^2 = A$.

Similarly, from $\frac{dy}{xz} = \frac{dz}{xy}$, we get $y^2 - z^2 = B$.

So the general solution is $\phi(x^2 - y^2, y^2 - z^2) = 0$.

3. Solve: $xzp + yzq = xy$.

Solution

The auxiliary equations are $\frac{dx}{xz} = \frac{dy}{yz} = \frac{dz}{xy}$.

From $\frac{dx}{xz} = \frac{dy}{yz}$, we get $\ln x = \ln y + \ln A \quad A = \frac{x}{y}$.

Again from the auxiliary equation, we get

$$ydx = xdy = zdz$$

$$ydx + xdy = 2zdz \quad d(xy) = d(z^2)$$

$$d(xy - z^2) = 0 \quad xy - z^2 = B$$

So the general solution is $\phi\left(\frac{x}{y}, xy - z^2\right) = 0$.

4. Solve: $p + q = 1$.

Solution

The auxiliary equations are $dx = dy = dz$.

From $dx = dy$, we get $A = x - y$ and from $dy = dz$, we get $B = y - z$.

So the general solution is $\phi(x - y, y - z) = 0$.

5. Solve: $x^2p + y^2q = z^2$.

Solution

The auxiliary equations are $\frac{dx}{x^2} = \frac{dy}{y^2} = \frac{dz}{z^2}$.

From $\frac{dx}{x^2} = \frac{dy}{y^2}$, we get $A = \frac{1}{x} - \frac{1}{y}$.

Again from $\frac{dy}{y^2} = \frac{dz}{z^2}$, we get $B = \frac{1}{y} - \frac{1}{z}$.

So the general solution is $\phi\left(\frac{1}{x} - \frac{1}{y}, \frac{1}{y} - \frac{1}{z}\right) = 0$.

6. Solve: $(mz - ny)p + (nx - lz)q = ly - mx$.

Solution

The auxiliary equation, are

$$\frac{dx}{mz - ny} = \frac{dy}{nx - lz} = \frac{dz}{ly - mx} = \frac{xdx + ydy + zdz}{0} = \frac{ldx + mdy + ndz}{0}.$$

Integrating, $xdx + ydy + zdz = 0$ and $ldx + mdy + ndz = 0$, we get

$$x^2 + y^2 + z^2 = a^2 \text{ and } lx + my + nz = b,$$

where a, b are arbitrary constants.

Thus the integral of the equation is

$$lx + my + nz = f(x^2 + y^2 + z^2).$$

where f is an arbitrary function.

7. Solve: $x^2(y - z)p + y^2(z - x)q = z^2(x - y)$.

Solution

The auxiliary equation are

$$\frac{dx}{x^2(y - z)} = \frac{dy}{y^2(z - x)} = \frac{dz}{z^2(x - y)} = k.$$

Then $$\frac{dx}{x} + \frac{dy}{y} + \frac{dz}{z} = k[x(y - z) + y(z - x) + z(x - y)] = 0,$$

and

$$(y + z)dx + (z + x)dy + (x + y)dz = k[x^2(y^2 - z^2) + y^2(z^2 - x^2) + z^2(x^2 - y^2)] = 0.$$

Integrating, we get

$$\ln x + \ln y + \ln z = A \Rightarrow xyz = A \text{ and } xy + yz + xz = B;$$

where A, B are arbitrary constants.

Thus the integral of the equation is

$$xyz = f(xy + yz + xz), \text{ where } f \text{ is an arbitrary function.}$$

8. Solve: $(x^2 - y^2 - z^2)p + 2xyq = 2xz$.

Solution

The auxiliary equations are

$$\frac{dx}{x^2 - y^2 - z^2} = \frac{dy}{2yx} = \frac{dz}{2xz}.$$

Form the last two equations, we get

$$\frac{dy}{y} = \frac{dz}{z}.$$

Integrating we get $\qquad y = Az.$

Again from the auxiliary equation we get

$$\frac{xdx + ydy + zdz}{x(x^2 - y^2 - z^2) + 2xy^2 + 2xz^2} = \frac{dy}{2yx} \text{ or, } \frac{d(x^2 + y^2 + z^2)}{x^2 + y^2 + z^2} = \frac{dy}{y}.$$

Integrating, we get $\quad x^2 + y^2 + z^2 = By.$

So the general solution of the given equation is

$$f\left(\frac{x^2 + y^2 + z^2}{y}, \frac{y}{z}\right) = 0,$$

where f is an arbitrary function.

9. Solve: $y^2p + x^2q = x^2y^2z^2$.

Solution

The auxiliary equation is

$$\frac{dx}{y^2} = \frac{dy}{x^2} = \frac{dz}{x^2y^2z^2}.$$

Taking first and second ratios, we get

$$x^2dx = y^2dy.$$

Integrating, we get $\qquad x^3 - y^3 = a.$

Now taking the multipliers as $x^2, y^2, -\frac{2}{z^2}$, we see that each of the ratios is equal to

$$\frac{x^2dx + y^2dy - \frac{2}{z^2}dz}{0} \qquad x^2dx + y^2dy - \frac{2}{z^2}dz = 0.$$

Integrating, we get $\qquad x^3 + y^3 + \frac{6}{z} = b.$

So the general solution is $\qquad \phi\left(x^3 - y^3, x^3 + y^3 + \frac{6}{z}\right) = 0,$

where ϕ is an arbitrary function.

10. Solve: $\frac{b-c}{a}yzdx + \frac{c-a}{b}xzdy + \frac{a-b}{c}xydz = 0.$

Solution

The auxiliary equation is

$$\frac{dx}{\frac{b-c}{a}yz} = \frac{dy}{\frac{c-a}{b}xz} = \frac{dz}{\frac{a-b}{c}xy}.$$

Taking ax, by, cz as multiplier we see that each of the ratios is equal to

$$\frac{axdx + bydy + czdz}{0} \qquad axdx + bydy + czdz = 0.$$

Integrating, we get $\quad ax^2 + by^2 + cz^2 = A.$

Again, taking multipliers as a^2x, b^2y, c^2z as multiplier we see that each of the ratios is equal to

$$\frac{a^2xdx + b^2ydy + c^2zdz}{0} \qquad a^2xdx + b^2ydy + c^2zdz = 0.$$

Integrating, we get $\quad a^2x^2 + b^2y^2 + c^2z^2 = B.$

So the general solution is

$$\phi(ax^2 + by^2 + cz^2, a^2x^2 + b^2y^2 + c^2z^2) = 0$$

where ϕ is an arbitrary function.

11. Solve: $p - 2q = 3z + \tan(y + 2x)$.

Solution

The auxiliary equation is

$$\frac{dx}{1} = \frac{dy}{-2} = \frac{dz}{3z + \tan(y + 2x)}.$$

Taking first two radios, we get $dy + 2dx = 0$. Integrating, we get $y + 2x = a$.

Now from first and third ratios, we get

$$\frac{dx}{1} = \frac{dz}{3z + \tan a}.$$

Integrating, we get $\quad x = \frac{1}{3}\ln(3z + \tan a) + \frac{1}{3}b$

or, $\quad 3x = \ln(3z + \tan a) + b = \ln(3z + \tan(y + 2x)) + b$

or, $\quad 3x - \ln(3z + \tan(y + 2x)) = \phi(a)$

or, $\quad 3x - \ln(3z + \tan(y + 2x)) = \phi(y + 2x)$, where ϕ is an arbitrary function.

12. Solve: $px + qy = z - a\sqrt{x^2 + y^2 + z^2}$.

Solution

The auxiliary equation is

$$\frac{dx}{x} = \frac{dy}{y} = \frac{dz}{z - a\sqrt{x^2 + y^2 + z^2}}.$$

From the first two ratios we get $\ln x = \ln y + \ln c$, or, $x = cy$.

Again, from the equation, we get

$$\frac{dx}{x} = \frac{dz}{z - a\sqrt{x^2 + y^2 + z^2}} = \frac{xdx + ydy + zdz}{x^2 + y^2 + z^2 - az\sqrt{x^2 + y^2 + z^2}}$$

or, $\quad \frac{dx}{x} = \frac{dz}{z - au} = \frac{du}{u - az}$ where $x^2 + y^2 + z^2 = u^2$

or, $\quad \frac{dx}{x} = \frac{dz + du}{z - au + u - az} = \frac{d(z + u)}{(1 - a)(u + z)}.$

Integrating, we get

$$(1-a)\ln x - \ln(z+u) = \ln b.$$

or, $$\frac{x^{1-a}}{z+u} = b$$

or, $$\frac{x^{1-a}}{z+\sqrt{x^2+y^2+z^2}} = b.$$

So the general solution of the given equation is

$$\phi\left(\frac{x^{1-a}}{z+\sqrt{x^2+y^2+z^2}}, \frac{x}{y}\right) = 0,$$

where ϕ is an arbitrary function.

13. Solve: $\dfrac{dx}{\cos(x+y)} = \dfrac{dy}{\sin(x+y)} = \dfrac{dz}{z+\frac{1}{z}}.$

Solution

From the given equation, we get

$$\frac{dx+dy}{\cos(x+y)+\sin(x+y)} = \frac{dx-dy}{\cos(x+y)-\sin(x+y)} = \frac{dz}{z+\frac{1}{z}}.$$

From the first two ratios of the equation, we get

$$\frac{\cos(x+y)-\sin(x+y)}{\cos(x+y)+\sin(x+y)} d(x+y) = d(x-y).$$

Integrating, we get

$$\ln\{\cos(x+y)+\sin(x+y)\} = x-y+k$$

or, $$[\cos(x+y)+\sin(x+y)]e^{y-x} = a.$$

Again, from $$\frac{dx+dy}{\cos(x+y)+\sin(x+y)} = \frac{dz}{z+\frac{1}{z}},$$

we get $$\frac{d(x+y)}{\sin\left(x+y+\frac{\pi}{4}\right)} = \sqrt{2}\frac{zdz}{z^2+1}.$$

Integrating, we get $$\frac{1}{\sqrt{2}}\ln(z^2+1) = \ln\tan\frac{1}{2}\left(x+y+\frac{\pi}{4}\right) + \ln B$$

or, $$(z^2+1)^{\frac{1}{\sqrt{2}}}\cot\frac{1}{2}\left(x+y+\frac{\pi}{4}\right) = B.$$

So the general solution is given by

$$\phi\left([\cos(x+y)+\sin(x+y)]e^{y-x}, (z^2+1)^{\frac{1}{\sqrt{2}}}\cot\frac{1}{2}\left(x+y+\frac{\pi}{4}\right)\right) = 0.$$

EXERCISE

- Solve the following differential equations:

1. $zp + x = 0$,
2. $px + qy = 3z$,
3. $2p + 3q = 1$,
4. $y^2zp - x^2zq = x^2y$,
5. $(y - z)p + (x - y)q = z - x$,
6. $(x^2 - y^2 - z^2)p + 2xyq = 2xz$,
7. $ap + bq + cz = 0$,
8. $p + q = z$,
9. $p + q = x + y + z$,
10. $(p + q)(x + y) = 1$,
11. $yq - xp = z$,
12. $yp - xq + x^2 - y^2 = 0$,
13. $yzp - xzq = xy$,
14. $zp + yq = x$,
15. $x^2p + y^2q = (x + y)z$,
16. $y^2p - xyq = x(z - 2y)$,
17. $(x^2 + 2y^2)p - xyq = xz$,
18. $yzp + 3xq = xy$,
19. $x(y - z)p + y(z - x)q = z(x - y)$,
20. $x^2(y - z)p + y^2(z - x)q = z^2(x - y)$,
21. $x(y^2 - z^2)p + y(z^2 - x^2)q = z(x^2 - y^2)$,
22. $x(y^2 - z^2)p - y(z^2 + x^2)q = z(x^2 + y^2)$,
23. $x(y^2 + z)p - y(x^2 + z)q = z(x^2 - y^2)$,
24. $p\tan x + q\tan y = \tan z$,
25. $y^2zp - x^2zq = xy^2$,
26. $(mz - ny)p + (nx - lz)q = ly - mx$,
27. $(x^2 - yz)p + (y^2 - xz)q = z^2 - xy$,
28. $x^2p + y^2q = z^2$,
29. $py + qx = xyz^2(x^2 - y^2)$,
30. $xp - yq = xy$,
31. $p + 3q = z + \cot(y - 3x)$,
32. $p - 2q = 3x^2 \sin(y + 2x)$,
33. $(x + y)(p - q) = z$,
34. $z(p - q) = z^2 + (x + y)^2$,
35. $(y^2 + z^2)p - xyq + xz = 0$,
36. $z(x + y)p + z(x - y)q = x^2 + y^2$,
37. $(y - xz)p + (x + yz)q = x^2 + y^2$,
38. $(y + xz)p - (x + yz)q = x^2 - y^2$,
39. $z(x + 2y)p - z(y + 2x)q + x^2 - y^2 = 0$,
40. $(1 + y)p + (1 + x)q = z$,
41. $(x + 2z)p + (4xz - y)q = 2x^2 + y$,
42. $x^2p + y^2q = (x + y)z$,
43. $x^2p + y^2q = x + y$,
44. $x(y^n - z^n)p + y(z^n - x^n)q = z(x^n - y^n)$.
45. Find the equation of all surfaces cutting the family of cones $x^2 + y^2 - a^2z^2 = 0$ orthogonally.

    [*Hints*: At any point (x, y, z) of the surface $z = f(x, y)$ direction ratios of the normal to the surface are $(p, q, -1)$ and those of the normal to the cone are $(x, y, -a^2z)$. Since, surface cuts the cone orthogonally, we have $px + qy + a^2z = 0$. Eliminating a^2 with the aid of the equation of the cone, we get

    $$z(px + qy) + x^2 + y^2 = 0,$$

    which is well known Lagrange's linear form. The Lagrange's auxiliary equation is $\frac{dx}{xz} = \frac{dy}{yz} = \frac{dz}{-(x^2 + y^2)}$, etc.]
46. Find the equation of all surfaces whose tangent planes cut off an intercept of constant length from z-axis.

    [*Hints*: At any point (x, y, z) of the surface $z = f(x, y)$ direction ratios of the normal to the surface are $(p, q, -1)$. So equation of the plane through the point $(0, 0, k)$ is $px + qy - z + k = 0$ which is well known Lagrange's linear form. The Lagrange's auxiliary equation is $\frac{dx}{x} = \frac{dy}{y} = \frac{dz}{z - k}$, etc.]

ANSWERS

1. $\phi(y, x^2 + z^2) = 0,$

2. $\phi\left(\frac{z}{x^3}, \frac{y}{x}\right) = 0,$

3. $\phi(x - 2z, 3x - 2y) = 0,$

4. $\phi(y^2 + z^2, x^3 + y^3) = 0,$

5. $\phi(x^2 + 2yz, x + y + z) = 0,$

6. $\phi\left(\frac{y}{z}, \frac{x^2 + y^2 + z^2}{y}\right) = 0,$

7. $z = e^{-\frac{cx}{a}}\phi(ay - bx)$, if $a \neq 0$; $z = e^{-\frac{cy}{b}}\phi(ay - bx)$, if $b \neq 0,$

8. $z = e^y\phi(y - x),$

9. $\phi\left(x - y, e^{-x}(x + y + z + 2)\right) = 0,$

10. $\phi\left(x - y, e^{-2z}(x + y)\right) = 0,$

11. $\phi(zx, yx) = 0,$

12. $\phi(y^2 + x^2, xy - z) = 0,$

13. $\phi(y^2 + z^2, x^2 + y^2) = 0,$

14. $\phi\left(\frac{x + z}{y}, x^2 - z^2\right) = 0,$

15. $\phi\left(\frac{1}{x} - \frac{1}{y}, \frac{z}{x + y}\right) = 0,$

16. $\phi(x^2 + y^2, zy - y^2) = 0,$

17. $\phi(yz, x^2y^2 + y^4) = 0,$

18. $\phi(x^2 - z^2, y^2 - 6z) = 0,$

19. $\phi(xyz, x + y + z) = 0,$

20. $\phi\left(xyz, \frac{1}{x} + \frac{1}{y} + \frac{1}{z}\right) = 0,$

21. $\phi(xyz, x^2 + y^2 + z^2) = 0,$

22. $\phi\left(\frac{x}{yz}, x^2 + y^2 + z^2\right) = 0,$

23. $\phi\left(y^2 + x^2 - 2z, xyz\right) = 0,$

24. $\phi\left(\frac{\sin z}{\sin y}, \frac{\sin x}{\sin y}\right) = 0,$

25. $\phi(x^2 - z^2, x^3 + y^3) = 0,$

26. $\phi(lx + my + nz, x^2 + y^2 + z^2) = 0,$

27. $\phi\left(\frac{x - y}{y - z}, \frac{y - z}{z - x}\right) = 0,$

28. $\frac{1}{y} - \frac{1}{x} = \phi\left(\frac{1}{x} - \frac{1}{z}\right),$

29. $y^2(x^2 - y^2) + \frac{2}{z} = \phi(x^2 - y^2),$

30. $xe^{-\frac{z}{xy}} = \phi(xy),$

31. $x - \ln|z + \cot(y - 3x)| = \phi(y - 3x),$

32. $x^3\sin(y + 2x) - z = \phi(y + 2x),$

33. $x - (x + y)\ln z = \phi(y + x),$

34. $e^{2y}[z^2 + (x + y)^2] = \phi(y + x),$

35. $\phi\left(\frac{y}{z}, x^2 + y^2 + z^2\right) = 0,$

36. $\phi(x^2 - y^2 - z^2, 2xy - z^2) = 0,$

37. $\phi(x^2 - y^2 + z^2, xy - z) = 0,$

38. $\phi(x^2 + y^2 - z^2, xy + z) = 0,$

39. $\phi(x^2 + y^2 + z^2, x^2 + y^2 + xy) = 0,$

40. $\phi\left((1 + x)^2 - (1 + y)^2, \frac{x + y + 2}{z}\right) = 0,$

41. $\phi(xy - z^2, x^2 - y - z) = 0,$

42. $\phi\left(\frac{1}{y}-\frac{1}{x}, \frac{xy}{z}\right)=0,$

43. $\phi(x^2y^2+y^4, yz)=0,$

44. $x^n+y^n+z^n=\phi(xyz),$

45. $\phi\left(\frac{x}{y}, x^2+y^2+z^2\right)=0,$

46. $\phi\left(\frac{y}{x}, \frac{z-k}{x}\right)=0.$

14.5 Integral Surfaces Passing Through a Given Curve

We have already seen that if $u(x, y, z)=a$ and $v(x, y, z)=b$ be two lineary independent solutions of the equation

$$\frac{dx}{P}=\frac{dy}{Q}=\frac{dz}{R},$$

then $\phi(u, v)=0$, where ϕ is an arbitrary function, represents the general solution of the equation

$$Pp+Qq=R.$$

Method 1: Let the parametric equation of the given curve be

$$x=x(t),\ y=y(t),\ z=z(t),$$

t being the parameter. Then from the solutions, we get

$$u(x(t), y(t), z(t))=a$$

and
$$v(x(t), y(t), z(t))=b.$$

Eliminating t from the above equations, we get a relation between a, b. Finally, replacing a, b by $u(x, y, z)$, $v(x, y, z)$ in the above relation of a, b we get the required equation of the integral surface.

Method 2: If the equations of the curves be

$$\phi(x, y, z)=0 \text{ and } \psi(x, y, z)=0.$$

then eliminating x, y, z from $\phi(x, y, z)=0$, $\psi(x, y, z)=0$, $u(x, y, z)=a$ and $v(x, y, z)=b$ we get a relation between a, b. Finally, replacing a, b by $u(x, y, z)$, $v(x, y, z)$ in the above relation of a, b, we get the required equation of the integral surface.

WORKED-OUT EXAMPLES

1. Find the integral surface of the linear partial differential equation

$$x(y^2+z^2)p+y(x^2+z^2)q=(x^2-y^2)z$$

which contains the straight line $x+y=0$, $z=1$.

Solution

The auxiliary equation is

$$\frac{dx}{x(y^2+z^2)}=\frac{dy}{y(x^2+z^2)}=\frac{dz}{(x^2-y^2)z}.$$

Taking multipliers as $\frac{1}{x}, \frac{1}{-y}, \frac{1}{z}$ we get that each ratio is equal to

$$\frac{\frac{dx}{x}-\frac{dy}{y}+\frac{dz}{z}}{y^2+z^2-(x^2+z^2)+x^2-y^2}, \text{ i.e., } \frac{dx}{x}-\frac{dy}{y}+\frac{dz}{z}=0.$$

Integrating, we get $\qquad \frac{xz}{y}=a. \qquad (1)$

Taking multipliers as $x, -y, -z$ we get that each ratio is equal to

$$\frac{xdx-ydy-zdz}{x^2(y^2+z^2)-y^2(x^2+z^2)-(x^2-y^2)z^2}, \text{ i.e., } xdx-ydy-zdz=0$$

Integrating, we get $\quad x^2-y^2-z^2=b. \qquad (2)$

Since the integral surface contains the line $x+y=0$, $z=1$, we put $x=t$, $y=-t$, $z=1$ in (1) and (2) and get

$$a=-1=b.$$

Now eliminating a, b we get

$$x^2-y^2-z^2=\frac{xy}{y}.$$

This is the required integral surface.

2. Find the integral surface of the linear partial differential equation

$$(y-z)p+(z--x)q=x-y$$

which contains the straight line $y=2x$, $z=0$.

Solution

The auxiliary equation is

$$\frac{dx}{y-z}=\frac{dy}{z-x}=\frac{dz}{x-y}.$$

Taking the multiliers as 1, 1, 1 we see that each of the ratios is equal to

$$\frac{dx+dy+dz}{y-z+z-x+x-y} \quad \text{or,} \quad dx+dy+dz=0.$$

Integrating we get $\quad x+y+z=a. \qquad (1)$

Again, taking the multipliers as x, y, z, we see that each of the ratios is equal to

$$\frac{xdx+ydy+zdz}{xy-xz+yz-yx+zx-zy}, \text{ or, } xdx+ydy+zdz=0.$$

Integrating, we get $\qquad x^2+y^2+z^2=b. \qquad (2)$

Since the integral surface contains the line $y=2x$, $z=0$, we put $x=t$, $y=2t$, $z=0$ in (1) and (2) and get

$$3t=a,\ 5t^2=b.$$

Eliminating t, we get

$$9b=5a^2.$$

Now eliminating a, b using (1) and (2), we get

$$9(x^2+y^2+z^2)=5(x+y+z)^2.$$

This is the required integral surface

3. Find the integral surface of the linear partial differential equation

$$x(y-z)p + y(z-x)q = z(x-y),$$

which contains the straight line $y = x$, $x = y = z$.

Solution

The auxiliary equation is

$$\frac{dx}{x(y-z)} = \frac{dy}{y(z-x)} = \frac{dz}{z(x-y)}.$$

Taking the multipliers as 1, 1, 1, we see that each of the ratios is equal to

$$\frac{dx+dy+dz}{xy-xz+yz-yx+zx-zy}, \text{ or, } dx+dy+dz=0.$$

Integrating, we get $\quad x + y + z = a. \quad$ (1)

Again, rewriting the equation as

$$\frac{\frac{dx}{x}}{y-z} = \frac{\frac{dy}{y}}{z-x} = \frac{\frac{dz}{z}}{x-y},$$

we see that each of the ratios is equal to

$$\frac{\frac{dx}{x}+\frac{dy}{y}+\frac{dz}{z}}{y-z+z-x+x-y} \text{ or, } \frac{dx}{x}+\frac{dy}{y}+\frac{dz}{z}=0.$$

Integrating, we get $\quad xyz = b. \quad$ (2)

Since the integral surface contains the line $y = x$, $x = y = z$, we put $x = y = z = t$ in (1) and (2) and get

$$3t = a,\ t^3 = b.$$

Eliminating t, we get $\quad a^3 = 27b.$

Now eliminating a, b using (1) and (2), we get

$$(x+y+z)^3 = 27xyz.$$

This is the required integral surface.

4. Find the integral surface of the linear partial differential equation

$$(x^2-yz)p+(y^2-zx)q = z^2-xy,$$

which contains the straight line $x = 1$, $y = 0$.

Solution

The auxiliary equation is

$$\frac{dx}{x^2-yz} = \frac{dy}{y^2-zx} = \frac{dz}{z^2-xy}.$$

From the given equation, we get

$$\frac{dx-dy}{x-y} = \frac{dy-dz}{y-z}.$$

Integrating, we get $\quad \dfrac{x-y}{y-z} = a. \quad$ (1)

Again, taking the multipliers as $y + z, x + z, x + y$ we see that each of the ratios is equal to

$$\frac{(y+z)dx+(x+z)dy+(x+y)dz}{(y+z)(x^2-yz)+(x+z)(y^2-zx)+(x+y)(z^2-xy)}$$

or, $\quad (y + z)dx + (x + z)dy + (x + y)dz = 0$

or, $\quad ydx + xdy + xdz + zdx + ydz + zdy = 0.$

Integrating, we get $\quad xy + yz + xz = b. \quad (2)$

Since the integral surface contains the line $y = 0, x = 1$. Using these in (1) and (2), we get

$$a = \frac{1}{-z} = \frac{1}{-b} \text{ or, } ab + 1 = 0.$$

Eliminating a, b from the above equation using (1) and (2), we get

$$(xy + yz + xz)(x - y) + y - z = 0.$$

This is the required integral surface.

5. Find the integral surface of the linear partial differential equation

$$xp + yq = z$$

which contains the straight line $y^2 = 4x, z = 1$.

Solution

The auxiliary equation is

$$\frac{dx}{x} = \frac{dy}{y} = \frac{dz}{z}.$$

Taking first two and last two ratios, we get

$$\frac{x}{y} = a \text{ and } \frac{y}{z} = b.$$

Using $y^2 = 4x, z = 1$ in the above results we get

$$b = 4a.$$

Eliminating a, b, we get $\quad \frac{y}{z} = 4\frac{x}{y}$ or, $y^2 = 4xz$.

This is the required integral surface.

6. Find the equation of the surface satisfying

$$4yzp + q + 2y = 0$$

and passing through $x + z = 3, y^2 + z^2 = 2$.

Solution

The auxiliary equation is $\quad \frac{dx}{4yz} = \frac{dy}{1} = \frac{dz}{-2y}.$

From the second and third ratios we get

$$2ydy + dz = 0.$$

Integrating, we get $\quad y^2 + z = a. \quad (1)$

Again, from the first and third ratios, we get

$$2zdz + dx = 0.$$

Integrating, we get $\quad z^2 + x = b. \quad (2)$

Adding (1) and (2), we get

$$y^2 + z + z^2 + x = a + b.$$

Using $x + z = 3$, $y^2 + z^2 = 2$, we get

$$a + b = 5.$$

Eliminating a, b, we get $\quad y^2 + z^2 + x + z = 5.$

This is the required integral surface.

EXERCISE

1. Find the integral surface of the linear partial differential equation $p + q = 1$, which passes through the curve $x = 0$, $y^2 = z$.
2. Find the integral surface of the linear partial differential equation $xp + yq = z$, which passes through the curve $x + y = 1$, $yz = 1$.
3. Find the integral surface of the linear partial differential equation $x(y^2 + z^2)p + y(x^2 + z^2)q = z(x^2 - y^2)$, which passes through the straight line $x + y = 0$, $z = 1$.
4. Find the integral surface of the linear partial differential equation $x(y^2 + z)p - y(x^2 + z)q = z(x^2 - y^2)$, which passes through the straight line $x + y = 0$, $z = 1$.
5. Find the integral surface of the linear partial differential equation $yzp + xzq = xy$, which passes through the curve $z^2 - y^2 = 1$, $x^2 - y^2 = 4$.
6. Find the integral surface of the linear partial differential equation $x^2p + y^2q + z^2 = 0$, which passes through the curve $xy = x + y$, $xz = 1$.
7. Find the integral surface of the linear partial differential equation $2y(z - 3)p + (2x - z)q = y(2x - 3)$, which passes through the curve $z = 0$, $x^2 + y^2 = 2x$.
8. Find the integral surface of the linear partial differential equation $(2xy - 1)p + (z - 2x^2)q = 2(x - yz)$ which passes through the line $x = 1$, $y = 0$.

ANSWERS

1. $(y - x)^2 = z - x$,
2. $(x + y)^2 = yz$,
3. $y(x^2 - y^2 - z^2) = xz$,
4. $x^2 + y^2 + 2xyz - 2z + 2 = 0$,
5. $x^2 - 2y^2 + z^2 = 5$,
6. $yz + 2xy + xz = 3xyz$,
7. $x^2 + y^2 - z^2 - 2x + 4z = 0$,
8. $x^2 + y^2 + z - zx - y = 1$.

14.6 Charpit's Method

Let a differential equation be given in the form

$$f(x, y, z, p, q) = 0. \tag{i}$$

We also have $\quad dz = pdx + qdy, \quad$ (ii)

Now we try to find a relation of the form

$$F(x, y, z, p, q) = 0 \tag{iii}$$

so that (ii) becomes integrable after putting the values of p, q obtained from (i) and (iii) in terms of x, y, z. Thus (iii) is such a relation that the values of p, q derived from it and the equation (i) will make (ii) integrable.

Thus z, p, q may be expressed as functions of x, y. Since, these values satisfy (i) and (iii), their derivatives with respect to x, y will vanish.

Now differentiating (i) and (iii) with respect to x, y, we get

$$\frac{\partial f}{\partial x}+\frac{\partial f}{\partial z}p+\frac{\partial f}{\partial p}\frac{\partial p}{\partial x}+\frac{\partial f}{\partial q}\frac{\partial q}{\partial x}=0 \tag{iv}$$

$$\frac{\partial F}{\partial x}+\frac{\partial F}{\partial z}p+\frac{\partial F}{\partial p}\frac{\partial p}{\partial x}+\frac{\partial F}{\partial q}\frac{\partial q}{\partial x}=0 \tag{v}$$

$$\frac{\partial f}{\partial y}+\frac{\partial f}{\partial z}q+\frac{\partial f}{\partial p}\frac{\partial p}{\partial y}+\frac{\partial f}{\partial q}\frac{\partial q}{\partial y}=0 \tag{vi}$$

and

$$\frac{\partial F}{\partial y}+\frac{\partial F}{\partial z}q+\frac{\partial F}{\partial p}\frac{\partial p}{\partial y}+\frac{\partial F}{\partial q}\frac{\partial q}{\partial y}=0 \tag{vii}$$

Eliminating $\frac{\partial p}{\partial x}$ from (iv) and (v), we get

$$\left(\frac{\partial f}{\partial x}+\frac{\partial f}{\partial z}p+\frac{\partial f}{\partial q}\frac{\partial q}{\partial x}\right)\frac{\partial F}{\partial p}-\left(\frac{\partial F}{\partial x}+\frac{\partial F}{\partial z}p+\frac{\partial F}{\partial q}\frac{\partial q}{\partial x}\right)\frac{\partial f}{\partial p}=0$$

or,

$$\frac{\partial f}{\partial x}\frac{\partial F}{\partial p}-\frac{\partial F}{\partial x}\frac{\partial f}{\partial p}+p\left(\frac{\partial f}{\partial z}\frac{\partial F}{\partial p}-\frac{\partial F}{\partial z}\frac{\partial f}{\partial p}\right)+\left(\frac{\partial f}{\partial q}\frac{\partial F}{\partial p}-\frac{\partial F}{\partial q}\frac{\partial f}{\partial p}\right)\frac{\partial q}{\partial x}=0. \tag{viii}$$

Again, eliminating $\frac{\partial q}{\partial y}$ from (vi) and (vii), we get

$$\frac{\partial f}{\partial y}\frac{\partial F}{\partial q}-\frac{\partial F}{\partial y}\frac{\partial f}{\partial q}+q\left(\frac{\partial f}{\partial z}\frac{\partial F}{\partial q}-\frac{\partial F}{\partial z}\frac{\partial f}{\partial q}\right)+\left(\frac{\partial f}{\partial p}\frac{\partial F}{\partial q}-\frac{\partial F}{\partial p}\frac{\partial f}{\partial q}\right)\frac{\partial p}{\partial y}=0. \tag{ix}$$

Adding (viii) and (ix) and using

$$\frac{\partial q}{\partial x}=\frac{\partial}{\partial x}\left(\frac{\partial z}{\partial y}\right)=\frac{\partial}{\partial y}\left(\frac{\partial z}{\partial x}\right)=\frac{\partial p}{\partial y}$$

and rearranging, we get

$$\left(\frac{\partial f}{\partial x}+p\frac{\partial f}{\partial z}\right)\frac{\partial F}{\partial p}+\left(\frac{\partial f}{\partial y}+q\frac{\partial f}{\partial z}\right)\frac{\partial F}{\partial q}-\left(p\frac{\partial f}{\partial p}+q\frac{\partial f}{\partial q}\right)\frac{\partial F}{\partial z}-\frac{\partial f}{\partial p}\frac{\partial F}{\partial x}-\frac{\partial f}{\partial q}\frac{\partial F}{\partial y}=0. \tag{x}$$

This is a linear equation, in which x, y, z, p, q are independent variables and F is the dependent variable.

Thus the auxiliary equations of the equation (i), which are known as *Charpit's auxiliary equations*, are

$$\frac{dp}{\frac{\partial f}{\partial x}+p\frac{\partial f}{\partial z}}=\frac{dq}{\frac{\partial f}{\partial y}+q\frac{\partial f}{\partial z}}=\frac{dz}{-p\frac{\partial f}{\partial p}-q\frac{\partial f}{\partial q}}=\frac{dx}{-\frac{\partial f}{\partial p}}=\frac{dy}{-\frac{\partial f}{\partial q}}=\frac{dF}{0}. \tag{xi}$$

Any integral of (i) will satisfy (x). Take an integral in the simplest form involving p or q or both for the equation (iii). Now solving (i), (iii) we obtain the values of p, q which make (ii) integrable.

This will give the complete integral of the equation (i) and from this, the general and singular solutions can be found.

The following steps are to be followed at the time of using Charpit's method:

(i) All the terms are to be placed in one side and denote the whole expression by f,

(ii) Write down the Charpit's auxiliary equation given by (xi),

(iii) Calculate the values of f_x, f_y, f_p, f_q using the expression of f and put these in the auxiliary equation,

(iv) Select any two ratios from the auxiliary equation which give the an equation involving p or q or both.

(v) This equation along with the given equation will give the values of p and q in terms of x, y, z.

(vi) These values of p, q are to be used in $dz = pdx + qdy$, which on integration gives the complete integral of the given equation.

It is to be mentioned in this connection that sometimes Charpit's auxiliary equation give $p = a$ and $q = b$. In such cases, the complete integral can be obtained by putting $p = a$, $q = b$ in the given equation (See example 3).

WORKED-OUT EXAMPLES

1. Solve by Charpit's method: $z = pq$.

Solution

Here $f(x, y, z, p, q) = z - pq$. Now the Charpit's auxiliary equations are:

$$\frac{dp}{\frac{\partial f}{\partial x}+p\frac{\partial f}{\partial z}}=\frac{dq}{\frac{\partial f}{\partial y}+q\frac{\partial f}{\partial z}}=\frac{dz}{-p\frac{\partial f}{\partial p}-q\frac{\partial f}{\partial q}}=\frac{dx}{-\frac{\partial f}{\partial p}}=\frac{dy}{-\frac{\partial f}{\partial q}}=\frac{dF}{0}$$

or, $$\frac{dp}{p}=\frac{dq}{q}=\frac{dz}{pq+pq}=\frac{dx}{q}=\frac{dy}{p}=\frac{dF}{0}.$$

Taking first two ratios, we get $\frac{dp}{p}=\frac{dq}{q}$. Integrating, we get

$$p = cq.$$

Using this in the given equation, we get

$$q=\sqrt{\frac{z}{c}}.$$

So, $$p=\sqrt{cz}.$$

Now from $$dz = pdx + qdy,$$

we get $$dz=\sqrt{cz}dx+\sqrt{\frac{z}{c}}dy.$$

or, $$\frac{dz}{\sqrt{z}}=\sqrt{c}dx+\frac{1}{\sqrt{c}}dy.$$

Integrating, we get $$2\sqrt{z}=\sqrt{c}x+\frac{1}{\sqrt{c}}y+d.$$

This is the complete integral.

2. Solve by Charpit's method: $zpq = p + q$.

Solution

Here $f(x, y, z, p, q) = zpq - p - q$. Now the Charpit's auxiliary equations are:

$$\frac{dp}{\frac{\partial f}{\partial x}+p\frac{\partial f}{\partial z}}=\frac{dq}{\frac{\partial f}{\partial y}+q\frac{\partial f}{\partial z}}=\frac{dz}{-p\frac{\partial f}{\partial p}-q\frac{\partial f}{\partial q}}=\frac{dx}{-\frac{\partial f}{\partial p}}=\frac{dy}{-\frac{\partial f}{\partial q}}=\frac{dF}{0}$$

or,
$$\frac{dp}{p^2q}=\frac{dq}{pq^2}=\frac{dz}{-zpq-zpq+p+q}=\frac{dx}{1-zq}=\frac{dy}{1-zp}=\frac{dF}{0}.$$

Taking first two ratios, we get $\frac{dp}{p}=\frac{dq}{q}$. Integrating, we get

$$p = cq.$$

Using this in the given equation, we get

$$q=\frac{c+1}{zc}.$$

So,
$$p=\frac{c+1}{z}.$$

Now from
$$dz = pdx + qdy,$$

we get
$$dz=\frac{c+1}{z}dx+\frac{c+1}{zc}dy.$$

or,
$$czdz = (c + 1)d\,(cx + y)$$

Integrating, we get
$$z^2=\frac{2(c+1)}{c}(cx+y)+d.$$

This is the complete integral.

3. Solve by Charpit's method: $z = px + qy + p^2 + q^2$.

Solution

Here $f(x, y, z, p, q) = px + qy + p^2 + q^2 - z$. Now the Charpit's auxiliary equations are

$$\frac{dp}{\frac{\partial f}{\partial x}+p\frac{\partial f}{\partial z}}=\frac{dq}{\frac{\partial f}{\partial y}+q\frac{\partial f}{\partial z}}=\frac{dz}{-p\frac{\partial f}{\partial p}-q\frac{\partial f}{\partial q}}=\frac{dx}{-\frac{\partial f}{\partial p}}=\frac{dy}{-\frac{\partial f}{\partial q}}=\frac{dF}{0}$$

or,
$$\frac{dp}{p-p}=\frac{dq}{q-q}=\frac{dz}{-p(x+2p)-q(y+2q)}=\frac{dx}{-(x+2p)}=\frac{dy}{-(y+2q)}=\frac{dF}{0}.$$

Taking first two ratios, we get $\frac{dp}{0}=\frac{dq}{0}$. Integrating, we get

$$p = a,\ q = b.$$

Using this in $\quad dz = pdx + qdy,$

we get $\quad dz = adx + bdy.$

Integrating we get $\quad z = ax + by + C.$

Using this in the given equation, we get $C = a^2 + b^2$.

Hence the complete integral is $z = ax + by + a^2 + b^2$.

4. Solve by Charpit's method: $2zx - px^2 - 2qxy + qp = 0$.

Solution

Here $f(x, y, z, p, q) = 2zx - px^2 - 2qxy + qp$. Now the Charpit's auxiliary equations are:

$$\frac{dp}{\frac{\partial f}{\partial x}+p\frac{\partial f}{\partial z}}=\frac{dq}{\frac{\partial f}{\partial y}+q\frac{\partial f}{\partial z}}=\frac{dz}{-p\frac{\partial f}{\partial p}-q\frac{\partial f}{\partial q}}=\frac{dx}{-\frac{\partial f}{\partial p}}=\frac{dy}{-\frac{\partial f}{\partial q}}=\frac{dF}{0}$$

$$\text{or, } \frac{dp}{2z-2qy}=\frac{dq}{-2qx+2qx}=\frac{dz}{px^2-pq+2qxy-pq}=\frac{dx}{x^2-q}=\frac{dy}{2xy-p}=\frac{dF}{0}.$$

From the second ratio we get $dq = 0$. Integrating, we get $q = c$.
Using this in the given equation we get

$$p=\frac{2x(z-cy)}{x^2-c}.$$

Now from $$dz = pdx + qdy,$$

we get $$dz=\frac{2x(z-cy)}{x^2-c}dx+cdy \text{ or, } \frac{d(z-cy)}{z-cy}=\frac{2xdx}{x^2-c}.$$

Integrating, we get $\ln(z-cy)=\ln(x^2-c)+\ln d$ or, $z=d(x^2-c)+cy$.
This is the complete integral.

5. Solve by Charpit's method: $pxy + pq + qy = yz$.

Solution

Here $f(x, y, z, p, q) = pxy + pq + qy - yz$. Now the Charpit's auxiliary equations are

$$\frac{dp}{\frac{\partial f}{\partial x}+p\frac{\partial f}{\partial z}}=\frac{dq}{\frac{\partial f}{\partial y}+q\frac{\partial f}{\partial z}}=\frac{dz}{-p\frac{\partial f}{\partial p}-q\frac{\partial f}{\partial q}}=\frac{dx}{-\frac{\partial f}{\partial p}}=\frac{dy}{-\frac{\partial f}{\partial q}}=\frac{dF}{0}$$

$$\text{or, } \frac{dp}{py-py}=\frac{dq}{px+q-z-qy}=\frac{dz}{-p(xy+q)-q(p+y)}=\frac{dx}{-(xy+q)}=\frac{dy}{-(p+y)}=\frac{dF}{0}.$$

Taking first two ratios, we get $dp = 0$. Integrating, we get $p = c$.
Using this in the given equation, we get

$$q=\frac{y(z-cx)}{y+c}.$$

Now from $$dz = pdx + qdy,$$

we get

$$dz=cdx+\frac{y(z-cx)}{y+c}dy \text{ or, } d(z-cx)=\frac{y(z-cx)}{y+c}dy \text{ or, } \frac{d(z-cx)}{z-cx}=\left(1-\frac{c}{y+c}\right)dy.$$

Integrating, we get

$$\ln(z-cx)=y-c\ln(y+c)+\ln d \text{ or, } \frac{(z-cx)(y+c)^c}{d}=e^y \text{ or, } z=cx+d(y+c)^{-c}e^y.$$

This is the complete integral.

6. Solve by Charpit's method: $(p^2 + q^2)y = qz$.

Solution

Here $f(x, y, z, p, q) = (p^2 + q^2)y - qz$. Now the Charpit's auxiliary equations are:

$$\frac{dp}{\frac{\partial f}{\partial x}+p\frac{\partial f}{\partial z}}=\frac{dq}{\frac{\partial f}{\partial y}+q\frac{\partial f}{\partial z}}=\frac{dz}{-p\frac{\partial f}{\partial p}-q\frac{\partial f}{\partial q}}=\frac{dx}{-\frac{\partial f}{\partial p}}=\frac{dy}{-\frac{\partial f}{\partial q}}=\frac{dF}{0}$$

or, $$\frac{dp}{-pq}=\frac{dq}{(p^2+q^2)-q^2}=\frac{dz}{-2(p^2+q^2)y+qz}=\frac{dx}{-2py}=\frac{dy}{-2qy+z}=\frac{dF}{0}.$$

Taking first two ratios, we get $pdp + qdq = 0$. Integrating, we get

$$p^2 + q^2 = c^2.$$

Using this in the given equation, we get $q=\frac{c^2y}{z}$.

So, $$p^2=c^2-\frac{c^4y^2}{z^2}=\frac{c^2}{z^2}(z^2-c^2y^2).$$

Now from $$dz = pdx + qdy,$$

we get $$dz=\frac{c}{z}\sqrt{z^2-c^2y^2}dx+\frac{c^2y}{z}dy$$

or, $$zdz-c^2ydy=c\sqrt{z^2-c^2y^2}dx$$

or, $$\frac{d(z^2-c^2y^2)}{\sqrt{z^2-c^2y^2}}=2cdx.$$

Integrating, we get $\sqrt{z^2-c^2y^2}=cx+d$ or, $z^2=c^2y^2+(cx+d)^2$.

This is the complete integral.

7. Solve by Charpit's method: $px + qy = qp$.

Solution

Here $f(x, y, z, p, q) = px + qy - pq$. Now the Charpit's auxiliary equations are:

$$\frac{dp}{\frac{\partial f}{\partial x}+p\frac{\partial f}{\partial z}}=\frac{dq}{\frac{\partial f}{\partial y}+q\frac{\partial f}{\partial z}}=\frac{dz}{-p\frac{\partial f}{\partial p}-q\frac{\partial f}{\partial q}}=\frac{dx}{-\frac{\partial f}{\partial p}}=\frac{dy}{-\frac{\partial f}{\partial q}}=\frac{dF}{0}$$

or, $$\frac{dp}{p}=\frac{dq}{q}=\frac{dz}{-p(x-q)-q(y-p)}=\frac{dx}{-(x-q)}=\frac{dy}{-(y-p)}=\frac{dF}{0}.$$

Taking first two ratios, we get $\frac{dp}{p}=\frac{dq}{q}$. Integrating, we get

$$\frac{p}{q}=c.$$

Using this in the given equation, we get

$$q=\frac{y+cx}{c}.$$

So, $p = y + cx.$

Now from $dz = pdx + qdy,$

we get

$$dz = (y + cx)dx + \frac{y + cx}{c}dy \text{ or, } cdz = (y + cx)d(y + cx).$$

Integrating, we get $z = \frac{(y+cx)^2}{2c} + d.$

This is the complete integral.

Now differentiating the complete integral with respect to d, we get $0 = 1$, which is absurd. Hence there is no singular solution.

EXERCISE

- Apply Charpit's method to solve the following equations:

1. $q = 3p^2,$
2. $px + qy + pq = 0,$
3. $(p^2 + q^2)x = pz,$
4. $(p^2 + q^2)y = qz,$
5. $p^2x + q^2y = z,$
6. $z^2(p^2z^2 + q^2) = 1,$
7. $p = (qy + z)^2,$
8. $q = (z + px)^2,$
9. $(p + q)(px + qy) = 1,$
10. $(x^2 - y^2)pq - (p^2 - q^2)xy = 1,$
11. $yzp^2 = q,$
12. $2x(z^2q^2 + 1) = pz,$
13. $2(z + px + qy) = yp^2,$
14. $z^2 = pqxy,$
15. $p + q = 3pq,$
16. $q = px + q^2,$
17. $px + 3qy = 2(z - x^2q^2).$

ANSWERS

1. $z = ax + 3a^2y + b,$
2. $az + \frac{1}{2}(ax + y)^2 = b,$
3. $z^2 = a^2x^2 + (ay + b)^2,$
4. $z^2 = a^2y^2 + (ax + b)^2,$
5. $\sqrt{(1 + a)z} = \sqrt{ax} + \sqrt{y} + b,$
6. $(a^2z^2 + 1)^3 = 9a^4(ax + y + b)^2,$
7. $yz = ax + 2\sqrt{ay} + b,$
8. $xz = ay + 2\sqrt{ax} + b,$
9. $z\sqrt{1 + a} = 2\sqrt{x + ay} + b,$
10. $z = \frac{a}{2}\ln(x^2 + y^2) + \frac{1}{a}\tan^{-1}\left(\frac{y}{x}\right) + b,$
11. $z^2 = 2ax + a^2y^2 + b,$
12. $z^2 = 2(a^2 + 1)x^2 + 2ay + b,$
13. $zy^3 = by^2 + axy - \frac{a^2}{4},$
14. $z = bx^a y^{\frac{1}{a}},$
15. $az = b - \frac{1}{2}(ax + y)^2,$
16. $z = a(1 - a)\ln x + ay + b,$
17. $z = a\left(a + \frac{y}{x}\right) + bx^2.$

Chapter 15

SERIES SOLUTION

15.1 Introduction

So far we have observed that in many cases, solutions of differential equations were obtained in terms of e^{mx}, $\sin x$, $\cos x$, $\ln x$, etc., each of which represents an infinite series. But it is found that the solution of some ordinary, homogeneous, linear, second order differential equations cannot be obtained using the method of solutions discussed so far. For such equations, we assume that it is possible to find a convergent series, whose terms are different powers of the independent variable as solutions.

Any second order, homogeneous, linear differential equation can be put as

$$P(x)\frac{d^2y}{dx^2} + Q(x)\frac{dy}{dx} + R(x)y = 0, \tag{i}$$

where $P(x)$, $Q(x)$, $R(x)$ are functions of x only. Let the above equation has a solution of the form

$$y = a_0x^{p_0} + a_1x^{p_1} + \ldots + a_kx^{p_k} \tag{ii}$$

which is either finite or a convergent series for some value(s) of x. We have to find the initial term, the relation between a_k's and the relation between p_k's.

If $y = u$ and $y = v$ be two linearly independent series solutions, finite or infinite, of the equation (i), then the general solution of the equation (i) will be

$$y = au + bv,$$

where a, b are arbitrary constants.

The form of the chosen series, as solution, depends on the nature of the point around which the solution is to be determined. In the next section we have discussed the nature of points in relation to the equation (i) as the nature of solution of the equation (i) depends on the nature of points.

15.2 Ordinary and Singular Points

A point $x = x_0$ (a finite value), is said to be an *ordinary point* of the differential equation given by (i), if $P(x_0) \neq 0$. On the other hand, if $P(x_0) = 0$, then the point $x = x_0$ is said to be a *singular point* of that differential equation. A differential equation may or may not have singular point.

Consider the equation, which is known as *Bessel's equation*:

$$x^2\frac{d^2y}{dx^2}+x\frac{dy}{dx}+(x^2-m^2)\,y=0. \qquad \text{(iii)}$$

Here, $P(0) = 0$, so $x = 0$ is a singular point of the equation. But the equation

$$\frac{d^2y}{dx^2}+x\frac{dy}{dx}+y=0$$

has no singular point.

A singular point $x = x_0$ of the differential equation (i) is said to be *regular singular point*, if both the limits

$$\lim_{x\to x_0}(x-x_0)\frac{Q(x)}{P(x)} \text{ and } \lim_{x\to x_0}(x-x_0)^2\frac{R(x)}{P(x)}$$

exist finitely.

On the other hand, if either of these limits is infinite, then $x = x_0$ is called an *irregular singular point*.

We have seen that $x = 0$ is a singular point of the equation (iii). Now we note that

$$\lim_{x\to 0} x\frac{Q(x)}{P(x)}=1 \text{ and } \lim_{x\to 0} x^2\frac{R(x)}{P(x)}=-m^2.$$

Hence, $x = 0$ is a regular singular point of the equation (15.2.1).

Now consider the equation

$$x^2\frac{d^2y}{dx^2}+\frac{dy}{dx}+y=0.$$

Here we see that $P(0) = 0$, but $\lim_{x\to 0} x\dfrac{Q(x)}{P(x)}$ does not exist finitely. Hence, $x = 0$ is an irregular singular point of the above equation.

15.3 Solution Around an Ordinary Point

If $x = x_0$ be an ordinary point of the differential equation (i), then we assume the general solution of the equation as a power series of the form

$$y=\Sigma_{n=0}^{\infty} c_n (x-x_0)^n \qquad \text{(iv)}$$

and calculate $\dfrac{dy}{dx}, \dfrac{d^2y}{dx^2}$. To determine the values of c_n, we put the values of $y, \dfrac{dy}{dx}, \dfrac{d^2y}{dx^2}$ in the equation (i) and compare the coefficients of different powers of $x - x_0$. This will give a relation between the coefficients, called reccurrance relation, which in turn will be helpful to determine the values of the coefficients in terms of c_0, c_1. Hence the general solution of the equation of (i) will be of the form

$$y = c_0 f(x) + c_1 g(x),$$

where $f(x)$, $g(x)$ are two independent solutions of the equation (i).

15.4 Solution Around a Regular Singular Point

If $x = x_0$ be a regular singular point of the differential equation (i), then rewriting the equation as

$$\frac{d^2y}{dx^2} = -\frac{Q(x)}{P(x)}\frac{dy}{dx} - \frac{R(x)}{P(x)}y,$$

we see that at the point $x = x_0$ both sides cannot be equal. Thus a power series of the form (i) will not satisfy the equationn if $x = x_0$ be a regular singular point of the differential equation. But if some powers of $x - x_0$ in the infinite series (i) be negative or fractional, then it can be a solution of the equation. This is the motivation of *Frobenius method* where, for a regular singular point $x = x_0$, the general solution is assumed as

$$y = \Sigma_{n=0}^{\infty} c_n(x - x_0)^{n+\alpha}, c_0 \neq 0, \quad \text{(v)}$$

where the constant α is to be determined. Since α can be negative or a fraction, the first term in the above series shows that the value of y, in general, is not finite at $x = x_0$. The method of solution is similar to that for an ordinary point, apart from that now α is to be determined.

Shifting the origin to the point $x = x_0$ i.e., putting $x = x_0 + z$ in the equation (i) and taking the general solution as

$$y = \Sigma_{n=0}^{\infty} c_n z^{n+\alpha}, c_0 \neq 0, \quad \text{(vi)}$$

we see that the equation (i) reduces to an identity.

Now equating to zero the coefficient of smallest power of z in the above identity, we get a quadratic equation in α, known as *indicial equation*. Roots of this equation will be of following types:

(i) Roots are unequal and do not differ by an integer,

(ii) Roots are unequal and differ by an integer making a coefficient of y infinite,

(iii) Roots are equal.

Next equate the coefficient of general power in the identity to zero, which gives a relation between the coefficients c_n, c_{n-2}, or c_n, c_{n-1}, known as *recurrence relation*. If the recurrence relation connects c_n, c_{n-2}, then the value of c_1 can be determined by equating to zero the coefficient of next highest power of z. On the other hand, if the recurrence relation connects c_n, c_{n-1}, then the above step may be skipped. Once the values of the coefficients are determined, the solution is obtained by putting these values in (vi).

The general solution of the equation (i) near a regular singular point $x = x_0$, i.e., $z = 0$ will be of the following forms:

Case I: If the roots k_1, k_2 of the indicial equation do not differ by an integer, then for these values we get two independent solutions u and v, and the general solution will be of the form

$$y = au + bv,$$

where a, b are arbitrary constants.

Case II: If the roots k_1, k_2 (say, $k_1 > k_2$) of the indicial equation differing by an integer and if one of the coefficients of y becomes infinite when $k = k_2$, we then change c_0 in y by $d_0(k - k_2)$. In this case, two independent solutions are obtained by putting $k = k_2$ in the modified form of y and $\frac{\partial y}{\partial k}$. Using $k = k_1$ in y, we get a constant multiple of the value of y obtained for $k = k_2$. Hence, we reject the solution obtained for $k = k_1$.

Case III: If the indicial equation has two equal roots, say k', k', we obtain two independent solutions by putting this value in y and $\frac{\partial y}{\partial k}$.

15.5 Solution for Large Values of Independent Variable

The concept of a regular singular point used so far is applicable if the point is finite. In some cases it may be required to know the nature of the singularity of a differential equation at infinity. To test the behavior of a solution of a differential equation at infinity, we put $z = \frac{1}{x}$ in the equation and then find the nature of the transformed equation, which is another differential equation having coefficients as polynomial in z, at $z = 0$.

WORKED-OUT EXAMPLES

1. Locate and classify the singular points of the equation

$$x^3(x-2)\frac{d^2y}{dx^2} - (x-2)\frac{dy}{dx} + 3xy = 0.$$

Solution

From $x^3(x - 2) = 0$, we get $x = 0, 2$. These are the singular points of the equation. Now

$$\lim_{x\to 2}(x-2)\left(-\frac{1}{x^3}\right) = 0 \text{ and } \lim_{x\to 2}(x-2)^2\left(\frac{3}{x^2(x-2)}\right) = 0.$$

Hence, $x = 2$ is a regular singular point of the equation. Again, we see that

$$\lim_{x\to 0} x\left(-\frac{1}{x^3}\right) = \infty \text{ and } \lim_{x\to 0} x^2\left(\frac{3}{x^2(x-2)}\right) = -\frac{3}{2}.$$

Hence, $x = 0$ is an irregular singular point of the equation.

2. Show that the point of infinity is a regular singular point of the equation

$$x^2y'' + (3x-1)\,y' + 3y = 0.$$

Solution

We put $z = \frac{1}{x}$, then

$$\frac{dy}{dx} = \frac{dy}{dz}\frac{dz}{dx} = -\frac{1}{x^2}\frac{dy}{dz} = -z^2\frac{dy}{dz}$$

and
$$\frac{d^2y}{dx^2} = -\frac{1}{x^2}\frac{d}{dz}\left(-z^2\frac{dy}{dz}\right) = -z^2\left[-z^2\frac{d^2y}{dz^2} - 2z\frac{dy}{dz}\right] = z^4\frac{d^2y}{dz^2} + 2z^3\frac{dy}{dz}.$$

Using these results in the given equation, we get

$$z^2\frac{d^2y}{dz^2}+2z\frac{dy}{dz}-\left(\frac{3}{z}-1\right)z^2\frac{dy}{dz}+3y=0 \text{ or, } z^2\frac{d^2y}{dz^2}+z(z-1)\frac{dy}{dz}+3y=0.$$

Clearly, $z=0$ is a singular point of the above equation. Now

$$\lim_{z\to 0} z\frac{z(z-1)}{z^2}=-1 \text{ and } \lim_{z\to 0} z^2\frac{z(z-1)}{z^2}=0.$$

Hence, $z=0$ is a regular singular point of the above equation, i.e., the point of infinity is a regular singular point of the given equation.

3. Solve the equation $(x^2+1)y''+xy'-xy=0$ near $x=0$.

Solution

Clearly, $x=0$ is an ordinary point of the equation. We assume the solution as

$$y=\sum_{n=0}^{\infty} c_n x^n. \tag{1}$$

Differentiating twice with respect to x, we get

$$y'=\sum_{n=0}^{\infty} nc_n x^{n-1} \text{ and } y''=\sum_{n=0}^{\infty} n(n-1)c_n x^{n-2}. \tag{2}$$

Using (1) and (2) in the given equation, we get

$$(x^2+1)\sum_{n=0}^{\infty} n(n-1)c_n x^{n-2}+x\sum_{n=0}^{\infty} nc_n x^{n-1}-x\sum_{n=0}^{\infty} c_n x^n=0$$

$$\text{or, } \sum_{n=0}^{\infty} n(n-1)c_n x^n+\sum_{n=0}^{\infty} n(n-1)c_n x^{n-2}+\sum_{n=0}^{\infty} nc_n x^n-\sum_{n=0}^{\infty} c_n x^{n+1}=0$$

$$\text{or, } \sum_{n=2}^{\infty} n(n-1)c_n x^n+\sum_{n=0}^{\infty}(n+2)(n+1)c_{n+2}x^n+\sum_{n=1}^{\infty} nc_n x^n-\sum_{n=1}^{\infty} c_{n-1}x^n=0$$

$$\text{or, } 2c_2+(6c_3+c_1-c_0)x+\sum_{n=2}^{\infty}[n(n-1)c_n+(n+2)(n+1)c_{n+2}+nc_n-c_{n-1}]x^n=0.$$

This is an identity in x, so we must have

$$c_2=0,\ 6c_3+c_1-c_0=0 \text{ and}$$

$$n(n-1)c_n+(n+2)(n+1)c_{n+2}+nc_n-c_{n-1}=0 \text{ for } n\geq 2.$$

From the last relation we get

$$c_{n+2}=\frac{c_{n-1}-n^2c_n}{(n+2)(n+1)},\ n\geq 2.$$

Putting $n=2$, we get $c_4=\dfrac{c_1-4c_2}{12}=\dfrac{c_1}{12}$.

Putting $n=3$, we get $c_5=\dfrac{c_2-9c_3}{20}=-\dfrac{9}{20}\dfrac{c_0-c_1}{6}=-\dfrac{3}{40}(c_0-c_1)$.

Proceeding in this way we can determine all the coefficients in terms of c_0, c_1. Thus the general solution is

$$y=c_0+c_1x+\frac{c_0-c_1}{6}x^3+\frac{c_1}{12}x^4-\frac{3}{40}(c_0-c_1)x^5+\frac{1}{180}(c_0-9c_1)x^6+\cdots.$$

$$=c_0\left(1+\frac{1}{6}x^3-\frac{3}{40}x^5+\frac{1}{180}x^6+\cdots.\right)+c_1x\left(1-\frac{1}{6}x^2+\frac{1}{12}x^3+\frac{3}{40}x^4-\frac{1}{20}x^5+\cdots\right)$$

4. Solve the equation $y'' + xy' + x^2y = 0$ near $x = 0$.

Solution

Clearly, $x = 0$ is an ordinary point of the equation. We assume the solution as

$$y = \sum_{n=0}^{\infty} c_n x^n. \tag{1}$$

Differentiating twice with respect to x, we get

$$y' = \sum_{n=0}^{\infty} nc_n x^{n-1} \text{ and } y'' = \sum_{n=0}^{\infty} n(n-1)c_n x^{n-2} \tag{2}$$

Using (1) and (2) in the given equation we get

$$\sum_{n=0}^{\infty} n(n-1)c_n x^{n-2} + x\sum_{n=0}^{\infty} nc_n x^{n-1} + x^2\sum_{n=0}^{\infty} c_n x^n = 0$$

or, $\sum_{n=2}^{\infty} n(n-1)c_n x^{n-2} + \sum_{n=1}^{\infty} nc_n x^n + \sum_{n=0}^{\infty} c_n x^{n+2} = 0$

or, $\sum_{n=0}^{\infty}(n+2)(n+1)c_{n+2}x^n + \sum_{n=1}^{\infty} nc_n x^n + \sum_{n=2}^{\infty} c_{n-2}x^n = 0$

or, $c_2 + (6c_3 + c_1)x + \sum_{n=2}^{\infty}[(n+2)(n+1)c_{n+2} + nc_n + c_{n-2}]x^n = 0.$

This is an identity in x, so we must have

$$c_2 = 0,\ 6c_3 + c_1 = 0 \text{ and}$$

$$(n+2)(n+1)c_{n+2} + nc_n + c_{n-2} = 0, \text{ for } n \geq 2.$$

From the last relation we get

$$c_{n+2} = -\frac{nc_n + c_{n-2}}{(n+2)(n+1)},\ n \geq 2.$$

Putting $n = 2$, we get $c_4 = -\dfrac{2c_2 + c_0}{12} = -\dfrac{c_0}{12}$.

Putting $n = 3$, we get $c_5 = -\dfrac{3c_3 + c_1}{20} = -\dfrac{c_1}{40}$.

Putting $n = 4$, we get $c_6 = -\dfrac{4c_4 + c_2}{30} = \dfrac{c_0}{90}$.

Proceeding in this way we can determine all the coefficients in terms of c_0, c_1. Thus the general solution is

$$y = c_0 + c_1x - \frac{c_1}{6}x^3 - \frac{c_0}{12}x^4 - \frac{c_1}{40}x^5 + \frac{c_0}{90}x^6 + \cdots$$

$$= c_0\left(1 - \frac{1}{12}x^4 + \frac{1}{90}x^6 + \cdots\right) + c_1x\left(1 - \frac{1}{6}x^2 - \frac{1}{40}x^4 + \cdots\right).$$

5. Solve the equation $2x^2y'' - xy' + (1 - x^2)y = 0$.

Solution

Here $x = 0$ is a regular singular point. We assume the solution of the equation as

$$y = \sum_{n=0}^{\infty} c_n x^{n+\alpha}. \tag{1}$$

Differentiating twice with respect to x we get

$$y' = \sum_{n=0}^{\infty}(n+\alpha)c_n x^{n+\alpha-1} \text{ and } y'' = \sum_{n=0}^{\infty}(n+\alpha)(n+\alpha-1)c_n x^{n+\alpha-2}. \quad (2)$$

Using (1) and (2) in the given equation, we get

$$2\sum_{n=0}^{\infty}(n+\alpha)(n+\alpha-1)c_n x^{n+\alpha} - \sum_{n=0}^{\infty}(n+\alpha)c_n x^{n+\alpha} + \sum_{n=0}^{\infty} c_n x^{n+\alpha} - \sum_{n=0}^{\infty} c_n x^{n+\alpha+2} = 0$$

or, $$[2\alpha(\alpha-1)c_0 - \alpha c_0 + c_0]x^{\alpha} + [2\alpha(\alpha+1)c_1 - (\alpha+1)c_1 + c_1]x^{\alpha+1} + \sum_{n=2}^{\infty}\left[2(n+\alpha)(n+\alpha-1)c_n - (n+\alpha)c_n + c_n - c_{n-2}\right]x^{n+\alpha} = 0.$$

This is an identity in x. Equating the coefficient of the term of the lowest power of x to zero, we get

$$2\alpha(\alpha-1)c_0 - \alpha c_0 + c_0 = 0$$

or, $$(2\alpha-1)(\alpha-1) = 0 \text{ as } c_0 \neq 0.$$

Two roots of the indicial equation are different and differ by not an integer.

From $$2\alpha(\alpha+1)c_1 - (\alpha+1)c_1 + c_1 = 0,$$

we see that $$c_1 = 0.$$

We also get $$2(n+\alpha)(n+\alpha-1)c_n - (n+\alpha)c_n + c_n - c_{n-2} = 0 \text{ for } n \geq 2.$$

From the last relation we get

$$c_n = \frac{c_{n-2}}{(2n+2\alpha-1)(n+\alpha-1)}, n \geq 2.$$

Since $c_1 = 0$, using the recurrence relation, we see that $c_{2k+1} = 0$ for $k = 1, 2, 3, \cdots$. Putting $n = 2, 4, 6, \ldots$ in the recurrence relation, we get

$$c_2 = \frac{c_0}{(2\alpha+3)(\alpha+1)}$$

$$c_4 = \frac{c_2}{(2\alpha+7)(\alpha+3)} = \frac{c_0}{(2\alpha+3)(2\alpha+7)(\alpha+1)(\alpha+3)}, \cdots$$

Thus $$y = x^{\alpha}\left[c_0 + \frac{c_0}{(2\alpha+3)(\alpha+1)}x^2 + \frac{c_0}{(2\alpha+3)(2\alpha+7)(\alpha+1)(\alpha+3)}x^4 + \cdots\right]$$

For $\alpha = 1, \frac{1}{2}$ two independent solutions. Hence the general solution is

$$y = A\sqrt{x}\left[1 + \frac{x^2}{2\cdot 3} + \frac{x^4}{2\cdot 3\cdot 4\cdot 7} + \cdots\right] + Bx\left[1 + \frac{x^2}{2\cdot 5} + \frac{x^4}{2\cdot 4\cdot 5\cdot 9} + \cdots\right].$$

6. Solve the equation $(2x + x^3)y'' - y' - 6xy = 0$.

Solution

Here $x = 0$ is a regular singular point. We assume the solution of the equation as

$$y = \sum_{n=0}^{\infty} c_n x^{n+\alpha}. \quad (1)$$

Differentiating twice with respect to x, we get

$$y' = \sum_{n=0}^{\infty}(n+\alpha)c_n x^{n+\alpha-1} \text{ and } y'' = \sum_{n=0}^{\infty}(n+\alpha)(n+\alpha-1)c_n x^{n+\alpha-2}. \quad (2)$$

Using (1) and (2) in the given equation, we get

$$2\sum_{n=0}^{\infty}(n+\alpha)(n+\alpha-1)c_n x^{n+\alpha-1} + \sum_{n=0}^{\infty}(n+\alpha)(n+\alpha-1)c_n x^{n+\alpha+1} -$$

$$\sum_{n=0}^{\infty}(n+\alpha)c_n x^{n+\alpha-1} - 6\sum_{n=0}^{\infty}c_n x^{n+\alpha+1} = 0$$

or, $\sum_{n=0}^{\infty}[2(n+\alpha)(n+\alpha-1)-(n+\alpha)]c_n x^{n+\alpha-1} +$

$$\sum_{n=1}^{\infty}(n+\alpha+2)(n+\alpha-3)c_n x^{n+\alpha+1} = 0$$

or, $\sum_{n=0}^{\infty}(n+\alpha)(2n+2\alpha-3)c_n x^{n+\alpha-1} + \sum_{n=0}^{\infty}\left[(n+\alpha)(n+\alpha-1)-6\right]c_n x^{n+\alpha+1} = 0.$

This is an identity in x. Equating the coefficient of the term of the lowest power of x to zero, we get

$$\alpha(2\alpha-3) = 0 \text{ as } c_0 \neq 0.$$

Two roots of the indicial equation are different and differ by not an integer.
Equating the coefficient of the term of the next highest power of x to zero, we get

$$(2\alpha-1)(\alpha+1)c_1 = 0 \Rightarrow c_1 = 0.$$

Equating the coefficient of $x^{n+\alpha-1}$ to zero, we get

$$(n+\alpha)(2n+2\alpha-3)c_n + (n+\alpha)(n+\alpha-5)c_{n-2} = 0$$

or, $c_n = -\dfrac{n+\alpha-5}{2n+2\alpha-3}c_{n-2}.$

Since $c_1 = 0$, using the recurrence relation we see that $c_{2k+1} = 0$ for $1,2,3,\cdots$.
Putting $n = 2,4,6,\cdots$ in the recurrence relation, we get

$$c_2 = -\frac{\alpha-3}{2\alpha+1}c_0$$

$$c_4 = -\frac{\alpha-1}{2\alpha+5}c_2 = \frac{(\alpha-1)(\alpha-3)}{(2\alpha+1)(2\alpha+5)}c_0, \cdots.$$

Thus $\quad y = x^{\alpha}\left[c_0 - \dfrac{\alpha-3}{2\alpha+1}c_0 x^2 + \dfrac{(\alpha-1)(\alpha-3)}{(2\alpha+1)(2\alpha+5)}c_0 x^4 + \cdots\right].$

For $\alpha = 0, \dfrac{3}{2}$ two independent solutions. Hence the general solution is

$$y = A\left[1+3x^2+\frac{3}{5}x^4+\cdots\right] + Bx^{3/2}\left[1+\frac{3}{8}x^2-\frac{3}{128}x^4+\cdots\right].$$

7. Solve the equation $x(1-x)y'' - 3xy' - y = 0$ near $x = 0$.

Solution

Here $x = 0$ is a regular singular point. We assume the solution of the equation as

$$y = \sum_{n=0}^{\infty}c_n x^{n+\alpha}. \quad (1)$$

Differentiating twice with respect to x we get

$$y' = \sum_{n=0}^{\infty}(n+\alpha)c_n x^{n+\alpha-1} \text{ and } y'' = \sum_{n=0}^{\infty}(n+\alpha)(n+\alpha-1)c_n x^{n+\alpha-2}. \quad (2)$$

Using (1) and (2) in the given equation we get

$$\sum_{n=0}^{\infty}(n+\alpha)(n+\alpha-1)c_n x^{n+\alpha-1} - \sum_{n=0}^{\infty}(n+\alpha)(n+\alpha-1)c_n x^{n+\alpha} - 3\sum_{n=0}^{\infty}(n+\alpha)c_n x^{n+\alpha} - \sum_{n=0}^{\infty} c_n x^{n+\alpha} = 0$$

or, $\sum_{n=0}^{\infty}(n+\alpha)(n+\alpha-1)\, c_n x^{n+\alpha-1} - \sum_{n=0}^{\infty}[(n+\alpha)(n+\alpha-1) + 3(n+\alpha)+1]c_n x^{n+\alpha} = 0$

or, $\sum_{n=0}^{\infty}(n+\alpha)(n+\alpha-1)c_n x^{n+\alpha-1} - \sum_{n=0}^{\infty}(n+\alpha+1)^2\, c_n x^{n+\alpha} = 0.$

This is an identity in x. Equating the coefficient of the term of the lowest power of x to zero, we get

$$\alpha(\alpha-1) = 0 \text{ as } c_0 \neq 0.$$

Two roots of the indicial equation are different and differ by an integer.

Equating the coefficient of $x^{n+\alpha-1}$ to zero, we get

$$(n+\alpha)(n+\alpha-1)c_n - (n+\alpha)^2 c_{n-1} = 0$$

or, $$c_n = \frac{n+\alpha}{n+\alpha-1}c_{n-1},\ n = 1, 2, 3, \cdots.$$

Thus $$c_1 = \frac{\alpha+1}{\alpha}c_0,$$

$$c_2 = \frac{\alpha+2}{\alpha+1}c_1 = \frac{\alpha+2}{\alpha}c_0,$$

$$c_3 = \frac{\alpha+3}{\alpha+2}c_2 = \frac{\alpha+3}{\alpha}c_0 \text{ and so on.}$$

Hence, $$y = x^{\alpha}\left[c_0 + \frac{\alpha+1}{\alpha}c_0 x + \frac{\alpha+2}{\alpha}c_0 x^2 + \frac{\alpha+3}{\alpha}c_0 x^3 + \cdots\right].$$

For $\alpha = 0$, the coefficients become infinite. To get rid of this difficulty, we put $c_0 = \alpha A$. So, the above series becomes

$$y = Ax^{\alpha}[\alpha + (\alpha+1)x + (\alpha+2)x^2 + (\alpha+3)x^3 + \cdots]. \quad (3)$$

For $\alpha = 0$, now we get

$$y = A[x + 2x^2 + 3x^3 + \cdots] = Au \text{ (say)}. \quad (4)$$

For the second independent solution, we differentiate (3) partially with respect to α and get

$$\frac{\partial y}{\partial \alpha} = Ax^{\alpha}\ln x[\alpha + (\alpha+1)x + (\alpha+2)x^2 + (\alpha+3)x^3 + \cdots] + Ax^{\alpha}[1 + x + x^2 + x^3 + \cdots].$$

Putting $\alpha = 0$ and changing A by B, we get

$$\left[\frac{\partial y}{\partial \alpha}\right]_{\alpha=0} = B\ln x[x+2x^2+3x^3+\cdots]+B[1+x+x^2+x^3+\cdots]$$

$$= Bu\ln x + B[1+x+x^2+x^3+\cdots] = Bv \text{ (say)}. \tag{5}$$

So the general solution is

$$y = Au + Bv,$$

where A, B are arbitrary constants and u, v are given by (4) and (5) respectively.

8. Solve the equation $x^2y''+xy'+(x^2-1)y=0$ (Bessel's equation of first order) near $x = 0$.

Solution

Here $x = 0$ is a regular singular point. We assume the solution of the equation as

$$y = \sum_{n=0}^{\infty} c_n x^{n+\alpha}. \tag{1}$$

Differentiating twice with respect to x we get

$$y' = \sum_{n=0}^{\infty}(n+\alpha)c_n x^{n+\alpha-1} \text{ and } y'' = \sum_{n=0}^{\infty}(n+\alpha)(n+\alpha-1)c_n x^{n+\alpha-2}. \tag{2}$$

Using (1) and (2) in the given equation, we get

$$\sum_{n=0}^{\infty}(n+\alpha)(n+\alpha-1)c_n x^{n+\alpha} + \sum_{n=0}^{\infty}(n+\alpha)c_n x^{n+\alpha} +$$

$$\sum_{n=0}^{\infty} c_n x^{n+\alpha+2} - \sum_{n=0}^{\infty} c_n x^{n+\alpha} = 0$$

or, $\sum_{n=0}^{\infty}[(n+\alpha)(n+\alpha-1)+(n+\alpha)-1]c_n x^{n+\alpha} + \sum_{n=0}^{\infty} c_n x^{n+\alpha+2} = 0$

or, $\sum_{n=0}^{\infty}(n+\alpha+1)(n+\alpha-1)c_n x^{n+\alpha} + \sum_{n=0}^{\infty} c_n x^{n+\alpha+2} = 0.$

This is an identity in x. Equating the coefficient of the term of the lowest power of x to zero, we get

$$(\alpha+1)(\alpha-1) = 0 \text{ as } c_0 \neq 0.$$

Two roots of the indicial equation are different and differ by an integer.

Equating the coefficient of the term of the next highest power of x to zero we get

$$\alpha(\alpha+2)c_1 = 0 \Rightarrow c_1 = 0.$$

Equating the coefficient of $x^{n+\alpha}$ to zero we get

$$(n+\alpha+1)(n+\alpha-1)c_n + c_{n-2} = 0 \text{ or, } c_n = -\frac{1}{(n+\alpha+1)(n+\alpha-1)}c_{n-2},\ n \geq 2.$$

Since $c_1 = 0$, the above recurrence relation gives that $c_{2k+1} = 0$ for $k = 1, 2, 3, \cdots$.

Now putting $n = 2, 4, 6, \cdots$, we get

$$c_2 = -\frac{1}{(\alpha+1)(\alpha+3)}c_0,$$

$$c_4 = -\frac{1}{(\alpha+3)(\alpha+5)}c_2 = \frac{1}{(\alpha+1)(\alpha+3)^2(\alpha+5)}c_0,$$

$$c_6 = -\frac{1}{(\alpha+5)(\alpha+7)}c_4 = -\frac{1}{(\alpha+1)(\alpha+3)^2(\alpha+5)^2(\alpha+7)}c_0 \text{ and so on.}$$

So, $$y = c_0 x^\alpha\left[1 - \frac{1}{(\alpha+1)(\alpha+3)}x^2 + \frac{1}{(\alpha+1)(\alpha+3)^2(\alpha+5)}x^4 - \frac{1}{(\alpha+1)(\alpha+3)^2(\alpha+5)^2(\alpha+7)}x^6 + \cdots\right].$$

For $\alpha = -1$, the coefficients become infinite. To get rid of this difficulty, we put $c_0 = (\alpha+1)A$. So, the above series becomes

$$y = Ax^\alpha\left[\alpha+1 - \frac{1}{\alpha+3}x^2 + \frac{1}{(\alpha+3)^2(\alpha+5)}x^4 - \frac{1}{(\alpha+3)^2(\alpha+5)^2(\alpha+7)}x^6 + \cdots\right]. \quad (3)$$

For the second independent solution, we differentiate (3) partially with respect to α and get

$$\frac{\partial y}{\partial \alpha} = Ax^\alpha \ln x\left[\alpha+1 - \frac{1}{\alpha+3}x^2 + \frac{1}{(\alpha+3)^2(\alpha+5)}x^4 - \frac{1}{(\alpha+3)^2(\alpha+5)^2(\alpha+7)}x^6 + \cdots\right]$$

$$+Ax^\alpha\left[1 + \frac{1}{(\alpha+3)^2}x^2 - \left\{\frac{2}{(\alpha+3)^3(\alpha+5)} + \frac{1}{(\alpha+3)^2(\alpha+5)^2}\right\}x^4 + \left\{\frac{1}{(\alpha+3)^2(\alpha+5)^2(\alpha+7)^2} + \frac{2}{(\alpha+3)^2(\alpha+5)^3(\alpha+7)} + \frac{2}{(\alpha+3)^3(\alpha+5)^2(\alpha+7)}\right\}x^6 + \cdots\right]. \quad (4)$$

Putting $\alpha = -1$ in (3) and (4) and changing A by B in (4) we get

$$[y]_{\alpha=-1} = \frac{A}{x}\left[-\frac{1}{2}x^2 + \frac{1}{2^2.4}x^4 - \frac{1}{2^2.4^2 6}x^6 + \cdots\right] = Au \text{ (say)}$$

and $$\left[\frac{\partial y}{\partial \alpha}\right]_{\alpha=-1} = \frac{B}{x}\ln x\left[-\frac{1}{2}x^2 + \frac{1}{2^2.4}x^4 - \frac{1}{2^2.4^2.6}x^6 + \cdots\right] +$$

$$\frac{B}{x}\left[1 + \frac{x^2}{2^2} - \frac{1}{2^2.4}\left(\frac{2}{2}+\frac{1}{4}\right)x^4 + \frac{1}{2^2.4^2.6}\left(\frac{2}{2}+\frac{2}{4}+\frac{1}{6}\right)x^6 + \cdots\right] = Bv \text{ (say)}.$$

So the general solution is $y = Au + Bv$, where A, B are arbitrary constants.

9. Solve the equation $xy'' + y' + xy = 0$ (Bessel's equation of zero order) near $x = 0$.

Solution

Here $x = 0$ is a regular singular point. We assume the solution of the equation as

$$y = \Sigma_{n=0}^{\infty} c_n x^{n+\alpha}. \quad (1)$$

Differentiating twice with respect to x, we get

$$y' = \Sigma_{n=0}^{\infty}(n+\alpha)c_n x^{n+\alpha-1} \text{ and } y'' = \Sigma_{n=0}^{\infty}(n+\alpha)(n+\alpha-1)c_n x^{n+\alpha-2}. \quad (2)$$

Using (1) and (2) in the given equation we get

$$\sum_{n=0}^{\infty}(n+\alpha)(n+\alpha-1)c_n x^{n+\alpha-1}+\sum_{n=0}^{\infty}(n+\alpha)c_n x^{n+\alpha-1}+\sum_{n=0}^{\infty}c_n x^{n+\alpha+1}=0$$

or, $\sum_{n=0}^{\infty}(n+\alpha)^2 c_n x^{n+\alpha-1}+\sum_{n=0}^{\infty}c_n x^{n+\alpha+1}=0.$

This is an identity in x. Equating the coefficient of the term of the lowest power of x to zero, we get

$$\alpha^2=0 \text{ as } c_0\neq 0.$$

Two roots of the indicial equation are equal.

Equating the coefficient of the term of the next highest power of x to zero, we get

$$(\alpha+1)^2 c_1=0\Rightarrow c_1=0.$$

Equating the coefficient of $x^{n+\alpha-1}$ to zero, we get

$$(n+\alpha)^2 c_n+c_{n-2}=0 \text{ or, } c_n=-\frac{1}{(n+\alpha)^2}c_{n-2},\ n\geq 2.$$

Since, $c_1=0$, the above recurrence relation gives that $c_{2k+1}=0$ for $k=1,2,3,\cdots$.

Now putting $n=2,4,6,\cdots$, we get

$$c_2=-\frac{1}{(\alpha+2)^2}c_0,\quad c_4=-\frac{1}{(\alpha+4)^2}c_2=\frac{1}{(\alpha+2)^2(\alpha+4)^2}c_0,$$

$$c_6=-\frac{1}{(\alpha+6)^2}c_4=-\frac{1}{(\alpha+2)^2(\alpha+4)^2(\alpha+6)^2}c_0 \text{ and so on.}$$

So,

$$y=c_0x^{\alpha}\left[1-\frac{1}{(\alpha+2)^2}x^2+\frac{1}{(\alpha+2)^2(\alpha+4)^2}x^4-\frac{1}{(\alpha+2)^2(\alpha+4)^2(\alpha+6)^2}x^6+\cdots\right]. \quad (3)$$

Putting $\alpha=0$ and $c_0=A$ we get

$$[y]_{\alpha=0}=A\left[1-\frac{1}{2^2}x^2+\frac{1}{2^2.4^2}x^4-\frac{1}{2^2.4^2.6^2}x^6+\cdots\right]=Au \text{ (say).}$$

For the second independent solution, we differentiate (3) partially with respect to α and get

$$\frac{\partial y}{\partial\alpha}=c_0x^{\alpha}\ln x\left[1-\frac{1}{(\alpha+2)^2}x^2+\frac{1}{(\alpha+2)^2(\alpha+4)^2}x^4-\frac{1}{(\alpha+2)^2(\alpha+4)^2(\alpha+6)^2}x^6+\cdots\right]+$$

$$c_0x^{\alpha}\left[\frac{2}{(\alpha+2)^3}x^2-\left\{\frac{2}{(\alpha+2)^3(\alpha+4)^2}+\frac{2}{(\alpha+4)^3(\alpha+2)^2}\right\}x^4+\left\{\frac{2}{(\alpha+2)^3(\alpha+4)^2(\alpha+6)^2}+\frac{2}{(\alpha+2)^2(\alpha+4)^3(\alpha+6)^2}+\frac{2}{(\alpha+2)^2(\alpha+4)^2(\alpha+6)^3}\right\}x^6+\cdots\right].$$

Putting $\alpha = 0$ and $c_0 = B$, we get

$$\left[\frac{\partial y}{\partial \alpha}\right]_{\alpha=0} B\ln x\left[1-\frac{1}{2^2}x^2+\frac{1}{2^2.4^2}x^4-\frac{1}{2^2.4^2.6^2}x^6+\cdots\right]$$

$$B\left[\frac{x^2}{2^2}-\left\{\frac{2}{2^3.4^2}+\frac{2}{4^3.2^2}\right\}x^4+\right.$$

$$\left.\left\{\frac{2}{2^3.4^2.6^2}+\frac{2}{2^2.4^3.6^2}+\frac{2}{2^2.4^2.6^3}\right\}x^6+\cdots\right]=Bv \text{ (say)}.$$

So the general solution is

$$y = Au + Bv,$$

where A, B are arbitrary constants.

10. Solve the equation $x(x-1)y''+(3x-1)y'+y=0$, near $x=0$.

Solution

Here $x=0$ is a regular singular point. We assume the solution of the equation as

$$y=\sum_{n=0}^{\infty} c_n x^{n+\alpha}. \tag{1}$$

Differentiating twice with respect to x, we get

$$y'=\sum_{n=0}^{\infty}(n+\alpha)c_n x^{n+\alpha-1} \text{ and } y''=\sum_{n=0}^{\infty}(n+\alpha)(n+\alpha-1)c_n x^{n+\alpha-2}. \tag{2}$$

Using (1) and (2) in the given equation, we get

$$\sum_{n=0}^{\infty}(n+\alpha)(n+\alpha-1)c_n x^{n+\alpha}-\sum_{n=0}^{\infty}(n+\alpha)(n+\alpha-1)c_n x^{n+\alpha-1}$$

$$+3\sum_{n=0}^{\infty}(n+\alpha)c_n x^{n+\alpha}-\sum_{n=0}^{\infty}(n+\alpha)c_n x^{n+\alpha-1}+\sum_{n=0}^{\infty}c_n x^{n+\alpha}=0$$

or, $\sum_{n=0}^{\infty}(n+\alpha+1)^2 c_n x^{n+\alpha}-\sum_{n=0}^{\infty}(n+\alpha)^2 c_n x^{n+\alpha-1}=0.$

This is an identity in x. Equating the coefficient of the term of the lowest power of x to zero, we get

$$\alpha^2=0 \text{ as } c_0\neq 0.$$

Two roots of the indicial equation are equal.

Equating the coefficient of the term of the next highest power of x to zero, we get

$$(\alpha+1)^2(c_0-c_1)=0 \Rightarrow c_1=c_0.$$

Equating the coefficient of $x^{n+\alpha-1}$ to zero we get

$$(n+\alpha)^2c_n-(n+\alpha)^2c_{n-1}=0 \text{ or, } c_n=c_{n-1}, n\geq 1.$$

Thus, $\quad c_0=c_1=\cdots=c_n=\cdots.$

So, $\quad y=c_0x^{\alpha}[1+x+x^2+x^3+x^4+\cdots]. \qquad (3)$

Putting $\alpha = 0$ and $c_0 = A$, we get

$$[y]_{\alpha=0}=A[1+x+x^2+x^3+x^4+\cdots\ldots]=A(1-x)^{-1}.$$

For the second independent solution, we differentiate (3) partially with respect to α and get

$$\frac{\partial y}{\partial \alpha} = c_0 x^{\alpha} \ln x (1-x)^{-1}.$$

Putting $\alpha = 0$ and $c_0 = B$, we get

$$\left[\frac{\partial y}{\partial \alpha}\right]_{\alpha=0} = B \ln x (1-x)^{-1}.$$

So the general solution is

$$y = A(1-x)^{-1} + B \ln x (1-x)^{-1},$$

where A, B are arbitrary constants.

EXERCISE

1. Locate and classify the singular points of the equations:

(i) $x(2-x)y'' + (2-x)y' - 2xy = 0$, (ii) $(x^2 - 3x + 2)y'' + xy' - 3y = 0$,

(iii) $(2x-1)xy'' - (x+1)y' + y = 0$, (iv) $x^3(x^2-1)y'' + 2x^4y' + 4y = 0$,

(v) $2y'' + xy' + 4y = 0$.

2. Solve the following equations near $x = 0$:

(i) $(1-x^2)y'' - xy' + 4y = 0$, (ii) $(1-x^2)y'' - 6xy' - 4y = 0$,

(iii) $y'' + x^2y' - 4xy = 0$, (iv) $(1-x^2)y'' + 2xy' - y = 0$.

3. Solve the equation $y'' + (x-1)^2 y' - 4(x-1)y = 0$, near $x = 1$.

4. Solve the equation $y'' + (x-2)y = 0$, near $x = 2$.

5. Solve the equation $x^2y'' - xy' + (x^2+1)y = 0$, near $x = 0$.

6. Solve the equation $(x - x^2)y'' + (1-5x)y' - 4y = 0$ near $x = 0$.

7. Solve the equation $x^2y'' + xy' + (x^2 - 4)y = 0$, near $x = 0$.

8. Solve the equation $(2x + x^3)y'' - y' - 6xy = 0$, near $x = 0$.

9. Solve the equation $9x(1-x)y'' - 12y' + 4y = 0$, near $x = 0$.

10. Solve the equation $(x - x^2)y'' + (1-x)y' - y = 0$, near $x = 0$.

11. Solve the equation $2x^2y'' - xy' + (1-x^2)y = 0$, near $x = 0$.

12. Solve the equation $2x^2(x-1)y'' + x(3x+1)y' - 2y = 0$, near $x = \infty$.

13. Solve the equation $x^2y'' + (3x-1)y' + y = 0$, near $x = \infty$.

14. Solve the equation $(x - x^2)y'' - 3y' + 2y = 0$, near $x = \infty$.

ANSWERS

1. (i) $x = 0, 2$ are regular singular points,
(ii) $x = 1, 2$ are regular singular points,
(iii) $x = 0, \frac{1}{2}$ are regular singular points,
(iv) $x = -1, 1$ are regular singular points and $x = 0$ is an irregular singular point,
(v) No singular point.

2. (i) $y = A(1-2x^2) + B(x - \frac{1}{2}x^3 - \frac{1}{8}x^5 + \cdots)$,

(ii) $y = A(1+2x^2+3x^4+4x^6+\cdots) + B\left(x + \frac{5}{3}x^3 + \frac{7}{3}x^5 + 3x^7 + \cdots\right)$,

(iii) $y = A\left(1 + \frac{2}{3}x^3 + \frac{1}{45}x^6 - \frac{1}{1620}x^9 + \cdots\right) + B\left(x + \frac{1}{4}x^4\right)$,

(iv) $y = A\left(1 + \frac{x^2}{2!} - \frac{x^4}{4!} + \cdots\right) + B\left(1 - \frac{x^3}{3!} - \frac{x^5}{5!} + \cdots\right)$.

3. $y = A\left[1 + \frac{2}{3}(x-1)^3 + \frac{1}{45}(x-1)^6 - \frac{1}{1620}(x-1)^9 + \cdots\right] + B\left[(x-1) + \frac{1}{4}(x-1)^4\right]$,

4. $y = A\left[1 - \frac{1}{3\cdot 2}(x-2)^3 + \frac{1}{3^2\cdot 2!\cdot 2\cdot 5}(x-2)^6 - \frac{1}{3^3\cdot 3!\cdot 2\cdot 5\cdot 8}(x-2)^9 + \cdots\ldots\right] +$

$B\left[(x-2) - \frac{1}{3\cdot 4}(x-2)^4 + \frac{1}{3^2\cdot 2!\cdot 4\cdot 7}(x-2)^7 - \frac{1}{3^3\cdot 3!\cdot 4\cdot 7\cdot 10}(x-2)^{10} + \cdots\right]$

5. $y = Ax\left[1 - \frac{x^2}{2^2} + \frac{x^4}{2^2\cdot 4^2} - \cdots\right] + Bx\ln x\left[1 - \frac{x^2}{2^2} + \frac{x^4}{2^2\cdot 4^2} - \cdots\right] +$

$Bx\left[\frac{x^2}{2^2} - \frac{x^4}{2^2\cdot 4^2}\left(1+\frac{1}{2}\right) + \cdots\right]$,

6. $y = A[1 + 2^2x + 3^2x^2 + 4^2x^3 + \cdots] + Bx\ln x[1 + 2^2x + 3^2x^2 + 4^2x^3 + \cdots] +$

$-2B[1\cdot 2x + 2\cdot 3x^2 + 3\cdot 4x^3 + \cdots]$,

7. $y = A\left[\frac{x^2}{16} - \frac{x^4}{192} + \cdots\right] + B\ln x\left[\frac{x^2}{16} - \frac{x^4}{192} + \cdots\right] + \frac{B}{x^2}\left[1 + \frac{x^2}{2^2} + \frac{x^4}{2^2\cdot 4^2} - \ldots\right]$,

8. $y = A\left[1 + 3x^2 + \frac{3}{5}x^4 + \cdots\right] + Bx^{3/2}\left[1 + \frac{3}{8}x^2 - \frac{3}{128}x^4 + \cdots\right]$

9. $y = A\left[1 + \frac{1}{3}x + \frac{1}{3}\cdot\frac{4}{6}x^2 + \frac{1}{3}\cdot\frac{4}{6}\cdot\frac{7}{9}x^3 + \cdots\right] +$

$Bx^{7/3}\left[1 + \frac{8}{10}x + \frac{8}{10}\cdot\frac{11}{13}x^2 + \frac{8}{10}\cdot\frac{11}{13}\cdot\frac{14}{16}x^3 + \cdots\right]$,

10. $y = A\left[1 + x + \frac{2}{4}x^2 + \frac{2}{4}\cdot\frac{5}{9}x^3 + \cdots\right] + B\ln x\left[1 + x + \frac{2}{4}x^2 + \frac{2}{4}\cdot\frac{5}{9}x^3 + \cdots\right]$
$-B[2x + x^2 + \cdots]$

11. $y = Ax\left[1 + \frac{x^2}{2\cdot 5} + \frac{x^4}{2\cdot 4\cdot 5\cdot 9} + \cdots\right] + B\sqrt{x}\left[1 + \frac{x^2}{2\cdot 3} + \frac{x^4}{2\cdot 4\cdot 3\cdot 7} + \cdots\right],$

12. $y = A\left[1 + \frac{2}{x} + \frac{7}{3x^2} + \frac{112}{45x^3} + \cdots\right] + \frac{B}{\sqrt{x}}\left[1 + \frac{4}{3x} + \frac{22}{15x^2} + \frac{484}{315x^3} + \cdots\right],$

13. $y = \frac{A}{x}\left[1 - \frac{1}{x} + \frac{1}{2x^2} - \frac{1}{6x^3} + \cdots\right] - \frac{B}{x}\ln x\left[1 - \frac{1}{x} + \frac{1}{2x^2} - \frac{1}{6x^3} + \cdots\right] + \frac{B}{x^2}\left[1 - \frac{3}{4x} + \cdots\right],$

14. $y = A(x^2 + 2x + 3) + \frac{B}{4}\left[\frac{4}{x} + \frac{5}{x^2} + \frac{6}{x^3} + \cdots\right].$

Laplace Transform

Chapter 1

THE LAPLACE TRANSFORM

1.1 Integral Transform

Equations arising in several problems of mathematical physics or applied mathematics, like Laplace's equation, Poisson's equation, are partial differential equations. In the past, the method of separation of variables was used for solving such differential equations. Later on integral transform became a useful tool to solve these equations.

Let $k(p, t)$ be a function of two variables p, t, where p is a parameter (may be real or complex) independent of t. The function $f(p)$ defined by the integral (assumed to be convergent)

$$f(p) = \int_{-\infty}^{\infty} k(p, t)\, F(t)\, dt$$

is called *integral transform* of the function $F(t)$ and is denoted by $T\,[F(t)]$. The function $k(p, t)$ is called *kernel* of the transformation.

1.2 Laplace Transform

If the kernel $k(p, t)$ is defined as $k(p, t) = 0$ for $t < 0$

$= e^{-pt}$ for $t \geq 0$.

Then $$f(p) = \int_{0}^{\infty} e^{-pt} F(t)\, dt$$

is called *Laplace transform* of the function $F(t)$ and is denoted by $L[F(t)]$.

The Laplace transform of $F(t)$ is said to exist if the above integral converges for some values of p, otherwise it does not exist. For example, if $f(t) = e^{t^2}$, then the integral is divergent for all values of p and if $f(t) = e^{t}$, then the integral is convergent for all values of p satisfying $p > 1$ only. So it is necessary to consider the sufficient conditions for the existence of Laplace transform of a function $f(t)$, which are:

If (a) $f(t)$ is integrable over any finite interval $[a, b]$, $0 < a < b$.

(b) there exists a real number c such that for any arbitrary $b > 0$,

$$\int_{b}^{\lambda} e^{-ct} f(t)\, dt \text{ tends to a finite limit as } \lambda \to \infty.$$

(c) for arbitrary $a > 0$,

$$\int_{\epsilon}^{a} |f(t)|\,dt \text{ tends to a finite limit as } \epsilon \to 0.$$

Then $L[F(t)]$ exists for $Re\, p \geq 0$.

1.3 Linearity Property of Laplace Transform

A transformation T is said to be linear if for every pair of functions $F_1(t)$, $F_2(t)$ and for every pair of constants a, b, we have

$$T[aF_1(t) + bF_2(t)] = aT[F_1(t)] + bT[F_2(t)].$$

THEOREM

The Laplace transform is a linear transform.

We have $L[F(t)] = \int_0^\infty e^{-pt} F(t)\,dt$. So,

$$T[aF_1(t) + bF_2(t)] = \int_0^\infty e^{-pt}[aF_1(t) + bF_2(t)]\,dt$$

$$= a\int_0^\infty e^{-pt} F_1(t)\,dt + b\int_0^\infty e^{-pt} F_2(t)\,dt$$

$$= aT[F_1(t)] + bT[F_2(t)].$$

Note: In general, $L[\Sigma_{i=1}^{n} c_i F_i(t)] = \Sigma_{i=1}^{n} c_i L[F_i(t)]$.

1.4 Piecewise Continuous Function

A function $F(t)$ is said to be piecewise continuous on a closed interval $[a, b]$, if it is defined on that interval and is such that the interval can be subdivided into a finite number of intervals, in each of which $F(t)$ is continuous and has finite right- and left-hand limits.

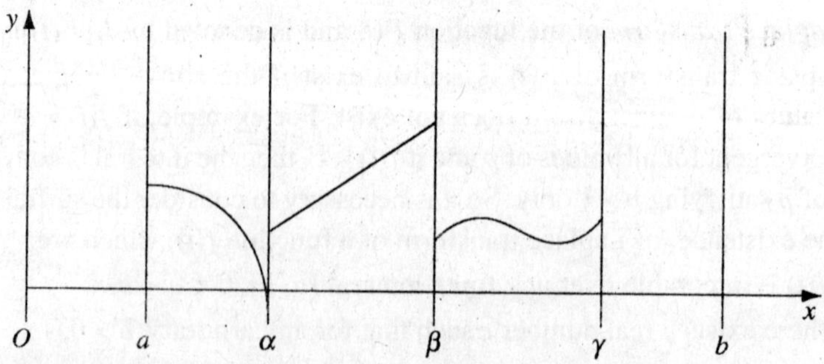

Fig.1.1: Piecewise continuous function

A piecewise continuous function is depicted in the above figure. The function is clearly discontinuous at the point α, β, γ, i.e., the function is not continuous in $[a, b]$, but in each of the subintervals (a, α), (α, β), (β, γ), the function is continuous and has finite right- and left-hand limits.

EXAMPLE

The function $f(x) = [x]$, where $[x]$ denotes the integer part of x, is piecewise continuous on $[0, 2]$.

1.5 Function of Exponential Order

A function $F(t)$ is said to be exponential order a as $t \to \infty$, if there exists a positive constant (real) M, a number a and a finite number t_0 such that $|F(t)| < Me^{at}$, for all $t > t_0$.

EXAMPLES

(i) The function $f(t) = e^{kt}$ is of exponential order at $t \to \infty$ for $k < a$, but not of exponential order at $t \to \infty$ for $k > a$.

(ii) The function $f(t) = t^2$ is of exponential order n, where n is a positive real number.

Note: If a function is of exponential order a as $t \to \infty$, then it is also of exponential order b for $b > a$.

1.6 Function of Class A

A function which is piecewise continuous on every finite interval in the range $t \geq 0$ and is of exponential order as $t \to \infty$ is known as "*a function of class A*".

1.7 Existence Theorem of Laplace Transform (Another Form)

If $F(t)$ is a function of class A, the Laplace transform exists for all $p > a$.

We have $L[F(t)] = \int_0^\infty e^{-pt} F(t)\,dt = \int_0^{t_0} e^{-pt} F(t)\,dt + \int_{t_0}^\infty e^{-pt} F(t)\,dt.$ (1.1)

The integral $\int_0^{t_0} e^{-pt} F(t)\,dt$ exists, since $F(t)$ is continuous on every interval $0 \leq t \leq t_0$.

Now, $\left| \int_{t_0}^\infty e^{-pt} F(t)\,dt \right| \leq \int_{t_0}^\infty \left| e^{-pt} F(t)\,dt \right| \leq \int_{t_0}^\infty e^{-pt} Me^{at}\,dt$, since $|F(t)| \leq Me^{at}$

$$= M\int_{t_0}^\infty e^{-(p-a)t}\,dt$$

$$= \frac{Me^{-(p-a)t_0}}{p-a},\ p > a$$

But, $\dfrac{Me^{-(p-a)t_0}}{p-a}$ can be made as small as we please by taking t_0 sufficiently large.

So, from eqn. (1.1), we conclude $L[F(t)]$ exists for all $p > a$.

Table of Laplace transform

	$F(t)$	$f(p)$
1.	t^n $n = 0, 1, 2, \cdots$	$\dfrac{n!}{p^{n+1}}$
2.	e^{at}	$\dfrac{1}{p-a}, p > a$
3.	$\sin at$	$\dfrac{a}{p^2+a^2}, p > 0$
4.	$\cos at$	$\dfrac{p}{p^2+a^2}, p > 0$
5.	$\sinh at$	$\dfrac{a}{p^2-a^2}, p > \lvert a\rvert$
6.	$\cosh at$	$\dfrac{p}{p^2-a^2}, p > \lvert a\rvert$

WORKED-OUT EXAMPLES

1. Show that the function $f(t) = t^2$ is of exponential order.

Solution

Using L'Hospital's rule, we see that

$$\lim_{t\to\infty} t^2 e^{-nt} = \lim_{t\to\infty} \frac{t^2}{e^{nt}} = \lim_{t\to\infty} \frac{2t}{ne^{nt}} = \lim_{t\to\infty} \frac{2}{n^2 e^{nt}} = 0, \text{ if } n > 0.$$

Hence the given function is of exponential order n, where n is any positive real number.

2. Show that the function $f(t) = e^{t^2}$ is not of exponential order.

Solution

Here, we see that

$$\lim_{t\to\infty} e^{t^2} e^{-nt} = \lim_{t\to\infty} e^{(t-n)t} = \infty, \text{ for all finite values of } n.$$

So there exists no value of a for which $e^{t^2} < Me^{at}$. Hence, the function $f(t) = e^{t^2}$ is not of exponential order.

3. Find the Laplace transforms of the following functions:

(i) $F(t) = 1$, (ii) $F(t) = t^n$, n being an integer,

(iii) $F(t) = e^{at}$, (iv) $F(t) = \sin at$,

(v) $F(t) = \cos at$, (vi) $F(t) = \cosh at$,

(vii) $F(t) = \sinh at$, (viii) $F(t) = t^n, -1 < n < 0$,

(ix) $F(t) = \sin \sqrt{t}$.

Solution

(i) $L[1] = \int_0^\infty e^{-pt}, 1\, dt = \frac{1}{p}, p > 0.$

(ii) $L[t^n] = \int_0^\infty e^{-pt} t^n dt = \frac{n}{p} \int_0^\infty e^{-pt} t^{n-1} dt = \frac{n}{p} L[t^{n-1}].$

Changing n by $n-1$ in the above result, we get $L[t^{n-1}] = \frac{n-1}{p} L[t^{n-2}]$.

Proceeding in this way, we get

$$L[t^n] = \frac{n}{p}\frac{n-1}{p} \cdots \frac{3}{p}\frac{2}{p}\frac{1}{p} L[1] = \frac{n!}{p^{n+1}}, p > 0.$$

(iii) $L[e^{at}] = \int_0^\infty e^{-pt} e^{at} dt = \int_0^\infty e^{-(p-a)t} dt = \frac{1}{p-a}, p > a.$

(iv) $L[\sin at] = \int_0^\infty e^{-pt} \sin at\, dt$

$$= \frac{e^{-pt}}{p^2 + a^2} [-p \sin at - a \cos at]_0^\infty = \frac{a}{p^2 + a^2}, p > 0.$$

(v) $L[\cos at] = \int_0^\infty e^{-pt} \cos at\, dt$

$$= \frac{e^{-pt}}{p^2 + a^2} [-p \cos at - a \sin at]_0^\infty$$

$$= \frac{p}{p^2 + a^2}, p > 0.$$

Alternative: Assuming the result $L[e^{at}] = \frac{1}{p-a}$ holds for complex number, we have

$$L[e^{iat}] = \frac{1}{p - ia} = \frac{p + ia}{p^2 + a^2} = \frac{p}{p^2 + a^2} + \frac{ia}{p^2 + a^2}.$$

Now $L[e^{iat}] = L[\cos at + i \sin at] = L[\cos at] + iL[\sin at]$. Comparing we get

$$L[\cos at] = \frac{p}{p^2 + a^2} \text{ and } L[\sin at] = \frac{a}{p^2 + a^2}.$$

(vi) $L[\cosh at] = \frac{1}{2} L[e^{at}] + \frac{1}{2} L[e^{at}] = \frac{1}{2}\frac{1}{p-a} + \frac{1}{2}\frac{1}{p+a} = \frac{p}{p^2 - a^2}, p > |a|.$

(vii) $L[\sinh at] = \frac{1}{2} L[e^{at}] + \frac{1}{2} L[e^{-at}] = \frac{1}{2}\frac{1}{p-a} + \frac{1}{2}\frac{1}{p+a} = \frac{p}{p^2 - a^2}, p > |a|.$

(viii) Here, it can be easily seen that $F(t) = t^n$, $-1 < n < 0$, is not a function of exponential order and also $F(t)$ is not a piecewise continuous function over every finite interval in the range $t \geq 0$. But $F(t)$ is integrable from 0 to any positive number t_0.

$$\text{Now } L[F(t)] = \int_0^\infty e^{-pt} t^n dt$$

$$= \frac{1}{p}\int_0^\infty e^{-x}\left(\frac{x}{p}\right)^n dx, \text{ (on putting } pt = x)$$

$$= \frac{1}{p^{n+1}}\int_0^\infty e^{-x}x^n dx = \frac{\Gamma(n+1)}{p^{n+1}}, \text{ if } p > 0, n > -1.$$

(ix) $$L\left[\sin\sqrt{t}\right] = L\left[t^{1/2} - \frac{t^{3/2}}{3!} + \frac{t^{5/2}}{5!} - \cdots\right]$$

$$= L\left[t^{1/2}\right] - \frac{1}{3!}L\left[t^{3/2}\right] + \frac{1}{5!}L\left[t^{5/2}\right] - \cdots$$

$$= \frac{\Gamma\left(\frac{3}{2}\right)}{p^{3/2}} - \frac{1}{3!}\frac{\Gamma\left(\frac{5}{2}\right)}{p^{5/2}} + \frac{1}{5!}\frac{\Gamma\left(\frac{7}{2}\right)}{p^{7/2}} - \cdots$$

$$= \frac{\frac{1}{2}\sqrt{\pi}}{p^{3/2}} - \frac{1}{3!}\frac{\frac{3}{2}\cdot\frac{1}{2}\sqrt{\pi}}{p^{5/2}} + \frac{1}{5!}\frac{\frac{5}{2}\cdot\frac{3}{2}\cdot\frac{1}{2}\sqrt{\pi}}{p^{7/2}} - \cdots$$

$$= \frac{\sqrt{\pi}}{2p^{3/2}}\left[1 - \frac{1}{4p} + \frac{1}{2!}\left(\frac{1}{4p}\right)^2 - \cdots\right] = \frac{\sqrt{\pi}}{2p^{3/2}}e^{-\frac{1}{4p}}.$$

4. Find $L[\cosh^2 t]$.

Solution

$$L[\cosh^2 t] = L\left[\tfrac{1}{2}(1 + \cosh 2t)\right]$$

$$= \frac{1}{2}L[1] + \frac{1}{2}L[\cosh 2t] = \frac{1}{2p} + \frac{1}{2}\frac{p}{p^2-4} = \frac{p^2-2}{p(p^2-4)}.$$

5. Find $L[F(t)]$ where $F(t) = \begin{cases} e^t, & 0 < t \leq 1 \\ t, & t > 1 \end{cases}$.

Solution

$$L[F(t)] = \int_0^\infty e^{-pt}F(t)\,dt = \int_0^1 e^{-pt}e^t dt + \int_1^\infty te^{-pt}dt$$

$$= \left[\frac{e^{-(p-1)t}}{p-1}\right]_0^1 + \left[-\frac{1}{p}te^{-pt} - \frac{1}{p^2}e^{-pt}\right]_1^\infty$$

$$= \frac{1-e^{-(p-1)}}{p-1} + \left(\frac{1}{p} + \frac{1}{p^2}\right)e^{-p}.$$

6. Find $L[F(t)]$ where $F(t) = \begin{cases} 1, & 0 < t \le 2 \\ t, & t > 2 \end{cases}$.

Solution

$$L[F(t)] = \int_0^\infty e^{-pt} F(t)\,dt = \int_0^2 e^{-pt}dt + \int_2^\infty te^{-pt}dt$$

$$= \left[-\frac{e^{-pt}}{p}\right]_0^2 + \left[-\frac{1}{p}te^{-pt} - \frac{1}{p^2}e^{-pt}\right]_2^\infty$$

$$= \frac{1-e^{-2p}}{p} + \left(\frac{2}{p} + \frac{1}{p^2}\right)e^{-2p}.$$

EXERCISE

1. Define Laplace transform of a function $F(t)$.
2. State and prove the existence theorem of Laplace transform.
3. Show that the function $\sin at$ is of exponential order.
4. Show that the function $\cos at$ is of exponential order.
5. Find the Laplace transform of the following functions:

(a) $3t - 5$, (b) $2t^2 + 8t - 5$, (c) $t^2 - e^{-t}$,

(d) $\frac{1}{\sqrt{\pi t}}$, (e) $a + bt + \frac{c}{\sqrt{t}}$, (f) $\sinh^2 at$,

(g) $\cosh^2 at$, (h) $(\sin t - \cos t)^2$, (i) $3\cosh 5t - 4\sinh 5t$,

(j) $(t^2+1)^3 - e^{3t}$, (k) $F(t) = \begin{cases} e^t & 0 < t < 5 \\ 3, & t \ge 5 \end{cases}$ (l) $F(t) = \begin{cases} 0 & 0 < t < 1 \\ (t-1)^2, & t \ge 1 \end{cases}$

(m) $F(t) = \begin{cases} \sin t, & 0 < t < \pi \\ 0, & \text{otherwise} \end{cases}$ (n) $F(t) = \frac{e^{at}-1}{a}$,

(o) $(t^3+1)^2$, (p) $4\cos^2 2t$, (q) $2e^{3t} - e^{-3t}$,

(r) $\sin 5t - \cos 3t$, (s) $\sin^2 at$, (t) $\cos^2 at$.

6. Show that $L\left[\frac{\cos\sqrt{t}}{\sqrt{t}}\right] = \sqrt{\frac{\pi}{p}}\,e^{-\frac{1}{4p}}$.
7. Show that $L\left[\sqrt{t} + \frac{1}{\sqrt{t}}\right] = \frac{(2p+1)\sqrt{\pi}}{2p^{3/2}}$.

ANSWERS

5.

(a) $\dfrac{3-5p}{p^2}, p>0,$ (b) $\dfrac{4+8p-5p^2}{p^3}, p>0,$ (c) $\dfrac{2+2p-p^3}{p^3(1+p)},$

(d) $\dfrac{1}{\sqrt{p}}, p>0,$ (e) $\dfrac{a}{p}+\dfrac{b}{p^2}+c\sqrt{\dfrac{\pi}{p}},$ (f) $\dfrac{2a^2}{p(p^2-4a^2)}, p>2a;$

(g) $\dfrac{p^2-2a^2}{p(p^2-4a^2)}, p>2a;$ (h) $\dfrac{p^2-2p+4}{p(p^2+4)}, p>0;$ (i) $\dfrac{3p-20}{p^2-25}, p>5;$

(j) $\dfrac{720}{p^7}+\dfrac{72}{p^5}+\dfrac{6}{p^3}-\dfrac{3}{p(p-3)}, p>3;$ (k) $\dfrac{1-e^{-5(p-1)}}{p-1}+\dfrac{3}{p}e^{-5p},$

(l) $\dfrac{2e^{-p}}{p^3},$ (m) $\dfrac{e^{-\pi p}+1}{p^2+1}, p>0$ (n) $\dfrac{1}{p(p-a)}, p>a$

(o) $\dfrac{720}{p^7}+\dfrac{12}{p^4}+\dfrac{1}{p}, p>0;$ (p) $\dfrac{2}{p}+\dfrac{2p}{p^2+16}, p>0,$ (q) $\dfrac{p+9}{p^2-9}, p>3;$

(r) $\dfrac{5}{p^2+25}-\dfrac{p}{p^2+9}, p>0;$ (s) $\dfrac{2a^2}{p(p^2+4a^2)}, p>0;$ (t) $\dfrac{p^2+2a^2}{p(p^2+4a^2)}, p>0.$

1.8 First Shifting Theorem

THEOREM

If $f(p)$ is the Laplace transform of $F(t)$, then $f(p-a)$ is the Laplace transform of $e^{at}F(t)$.

Proof: By definition,

$$f(p) = L[F(t)] = \int_0^{\infty} e^{-pt} F(t)\,dt.$$

So,
$$f(p-a) = \int_0^{\infty} e^{-(p-a)t} F(t)\,dt = \int_0^{\infty} e^{-pt}[e^{at}F(t)]\,dt = L[e^{at}F(t)].$$

1.9 Second Shifting Theorem

THEOREM

If $L[F(t)] = f(p)$ and $G(t) = F(t-a), t>a$
$\qquad = 0, t<a$

then $L[G(t)] = e^{-ap}L[F(t)]$.

Proof: By definition,

$$L[G(t)] = \int_0^\infty e^{-pt} G(t)\,dt$$

$$= \int_0^a e^{-pt} G(t)\,dt + \int_a^\infty e^{-pt} G(t)\,dt$$

$$= \int_a^\infty e^{-pt} F(t-a)\,dt$$

$$= \int_0^\infty e^{-(x+a)p} F(x)\,dx, \text{ on putting } t = a + x$$

$$= e^{-pa} \int_0^\infty e^{-px} F(x)\,dx$$

$$= e^{-pa} \int_0^\infty e^{-pt} F(t)\,dt$$

$$= e^{-pa} L[F(t)].$$

1.10 Change of Scale Property

THEOREM

If $L[F(t)] = f(p)$, then $L[F(at)] = \frac{1}{a} f\left(\frac{p}{a}\right)$.

Proof: By definition,

$$L[F(at)] = \int_0^\infty e^{-pt} F(at)\,dt$$

$$= \frac{1}{a} \int_0^\infty e^{-\frac{p}{a}x} F(x)\,dx, \text{ on putting } t = \frac{x}{a}$$

$$= \frac{1}{a} \int_0^\infty e^{-\frac{p}{a}t} F(t)\,dt$$

$$= \frac{1}{a} f\left(\frac{p}{a}\right).$$

WORKED-OUT EXAMPLES

1. Find Laplace transform of the following functions:

(a) $t^4 e^{-2t}$ (b) e^{-2t} sinh 3t, (c) $e^{-3t} \sin 4t$,

(d) $e^t \sin^2 t$, (e) $e^t \cos^2 t$, (f) $\cos 6t$,

(g) $e^{-3t}(3\cos 6t - 5\sin 6t)$, (h) $\frac{e^{-3t}}{\sqrt{t}}$, (i) $t^{5/2} e^{2t}$.

Solutions

(a) $L[t^4]=\dfrac{24}{p^5}$, so by first shifting theorem, $L[t^4e^{-2t}]=\dfrac{24}{(p+2)^5}$,

(b) $L[\sinh 3t]=\dfrac{3}{p^2-9}$, so by first shifting theorem,

$$L[e^{-2t}\sinh 3t]=\frac{3}{(p+2)^2-9}=\frac{3}{p^2+4p-9},$$

(c) $L[\sin 4t]=\dfrac{4}{p^2+16}$, so by first shifting theorem,

$$L[e^{-3t}\sin 4t]=\frac{4}{(p+3)^2+16}=\frac{4}{p^2+6p+25},$$

(d) $L[\sin^2 t]=L\left[\dfrac{1-\cos 2t}{2}\right]=\tfrac{1}{2}L[1]-\tfrac{1}{2}L[\cos 2t]=\dfrac{1}{2p}-\dfrac{p}{2(p^2+4)}=\dfrac{2}{p(p^2+4)}$,

so the first shifting theorem, $L[e^t\sin^2 t]=\dfrac{2}{(p-1)[(p-1)^2+4]}=\dfrac{2}{(p-1)(p^2-2p+5)}$.

(e) $L[\cos^2 t]=L\left[\dfrac{1+\cos 2t}{2}\right]=\tfrac{1}{2}L[1]+\tfrac{1}{2}L[\cos 2t]=\dfrac{1}{2p}+\dfrac{p}{2(p^2+4)}=\dfrac{p^2+2}{p(p^2+4)}$,

$$L[e^t\cos^2 t]=\frac{(p-1)^2+2}{(p-1)[(p-1)^2+4]}=\frac{p^2-2p+3}{(p-1)(p^2-2p+5)}.$$

(f) $L[\cos t]=\dfrac{p}{p^2+1}$, so by change of scale property,

$$L[\cos 6t]=\frac{1}{6}\frac{p/6}{\left(\frac{p}{6}\right)^2+1}=\frac{p}{p^2+36}.$$

(g) $L[3\cos 6t-5\sin 6t]=3L[\cos 6t]-5L[\sin 6t]=\dfrac{3p}{p^2+36}-\dfrac{30}{p^2+36}=\dfrac{3p-30}{p^2+36}$,

so by first shifting theorem, $L[e^{-3t}(3\cos 6t-5\sin 6t)]=\dfrac{3(p+3)-30}{(p+3)^2+36}=\dfrac{3(p-7)}{p^2+6p+45}$.

(h) $L\left[\dfrac{1}{\sqrt{t}}\right]=\dfrac{\Gamma\left(-\frac{1}{2}+1\right)}{p^{-\frac{1}{2}+1}}=\dfrac{\Gamma\left(\frac{1}{2}\right)}{p^{\frac{1}{2}}}=\sqrt{\dfrac{\pi}{p}}$. So, by first shifting theorem,

$$L\left[\frac{e^{-3t}}{\sqrt{t}}\right]=\sqrt{\frac{\pi}{p+3}}.$$

(i) $L\left[t^{\frac{5}{2}}\right]=\dfrac{\Gamma\left(\frac{5}{2}+1\right)}{p^{\frac{5}{2}+1}}=\dfrac{\Gamma\left(\frac{7}{2}\right)}{p^{\frac{7}{2}}}=\dfrac{5}{2}\dfrac{3}{2}\dfrac{1}{2}\Gamma\left(\dfrac{1}{2}\right)\dfrac{1}{p^{\frac{7}{2}}}=\dfrac{15\sqrt{\pi}}{8p^{\frac{7}{2}}}$. So, by first shifting theorem, $L\left[t^{5/2}e^{2t}\right]=\dfrac{15\sqrt{\pi}}{8(p-2)^{7/2}}$.

2. Find the Laplace transform of the function $F(t)$, where

$$F(t)=\begin{cases}\sin\left(t-\dfrac{\pi}{3}\right), & t\geq\dfrac{\pi}{3}\\ 0, & t<\dfrac{\pi}{3}.\end{cases}$$

Solution

Let $\varphi(t)=\sin t$, then $F(t)=\begin{cases}\varphi\left(t-\dfrac{\pi}{3}\right), & t\geq\dfrac{\pi}{3}\\ 0, & t<\dfrac{\pi}{3}.\end{cases}$

Now, $L[\varphi(t)]=\dfrac{1}{p^2+1}=f(p)$ (say). So by second shifting theorem,

$$L[F(t)]=e^{-\frac{\pi p}{3}}f(p)=e^{-\frac{\pi p}{3}}\frac{1}{p^2+1},\ p>0.$$

EXERCISE

1. Prove that if $f(p)$ is the Laplace transform of $F(t)$, then $f(p-a)$ is the Laplace transform of $e^{at}F(t)$.

2. Prove that if $L[F(t)]=f(p)$ and $G(t)=F(t-a), t>a$
$\qquad\qquad\qquad\qquad\qquad\qquad = 0, t<a,$

then $L[G(t)]=e^{-ap}L[F(t)]$.

3. Prove that if $L[F(t)]=f(s)$, then $L[F(at)]=\dfrac{1}{a}f\left(\dfrac{s}{a}\right)$.

Hence, if $L[F(t)]=\dfrac{p^2-p+1}{(2p+1)^2(p-1)}$, find $L[F(2t)]$ applying the above property.

4. Find the Laplace transform of the following functions:

(a) t^3e^{-2t}, (b) $e^t\cos^2 t$, (c) $e^{3t}\sin^2 t$,

(d) $e^{-4t}\sin 2t$, (e) $e^{-t}\sin t$, (f) $e^{-t}\cos t$,

(g) $e^{-t}\cosh 2t$, (h) $e^{-2t}\sinh 3t$, (i) $e^{-4t}\cosh 2t$,

(j) $e^{-t}[\sin 2t-5\cosh 2t]$, (k) $e^{2t}[3\sin 4t-4\cos 4t]$,

(l) $e^{-4t}[4\sin 2t + 6\cos 2t]$, (m) $(t+3)^2 e^t$, (n) $\cosh 2t \cos 2t$,

(o) $\sinh 2t \cos 2t$.

5. Find the Laplace transform of the function $F(t)$, where

$$F(t) = \begin{cases} \cos\left(t - \frac{\pi}{3}\right), & t \geq \frac{\pi}{3} \\ 0, & t < \frac{\pi}{3}. \end{cases}$$

6. Find the Laplace transform of the function $F(t)$, where

$$F(t) = \begin{cases} \sinh(t - \pi), & t \geq \pi \\ 0, & t < \pi. \end{cases}$$

7. Find the Laplace transform of the function $F(t)$, where

$$F(t) = \begin{cases} \cosh\left(t - \frac{\pi}{2}\right), & t \geq \frac{\pi}{2} \\ 0, & t < \frac{\pi}{2}. \end{cases}$$

8. Find the Laplace transform of the function $F(t)$, where

$$F(t) = \begin{cases} \cos t, & 0 < t < \pi \\ \sin t, & t \geq \pi. \end{cases}$$

ANSWERS

3. $\dfrac{p^2 - 2p + 4}{4(p+1)^2(p-2)}$

4. (a) $\dfrac{6}{(p+2)^4}$, (b) $\dfrac{p^2 - 2p + 3}{(p-1)(p^2 - 2p + 5)}$, (c) $\dfrac{2}{(p-3)(p^2 - 6p + 13)}$,

(d) $\dfrac{2}{p^2 + 8p + 20}$, (e) $\dfrac{1}{p^2 + 2p + 2}$, (f) $\dfrac{p+1}{p^2 + 2p + 2}$

(g) $\dfrac{p+1}{p^2 + 2p - 3}$, (h) $\dfrac{3}{p^2 + 4p - 5}$, (i) $\dfrac{p+4}{p^2 + 8p + 12}$,

(j) $\dfrac{2}{p^2 + 2p + 5} - \dfrac{5(p+1)}{p^2 + 2p - 3}$, (k) $\dfrac{4(5-p)}{p^2 - 4p + 20}$, (l) $\dfrac{6p + 32}{p^2 + 8p + 20}$,

(m) $\dfrac{9p^2 - 12p + 5}{(p-1)^3}$, (n) $\dfrac{p^3}{p^4 + 64}$, (o) $\dfrac{2(p^2 - 8)}{p^4 + 64}$.

5. $\dfrac{pe^{-p\pi/3}}{p^2 + 1}, p > 0.$ **6.** $\dfrac{e^{-p\pi}}{p^2 - 1}, p > 0.$ **7.** $\dfrac{pe^{-p\pi/2}}{p^2 - 1}, p > 0.$

8. $\dfrac{p + (p-1)e^{-p\pi}}{p^2 + 1}, p > 0.$

Chapter 2

Laplace Transform of Derivatives

2.1 Introduction

In the previous chapter, we have discussed about the Laplace transform of a function. In a differential equation the unknown function occurs in the form of derivatives or differentials. So, before employing Laplace transform to solve a differential equation it is necessary to know that Laplace transform of the derivatives of different order of a function.

Theorem

Let $F(t)$ be continuous function for all $t \geq 0$ and be of exponential order a as $t \to \infty$ and if $F'(t)$ is of class A, then the Laplace transform of $F'(t)$ exists when $p > a$, and

$$L[F'(t)] = pL[F(t)] - F(0).$$

Proof: If $F'(t)$ is continuous for $t \geq 0$, then

$$L[F'(t)] = \int_0^\infty e^{-pt} F'(t)\,dt = \lim_{t\to\infty} e^{-pt} F(t) - F(0) + p\int_0^\infty e^{-pt} F(t)\,dt.$$

Now, $|F(t)| < Me^{at}$, for all $t \geq 0$ and for some constants a and M.

We have

$$|e^{-pt} F(t)| \leq Me^{-(p-a)t} \to 0 \text{ as } t \to \infty, \text{ if } p > a. \text{ So, } \lim_{t\to\infty} e^{-pt} F(t) = 0 \text{ for } p > a.$$

Therefore, $L[F'(t)] = pL[F(t)] - F(0)$.

In case, $F'(t)$ is merely a piecewise continuous function, the integral $\int_0^\infty e^{-pt} F'(t)\,dt$ may be broken as the sum of integrals in different ranges from 0 to ∞ such that $F'(t)$ is continuous in each sub intervals, then proceeding as before we shall get result.

Note: If $F(t)$ is discontinuous at $t = a$, then

$$L[F'(t)] = pL[F(t)] - F(0) - e^{-ap}[F(a+0) - F(a-0)].$$

THEOREM

Let $F(t)$ and its derivatives $F'(t), F''(t), \ldots F^{n-1}(t)$ be continuous function for all $t \geq 0$ and be of exponential order a as $t \to \infty$ and if $F^n(t)$ is of class A, then the Laplace transform of $F^n(t)$ exists when $p > a$, and is given by

$$L[F^n(t)] = p^n L[F(t)] - p^{n-1}F(0) - p^{n-2}F'(0) - p^{n-3}F''(0) - \ldots - F^{n-1}(0)$$

Proof: From the above theorem we get

$$L[F'(t)] = pL[F(t)] - F(0)$$

Applying the result on $F''(t)$ we get,

$$\begin{aligned} L[F''(t)] &= pL[F'(t)] - F'(0) \\ &= p[pL[F(t)] - F(0)] - F'(0) \\ &= p^2 L[F(t)] - pF(0) - F'(0). \end{aligned}$$

Again, applying the result on $F'''(t)$ we get,

$$\begin{aligned} L[F'''(t)] = pL[F''(t)] - F''(0) &= p[p^2 L[F(t)] - pF(0) - F'(0)] - F''(0) \\ &= p^3 L[F(t)] - p^2 F(0) - pF'(0) - F''(0). \end{aligned}$$

Proceeding, similarly we get,

$$L[F^n(t)] = p^n L[F(t)] - p^{n-1}F(0) - p^{n-2}F'(0) - p^{n-3}F''(0) - \ldots - F^{n-1}(0).$$

2.2 Initial Value Problem

THEOREM

Let $F(t)$ be continuous function for all $t \geq 0$ and be of exponential order as $t \to \infty$ and if $F'(t)$ is of class A, then

$$\lim_{t \to 0} F(t) = \lim_{p \to \infty} pL[F(t)].$$

Proof: We have $L[F'(t)] = pL[F(t)] - F(0)$ [2.1]

Since, $F'(t)$ is of class A, $\lim_{p \to \infty} \int_0^\infty e^{-pt} F'(t)\, dt = 0.$

Taking limit $p \to \infty$ in (2.1) we have,

$$\lim_{p \to \infty} pL[F(t)] - F(0) = 0$$

$$\Rightarrow \lim_{p \to \infty} pL[F(t)] = F(0) = \lim_{t \to 0} F(t)$$

If $F(t)$ is discontinuous at $t = 0$, but $\lim_{t \to 0} F(t)$ exists, the above results still holds.

2.3 Final Value Problem

THEOREM

Let $F(t)$ be continuous function for all $t \geq 0$ and be of exponential order as $t \to \infty$ and if $F'(t)$ is of class A, then

$$\lim_{t\to\infty} F(t) = \lim_{p\to 0} pL[F(t)].$$

Proof: We have $$L[F'(t)] = pL[F(t)] - F(0) \qquad [2.2]$$

Since $F'(t)$ is of class A, $\lim_{p\to 0} \int_0^\infty e^{-pt} F'(t)\,dt = \lim_{p\to 0} pL[F(t)] - F(0)$

$$\Rightarrow \int_0^\infty F'(t)\,dt = \lim_{p\to 0} pL[F(t)] - F(0)$$

$$\Rightarrow \lim_{t\to\infty} F(t) - F(0) = \lim_{p\to 0} pL[F(t)] - F(0)$$

$$\Rightarrow \lim_{t\to\infty} F(t) = \lim_{p\to 0} pL[F(t)]$$

If $F(t)$ is discontinuous at $t = \infty$, but $\lim_{t\to\infty} F(t)$ exists, the above results still holds.

THEOREM

If $F(t)$ is a function of class A and if $L[F(t)] = f(p)$ then $L[tF(t)] = -f'(p)$.

Proof: We know that $f(p) = \int_0^\infty e^{-pt} F(t)\,dt$.

So $f'(p) = \frac{d}{dp}\int_0^\infty e^{-pt} F(t)\,dt = -\int_0^\infty te^{-pt} F(t)\,dt = -L[tF(t)]$

THEOREM

If $F(t)$ is a function of class A and if $L[F(t)] = f(p)$ then

$$L[t^n F(t)] = (-1)^n \frac{d^n}{dp^n} f(p), n = 1, 2, 3, \ldots.$$

Proof: From the above theorem it is clear that the theorem is true for $n = 1$.
Now assume that the theorem is true for $n = s$, i.e.,

$$L[t^s F(t)] = (-1)^s \frac{d^s}{dp^s} f(p),$$

i.e., $$\int_0^\infty t^s e^{-pt} F(t)\,dt = (-1)^s \frac{d^s}{dp^s} f(p).$$

Now differentiating both sides with respect to p, we get,

$$\frac{d}{dp}\int_0^{\infty} t^s e^{-pt} F(t)\,dt = (-1)^s \frac{d^{s+1}}{dp^{s+1}} f(p)$$

$$\Rightarrow \int_0^{\infty} \frac{\partial}{\partial p}[t^s e^{-pt} F(t)]dt = (-1)^s \frac{d^{s+1}}{dp^{s+1}} f(p)$$

$$\Rightarrow -\int_0^{\infty} t^{s+1} e^{-pt} F(t)\,dt = (-1)^s \frac{d^{s+1}}{dp^{s+1}} f(p)$$

$$\Rightarrow L[t^{s+1} F(t)] = (1-)^{s+1} \frac{d^{s+1}}{dp^{s+1}} f(p)$$

Which shows that if the theorem is true for $n = s$, then it is true for $n = s + 1$ and we have already seen that the theorem is true for $n = 1$, hence it is true for $n = 2$, $n = 3$ and so on. So the theorem is true for all positive integral values of n.

THEOREM

If $L[F(t)] = f(p)$, then $L[t^{-1}F(t)] = \int_p^{\infty} f(x)dx$, provided, $\lim_{t\to 0} \frac{F(t)}{t}$ exists.

Proof: Let $tG(t) = F(t)$, then $L[F(t)] = L[tG(t)] = -\frac{d}{dp} L[G(t)]$

$$\Rightarrow f(p) = -\frac{d}{dp} L[G(t)]$$

Now integrating both sides with respect to p, taking limit p to ∞ for p, we get

$$-L[G(t)]_p^{\infty} = \int_p^{\infty} f(p)\,dp$$

$$\Rightarrow -\lim_{p\to\infty} L[G(t)] + L[G(t)] = \int_p^{\infty} f(p)\,dp$$

$$\Rightarrow L[G(t)] = \int_p^{\infty} f(p)\,dp$$

$$\Rightarrow L\left[\frac{F(t)}{t}\right] = \int_p^{\infty} f(p)\,dp$$

Alternative method: Given that $f(p) = \int_0^{\infty} e^{-pt} F(t)\,dt$.

Integrating both sides with respect to p, taking limit p to ∞ we get

$$\int_p^{\infty} f(p)\,dp = \int_p^{\infty}\left[\int_0^{\infty} e^{-pt} F(t)\,dt\right]dp.$$

Since t, p are independent, we can write,

$$\int_p^\infty f(p)\,dp = \int_0^\infty [F(t)\int_p^\infty e^{-pt}dp]dt$$

$$= \int_0^\infty \left[F(t)\left\{ -\frac{1}{t}e^{-pt} \right\}_p^\infty \right] dt$$

$$= \int_0^\infty \frac{F(t)}{t} e^{-pt} dt = L\left[\frac{F(t)}{t}\right].$$

Repeating the process we find that, for any positive integer n,

$$L\left[\frac{F(t)}{t^n}\right] = \int_p^\infty dp_n \int_{p_n}^\infty dp_{n-1} \cdots\cdots \int_{p_3}^\infty dp_2 \int_{p_2}^\infty f(p_1)\,dp_1.$$

THEOREM

If $L[F(t)] = f(p)$, then $L\left[\int_0^t F(u)\,du\right] = \dfrac{f(p)}{p}$.

Proof: Let $G(t) = \int_0^t F(u)\,du$. Then $G(0) = 0$ and $G'(t) = F(t)$. Now using the formula

$$L[G'(t)] = pL[G(t)] - G(0),$$

we get
$$f(p) = L[F(t)] = pL\left[\int_0^t F(u)\,du\right]$$

or,
$$L\left[\int_0^t F(u)\,du\right] = \frac{f(p)}{p}.$$

2.4 Periodic Functions

THEOREM

Let $F(t)$ be a periodic function of period $T > 0$, i.e., $F(t + 2T) = F(t + T) = F(t)$, etc.

Then
$$L[F(t)] = \frac{\int_0^T e^{-pt} F(t)\,dt}{1 - e^{-pT}}.$$

Proof: We have

$$L[F(t)] = \int_0^\infty e^{-pt} F(t)\,dt = \int_0^T e^{-pt} F(t)\,dt + \int_T^{2T} e^{-pt} F(t)\,dt + \int_{2T}^{3T} e^{-pt} F(t)\,dt + \cdots$$

$$= \int_0^T e^{-pt} F(t)\,dt + \int_0^T e^{-p(u+T)} F(u+T)\,du + \int_0^T e^{-p(u+2T)} F(u+2T)\,du + \cdots$$

$$= \int_0^T e^{-pt}F(t)\,dt + e^{-pT}\int_0^T e^{-pu}F(u)\,du + e^{-2pT}\int_0^T e^{-pu}F(u)\,du + \ldots$$

$$= (1 + e^{-pT} + e^{-2pT} + \cdots)\int_0^T e^{-pt}F(t)\,dt$$

$$= \frac{\int_0^T e^{-pt}F(t)\,dt}{1 - e^{-pT}}.$$

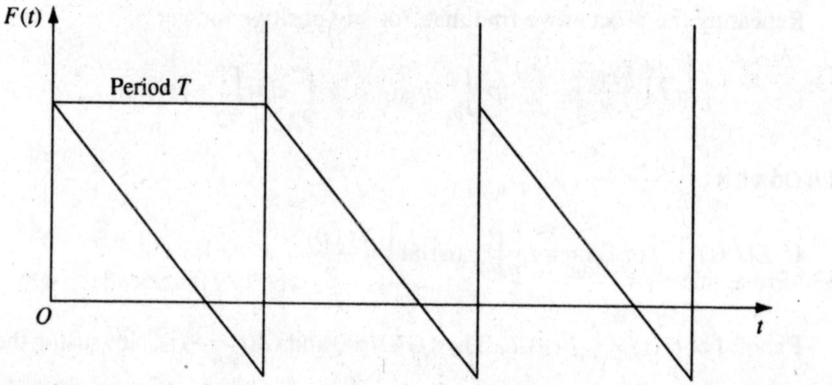

Fig.2.1 Periodic function

WORKED-OUT EXAMPLES

1. Find Laplace transform of the functions:

(a) $t\cos at$, (b) t^2e^{at}, (c) $t\cosh 3t$,

(d) $t^2\sin at$, (e) $t\sin^2 t$.

2. Show that $L\left(\frac{e^{-at} - e^{-bt}}{t}\right) = \ln\left(\frac{p+b}{p+a}\right)$. Hence, show that $\int_0^\infty \frac{e^{-at} - e^{-bt}}{t}\,dt = \ln\left(\frac{b}{a}\right)$.

3. Find the Laplace transform of $\frac{\sin at}{t}$.

4. Show that Laplace transform of $\frac{\cos at}{t}$ does not exist.

5. Prove that $\int_0^\infty \frac{e^{-t}\sin t}{t}\,dt = \frac{\pi}{4}$.

6. Find the Laplace transform of the function $F(t-a)$, where

$$F(t-a) = \begin{cases} 1, & t \ge a \\ 0, & t < a. \end{cases}$$

7. If $F(t) = t, 0 < t < 1$ and $F(t + 1) = F(t)$, find $L[F(t)]$.

8. Find $L\left[\dfrac{\cos at - \cos bt}{t}\right]$.

9. Using the formula for Laplace transform of derivatives, prove that $L[\sin at] = \dfrac{a}{p^2 + a^2}$.

10. Show that $L\left[\dfrac{\cos\sqrt{t}}{\sqrt{t}}\right] = \sqrt{\dfrac{\pi}{p}}\, e^{-\frac{1}{4p}}$.

11. Find $L\left[\int_0^t \dfrac{\sin x}{x}\, dx\right]$.

12. Find $L\left[\int_0^t \left(\dfrac{1 - e^{-2x}}{x}\right) dx\right]$.

13. Prove that $L\left[\dfrac{1 - \cos t}{t^2}\right] = \dfrac{\pi}{2} + \dfrac{1}{2} p \ln\left(\dfrac{p^2}{p^2 + 1}\right) - \tan^{-1} p$. Hence deduce that

$$\int_0^\infty \frac{\sin^2 t}{t^2}\, dt = \frac{\pi}{2}.$$

Solutions

1. (a) We know that $L[\cos at] = \dfrac{p}{p^2 + a^2}$, so

$$L[t \cos at] = -\frac{d}{dp}\left[\frac{p}{p^2 + a^2}\right] = \frac{p^2 - a^2}{(p^2 + a^2)^2}.$$

(b) We have $L[e^{at}] = \dfrac{1}{p - a}$, so

$$L[t^2 e^{at}] = \frac{d^2}{dp^2}\left[\frac{1}{p - a}\right] = -\frac{d}{dp}\left[\frac{1}{(p - a)^2}\right] = \frac{2}{(p - a)^3}.$$

(c) We know that $L[\cosh 3t] = \dfrac{p}{p^2 - 9}$.

So, $L[t \cosh 3t] = -\dfrac{d}{dp}\left[\dfrac{p}{p^2 - 9}\right] = \dfrac{p^2 + 9}{(p^2 - 9)^2}$.

(d) We have $L[\sin at] = \dfrac{a}{p^2 + a^2}$.

So, $L[t^2 \sin at] = \dfrac{d^2}{dp^2}\left[\dfrac{a}{p^2 + a^2}\right] = \dfrac{d}{dp}\left[\dfrac{-2ap}{(p^2 + a^2)^2}\right] = \dfrac{2a(3p^2 - a^2)}{(p^2 + a^2)^3}$.

(e) We know that $L[\cos 2t] = \dfrac{p}{p^2+4}$.

So, $L[t\cos 2t] = -\dfrac{d}{dp}\left[\dfrac{p}{p^2+4}\right] = \dfrac{p^2-4}{(p^2+4)^2}$.

Now, $L[t\sin^2 t] = \dfrac{1}{2}L[t(1-\cos 2t)]$

$$= \frac{1}{2}L[t] - \frac{1}{2}L[t\cos 2t] = \frac{1}{2p^2} - \frac{p^2-4}{2(p^2+4)^2}$$

$$= \frac{6p^2+8}{p^2(p^2+4)^2}.$$

2. Let $F(t) = e^{-at} - e^{-bt}$, then $f(p) = L[F(t)] = \dfrac{1}{p+a} - \dfrac{1}{p+b}$.

Now using the formula,

$$L\left[\frac{F(t)}{t}\right] = \int_p^\infty f(p)\,dp = \int_p^\infty \left[\frac{1}{p+a} - \frac{1}{p+b}\right]dp = \ln\left(\frac{p+b}{p+a}\right)$$

i.e., $$\int_0^\infty e^{-pt}\frac{e^{-at}-e^{-bt}}{t}\,dt = \ln\frac{b+p}{a+p}.$$

Now making $p \to 0$, we get $\displaystyle\int_0^\infty \frac{e^{-at}-e^{-bt}}{t}\,dt = \ln\left(\frac{b}{a}\right)$.

3. We have $L[\sin at] = \dfrac{a}{p^2+a^2} = f(p)$ (say). Then

$$L\left[\frac{\sin at}{t}\right] = \int_p^\infty f(x)\,dx = \int_p^\infty \frac{x}{x^2+a^2}\,dx = \left[\tan^{-1}\left(\frac{x}{a}\right)\right]_p^\infty$$

$$= \frac{\pi}{2} - \tan^{-1}\left(\frac{p}{a}\right) = \cot^{-1}\left(\frac{p}{a}\right).$$

4. We have $L[\cos at] = \dfrac{a}{p^2+a^2} = f(p)$ (say). Then

$$L\left[\frac{\sin at}{t}\right] = \int_p^\infty f(x)\,dx = \int_p^\infty \frac{x}{x^2+a^2}\,dx = \left[\frac{1}{2}\ln(x^2+a^2)\right]_p^\infty$$

$= \dfrac{1}{2}\lim_{x\to\infty} \ln(x^2+a^2) - \ln(p^2+a^2)$, which does not exist.

5. We have $L[\sin t] = \dfrac{1}{p^2+1} = f(p)$ (say). Then

$$L\left[\frac{\sin at}{t}\right] = \int_p^\infty f(x)\,dx = \int_p^\infty \frac{x}{x^2+1}\,dx = \left[\tan^{-1} x\right]_p^\infty = \frac{\pi}{2} - \tan^{-1} p = \cot^{-1} p.$$

i.e. $$\int_0^\infty e^{-pt}\frac{\sin t}{t}\,dt = \cot^{-1} p.$$

Putting $p = 1$, we get $\displaystyle\int_0^\infty e^{-t}\frac{\sin t}{t}\,dt = \frac{\pi}{4}.$

6. $$L[F(t-a)] = \int_0^\infty e^{-pt}F(t-a)\,dt = \int_a^\infty e^{-pt}\,dt = -\left[\frac{e^{-pt}}{p}\right]_a^\infty = \frac{e^{-pa}}{p}.$$

7. Using the formula of Laplace transform of periodic function

$$L[F(t)] = \frac{\int_0^T e^{-pt}F(t)\,dt}{1-e^{-pT}},$$

we get $$L[F(t)] = \frac{\int_0^1 te^{-pt}\,dt}{1-e^{-p}} = \frac{-\left[\dfrac{te^{-pt}}{p}\right]_0^1 - \left[\dfrac{e^{-pt}}{p^2}\right]_0^1}{1-e^{-p}}$$

$$= \frac{-pe^{-p} + (1-e^{-p})}{p^2(1-e^{-p})} = \frac{1-(p+1)e^{-p}}{p^2(1-e^{-p})}.$$

8. Let $L(t) = \cos at - \cos bt$, then

$$f(p) = L[F(t)] = \frac{p}{p^2+a^2} - \frac{p}{p^2+b^2}.$$

Now using the formula

$$L\left[\frac{F(t)}{t}\right] = \int_p^\infty f(x)\,dx,$$

we get $$L\left[\frac{\cos at - \cos bt}{t}\right] = \int_p^\infty \left[\frac{x}{x^2+a^2} - \frac{x}{x^2+b^2}\right]dx$$

$$= -\frac{1}{2}\ln\left(\frac{p^2+a^2}{p^2+b^2}\right)$$

$$= \frac{1}{2}\ln\left(\frac{p^2+b^2}{p^2+a^2}\right).$$

9. Let $F(t) = \sin at$, then $F'(t) = a\cos at$, $F''(t) = -a^2 \sin at$. Now using the formula

$$L[F''(t)] = p^2 L[F(t)] - pF(0) - F'(0),$$

we get
$$-a^2 L[\sin at] = p^2 L[\sin at] - 0 - a$$

or,
$$(p^2 + a^2) L[\sin at] = a$$

or,
$$L[\sin at] = \frac{a}{p^2 + a^2}.$$

10. Let $F(t) = \sin\sqrt{t}$, then $F(0) = 0$ and $F'(t) = \dfrac{\cos\sqrt{t}}{2\sqrt{t}}$.

Now using the result $L[F'(t)] = pL[F(t)] - F(0)$,

we get $L\left[\dfrac{\cos\sqrt{t}}{2\sqrt{t}}\right] = pL[\sin\sqrt{t}] - 0 = p\dfrac{\sqrt{\pi}}{2p^{3/2}} e^{-1/4p}$

or, $L\left[\dfrac{\cos\sqrt{t}}{\sqrt{t}}\right] = \sqrt{\dfrac{\pi}{p}} e^{-1/4p}.$

11. We have $L[\sin t] = \dfrac{1}{p^2 + 1}$.

Now $L\left[\dfrac{\sin t}{t}\right] = \displaystyle\int_p^\infty L[\sin t]\,dp$

$$= \int_p^\infty \frac{1}{p^2+1} dp = \left[\tan^{-1} p\right]_p^\infty = \frac{\pi}{2} - \tan^{-1} p = \tan^{-1}\frac{1}{p}.$$

So,
$$L\left[\int_0^t \frac{\sin x}{x} dx\right] = \frac{1}{p}\tan^{-1}\frac{1}{p}.$$

12. We have $L[1 - e^{-2t}] = \dfrac{1}{p} - \dfrac{1}{p+2}$.

Now
$$L\left[\frac{1-e^{-2t}}{t}\right] = \int_p^\infty \left[\frac{1}{p} - \frac{1}{p+2}\right] dp$$

$$= [\ln p - \ln(p+2)]_p^\infty$$

$$= \left[\ln\left(\frac{p+2}{p}\right)\right]_\infty^p = \ln\left(\frac{p+2}{p}\right).$$

So,
$$L\left[\int_0^t \left(\frac{1-e^{-2x}}{x}\right) dx\right] = \frac{1}{p}\ln\left(\frac{p+2}{p}\right).$$

13. We have $L[1-\cos t]=\dfrac{1}{p}-\dfrac{p}{p^2+1}$

So, $L\left[\dfrac{1-\cos t}{t}\right]=\int_p^\infty \dfrac{1}{p}dp-\int_p^\infty \dfrac{p}{p^2+1}dp$

$$=\left[\ln p-\frac{1}{2}\ln(p^2+1)\right]_p^\infty$$

$$=\frac{1}{2}\left[\ln\frac{p^2}{p^2+1}\right]_p^\infty=\frac{1}{2}\ln\left(\frac{p^2+1}{p^2}\right).$$

Hence, $L\left[\dfrac{1-\cos t}{t^2}\right]=\dfrac{1}{2}\int_p^\infty \ln(p^2+1)\,dp-\int_p^\infty \ln p\,dp$

$$=\frac{1}{2}[p\ln(p^2+1)]_p^\infty-\int_p^\infty \frac{p^2}{p^2+1}dp-[p\ln p]_p^\infty+\int_p^\infty dp$$

$$=\frac{1}{2}\left[p\ln\left(\frac{p^2+1}{p^2}\right)\right]_p^\infty+\int_p^\infty \frac{p^2}{p^2+1}dp$$

$$=\frac{1}{2}p\ln\left(\frac{p^2}{p^2+1}\right)+\frac{\pi}{2}-\tan^{-1}p.$$

Now, taking limit $p\to 0$, we get

$$\int_0^\infty \frac{1-\cos t}{t^2}dt=\frac{\pi}{2}\quad\text{or,}\quad\int_0^\infty \frac{\sin^2 t}{t^2}dt=\frac{\pi}{2}.$$

EXERCISE

1. (a) State the conditions of existence of the Laplace transform of a derivative of a function $f(t)$ and prove that

$$L[F'(t)]=pL[F(t)]-F(0).$$

(b) If $f(t)$ be continuous for all $t\ge 0$ and be of exponential order as $t\to\infty$ and $f'(t)$ is of class A, then prove the initial and final value theorem of Laplace theorem.

2. Prove that

(i) $L[F''(t)]=p^2L[F(t)]-pL(0)-F'(0)$,

(ii) $L[F'''(t)]=p^3L[F(t)]-p^2F(0)-pF'(0)-F''(0)$

(iii) $L[F^n(t)]=p^nL[F(t)]-p^{n-1}F(0)-p^{n-2}F'(0)\cdots-pF^{n-2}(0)-F^{n-1}(0)$.

3. Find:

(a) $L[(1+te^{-1})^3]$, (b) $L[te^{-t}\cos t]$, (c) $L[te^t \sin t]$,

(d) $L[\sin at - at\cos at]$, (e) $L[t^n e^{at}]$, (f) $L[t\{3\sin 2t - 2\cos 2t\}]$,

(g) $L[(t^2 - 3t + 2)\sin 3t]$, (h) $L[t^2 \cos at]$, (i) $L[t\cos^2 t]$,

(j) $L[t\cosh 3t]$, (k) $L[t\sinh 2t]$, (l) $L[t^3 \cos t]$.

4. Show that $\int_0^\infty \frac{\sin t}{t}\,dt = \frac{\pi}{2}$.

5. If $F(t) = t^2, 0 < t < 2$ and $F(t+2) = F(t)$, find $L[F(t)]$.

6. If $F(t) = 3t,\ 0 < t < 2$

$= 6,\ 2 < t < 4$

and $F(t+4) = F(t)$, find $L[F(t)]$.

7. Find $L[F(t)]$ where $F(t) = \sin t,\ 0 < t < \pi$

$= 0,\ \ \pi < t < 2\pi$

and $F(t)$ is a periodic function of period 2π.

8. Show that $\int_0^\infty \frac{e^{4t} - e^{6t}}{t} = \ln\frac{3}{2}$.

9. Show that $\int_0^\infty \frac{e^{-3t} - e^{-6t}}{t} = \ln 2$.

10. Show that $\int_0^\infty te^{-3t}\sin t\,dt = \frac{3}{50}$.

11. Show that $\int_0^\infty te^{-2t}\cos t\,dt = \frac{3}{25}$.

12. Show that $\int_0^\infty t^3 e^{-t}\sin t\,dt = 0$.

13. Find $L\left[\frac{\sin at - \sin bt}{t}\right]$.

14. Find $L\left[\frac{\sinh t}{t}\right]$.

15. Find $L\left[\frac{\sin^2 t}{t}\right]$.

16. Show that $\int_0^\infty e^{-t}\frac{\sin^2 t}{t}dt = \frac{1}{4}\ln 5.$

17. If $L[F(t)] = f(p)$, show that $\int_0^\infty \frac{F(t)}{t}dt = \int_0^\infty f(p)dp$, assuming that the integrals converge.

18. Show that if $f(t)$ be a periodic function of period T, then $L[F(t)] = \frac{\int_0^T e^{-pt}F(t)dt}{1-e^{-pT}}$.

Hence deduce that $L[\cos t] = \frac{p}{p^2+1}$.

19. Find the Laplace transform of $F(t)$, where

$$F(t) = \begin{cases} \frac{t}{T}, & 0 < t < T \\ 1, & t > T. \end{cases}$$

20. Find the Laplace transform of $F(t)$, where

$$F(t) = \begin{cases} \sin \omega t, & 0 < t < \frac{\pi}{\omega} \\ 0, & \frac{\pi}{\omega} < t < \frac{2\pi}{\omega}. \end{cases}$$

21. Define Laplace transform, $L\{f(t)\}$. If $L\{f(t)\} = F(s)$, prove that

(i) $L[t^2 f(t)] = \frac{d^2}{ds^2}F(s)$, (ii) $L[t^n f(t)] = (-1)^n \frac{d^n}{ds^n}F(s)$.

22. Using the formula for Laplace tranfrom of derivatives, prove that $L[\cos at] = \frac{p}{p^2+a^2}$.

23. Show that

(i) $L[\sinh at \sin at] = \frac{2a^2 p}{p^4+4a^4}$ and (ii) $L[\sinh at \cos at] = \frac{a(p^2-2a^2)}{p^4+4a^4}$.

[*Hints*: Consider the Laplace transform of $f(t) = e^{iat}\sinh at$.]

24. Evaluate: $L\left[\int_0^t \left(\frac{e^{-x}-e^{-2x}}{x}\right)dx\right]$.

25. Evaluate: $L\left[\int_0^t \frac{2(1-\cos x)}{x}dx\right]$.

26. Evaluate: $L\left[\int_0^t \frac{2(1-\cosh x)}{x}dx\right]$.

27. Verify that $L\left[\int_0^t (e^{-x}-x)dx\right] = \frac{1}{p}L[e^{-t}-t]$.

ANSWERS

3. (a) $\frac{1}{p}+\frac{3}{(p+1)^2}+\frac{6}{(p+2)^3}+\frac{6}{(p+3)^4}$, (b) $\frac{p(p+2)}{(p^2+2p+2)^2}$,

(c) $-\frac{2(p-1)}{(p^2-2p+2)^2}$, (d) $\frac{2a^3}{(p^2+a^2)^2}$, (e) $\frac{n!}{(p-a)^{n+1}}, p>a$,

(f) $\frac{8+12p-2p^2}{(p^2+4)^2}$, (g) $\frac{6p^4-18p^3+126p^2-162p+432}{(p^2+9)^3}$,

(h) $\frac{2p(p^2-3a^2)}{(p^2+a^2)^3}$, (i) $\frac{p^4+2p^2+8}{p^2(p^2+4)^2}$, (j) $\frac{p^2+9}{(p^2-9)^2}$,

(k) $\frac{4p}{(p^2-4)^2}$, (l) $\frac{6(p^4-6p^2+1)}{(p^2+1)^4}$.

5. $\frac{2-(4p^2+4p+2)e^{-2p}}{p^3(1-e^{-2p})}$. **6.** $\frac{3[1-(1+2pe^{-2p})e^{-2p}}{p^2(1-e^{-4p})}$.

7. $\frac{-1}{(p^2+1)(1-e^{-p\pi})}$. **13.** $\tan^{-1}\left\{\frac{p(a-b)}{p^2+ab}\right\}$. **14.** $\frac{1}{2}\ln\left(\frac{p+1}{p-1}\right)$.

15. $\frac{1}{4}\ln\left(\frac{p^2+4}{p^2}\right)$. **19.** $-\frac{e^{-pT}-1}{p^2T}$. **20.** $\frac{\omega\left(1+e^{-\frac{pT}{\omega}}\right)}{(p^2+\omega^2)\left(1-e^{-\frac{2p\pi}{\omega}}\right)}$.

24. $\frac{1}{p}\ln\left(\frac{p+2}{p+1}\right)$. **25.** $\frac{1}{p}\ln\left(1+\frac{1}{p^2}\right)$. **26.** $\frac{1}{p}\ln\left(1-\frac{1}{p^2}\right)$.

Chapter 3

INVERSE LAPLACE TRANSFORM

3.1 Introduction

When an integral transform is used to solve an ordinary or a partial differential equation we get a differential equation or a result involving the transformed variable, solving which we have to get result in the original variable. So an integral transform will be meaningful if its inverse transform exists.

If $f(p)$ is the Laplace transform of a function $F(t)$, $F(t)$ is called *inverse Laplace transform of* $f(p)$ and is written as $F(t) = L^{-1}[f(p)]$.

As the direct determination of inverse transform of a given function is a difficult task, we will calculate the inverse Laplace transform of a function using the results of Laplace transforms of different functions discussed in chapter 2.

3.2 Linearity Property

THEOREM

If $f_1(p), f_2(p)$ be Laplace transform of the functions $F_1(t), F_2(t)$ respectively and c_1, c_2 be two constants, then

$$L^{-1}[c_1 f_1(p) + c_2 f_2(p)] = c_1 L^{-1}[f_1(p)] + c_2 L^{-1}[f_2(p)] = c_1 F_1(t) + c_2 F_2(t).$$

Proof: We have $L[c_1 F_1(t) + c_2 F_2(t)] = c_1 L[F_1(t)] + c_2 L[F_2(t)] = c_1 f_1(p) + c_2 f_2(p)$.

Taking inverse of both sides,

$$L^{-1}[c_1 f_1(p) + c_2 f_2(p)] = c_1 F_1(t) + c_2 F_2(t) = c_1 L^{-1}[f_1(p)] + c_2 L^{-1}[f_2(p)].$$

Note: In general, $L^{-1}\left[\Sigma_{k=1}^{n} c_k f_k(p)\right] = \Sigma_{k=1}^{n} c_k L^{-1}[f_k(p)]$.

WORKED-OUT EXAMPLES

- Show that

1. $L^{-1}\left(\frac{1}{p}\right)=1.$ **2.** $L^{-1}\left(\frac{1}{p-a}\right)=e^{at}.$

3. $L^{-1}\left(\frac{1}{p^2-a^2}\right)=\frac{1}{a}\sinh at.$ **4.** $L^{-1}\left(\frac{p}{p^2-a^2}\right)=\cosh at.$

5. $L^{-1}\left(\frac{1}{p^2+a^2}\right)=\frac{1}{a}\sin at.$ **6.** $L^{-1}\left(\frac{p}{p^2+a^2}\right)=\cos at.$

7. $L^{-1}\left(\frac{1}{p^{n+1}}\right)=\frac{t^n}{n!}.$

- Find the inverse Laplace transform of the following functions:

8. $\frac{2p+1}{p(p+1)}.$ **9.** $\frac{3p-8}{4p^2+25}.$ **10.** $\frac{2p-5}{p^2-9}.$

11. $\frac{3p}{p^2+2}-\frac{4p}{p^2-16}.$ **12.** $\frac{6}{2p-3}+\frac{4p+9}{9p^2-16}.$ **13.** $\frac{3}{p^2-3}+\frac{3p+4}{p^2}.$

Solutions

1. Since $L[1]=\frac{1}{p}$, $L^{-1}\left[\frac{1}{p}\right]=1.$

2. Since $L[e^{at}]=\frac{1}{p-a}$, $L^{-1}\left[\frac{1}{p-a}\right]=e^{at}.$

3. Since $L[\sinh at]=\frac{a}{p^2-a^2}$, $L^{-1}\left[\frac{1}{p^2-a^2}\right]=\frac{1}{a}\sinh at.$

4. Since $L[\cosh at]=\frac{p}{p^2-a^2}$, $L^{-1}\left[\frac{p}{p^2-a^2}\right]=\frac{1}{a}\cosh at.$

5. Since $L[\sin at]=\frac{a}{p^2+a^2}$, $L^{-1}\left[\frac{1}{p^2+a^2}\right]=\frac{1}{a}\sin at.$

6. Since $L[\cos at]=\frac{p}{p^2+a^2}$, $L^{-1}\left[\frac{1}{p^2+a^2}\right]=\cos at.$

7. Since $L[t^n]=\frac{n!}{p^{n+1}}$, $L^{-1}\left[\frac{1}{p^{n+1}}\right]=\frac{t^n}{n!}.$

8. $L^{-1}\left[\frac{2p+1}{p(p+1)}\right]=L^{-1}\left[\frac{1}{p}+\frac{1}{p+1}\right]=L^{-1}\left[\frac{1}{p}\right]+L^{-1}\left[\frac{1}{p+1}\right]=1+e^{-t}.$

9. $L^{-1}\left[\frac{3p-8}{4p^2+25}\right]=3L^{-1}\left[\frac{p}{4p^2+25}\right]-8L^{-1}\left[\frac{1}{4p^2+25}\right]=\frac{3}{4}\cos\left(\frac{5t}{2}\right)-\frac{4}{5}\sin\left(\frac{5t}{2}\right).$

10. $L^{-1}\left[\frac{2p-5}{p^2-9}\right]=2L^{-1}\left[\frac{p}{p^2-9}\right]-5L^{-1}\left[\frac{1}{p^2-9}\right]=2\cosh 3t-\frac{5}{3}\sinh 3t.$

11. $L^{-1}\left[\frac{3p}{p^2+2}-\frac{4p}{p^2-16}\right]=3L^{-1}\left[\frac{p}{p^2+2}\right]-4L^{-1}\left[\frac{p}{p^2-16}\right]=3\cos\sqrt{2}t-4\cosh 4t.$

12. $L^{-1}\left[\frac{6}{2p-3}+\frac{4p+9}{9p^2-16}\right]=3L^{-1}\left[\frac{1}{p-3/2}\right]+\frac{4}{9}L^{-1}\left[\frac{p}{p^2-16/9}\right]+L^{-1}\left[\frac{1}{p^2-16/9}\right]$

$$=3e^{3t/2}+\frac{4}{9}\cosh\left(\frac{4t}{3}\right)+\frac{3}{4}\sinh\left(\frac{4t}{3}\right).$$

13. $L^{-1}\left[\frac{3}{p^2-3}+\frac{3p+4}{p^2}\right]=3L^{-1}\left[\frac{1}{p^2-3}\right]+3L^{-1}\left[\frac{1}{p}\right]+4L^{-1}\left[\frac{1}{p^2}\right]$

$$=\sqrt{3}\sinh\sqrt{3}t+3+4t.$$

Table of Inverse Laplace Transform

	$F(p)$	$L^{-1}[f(p)]=F(t)$
1.	$\frac{1}{p^{n+1}}, n=0,1,2,\ldots$	$\frac{t^n}{n!}$
2.	$\frac{1}{p-a}$	e^{at}
3.	$\frac{1}{p^2+a^2}$	$\frac{1}{a}\sin at$
4.	$\frac{p}{p^2+a^2}$	$\cos at$
5.	$\frac{1}{p^2-a^2}$	$\frac{1}{a}\sinh at$
6.	$\frac{p}{p^2-a^2}$	$\cosh at$

3.3 First Shifting Theorem

THEOREM

If $L^{-1}[f(p)] = F(t)$, then $L^{-1}[f(p-a)] = e^{at}F(t) = e^{at}L^{-1}[f(p)]$.

Proof: We have $f(p) = \int_0^\infty e^{-pt}F(t)dt$.

So, $f(p-a) = \int_0^\infty e^{-(p-a)t}F(t)dt = \int_0^\infty e^{-pt}[e^{at}F(t)]dt = L[e^{at}F(t)]$.

Hence, $L^{-1}[f(p-a)] = e^{at}F(t)$.

3.4 Second Shifting Theorem

THEOREM

If $L^{-1}[f(p)] = F(t)$, then $L^{-1}[e^{-ap}f(p)] = G(t)$, where

$$G(t) = \begin{cases} F(t-a), & t > a \\ 0, & t < a. \end{cases}$$

Proof: We have $f(p) = \int_0^\infty e^{-pt}F(t)dt$. So,

$$e^{-ap}f(p) = \int_0^\infty e^{-p(t+a)}F(t)dt$$

$$= \int_0^\infty e^{-px}F(x-a)dx$$

$$= \int_0^a e^{-pt}(0)dx + \int_0^\infty e^{-pt}F(t-a)dt$$

$$= \int_0^\infty e^{-pt}G(t)dt$$

$$= L[G(t)]$$

where $G(t) = \begin{cases} F(t-a), & t > a \\ 0, & t < a. \end{cases}$

3.5 Change of Scale Property

THEOREM

If $L^{-1}[f(p)] = F(t)$, then $L^{-1}[f(ap)] = \frac{1}{a}F\left(\frac{t}{a}\right)$.

Proof: We have $f(p) = \int_0^\infty e^{-pt}F(t)\,dt$.

So, $f(ap) = \int_0^\infty e^{-apt}F(t)\,dt = \frac{1}{a}\int_0^\infty e^{-px}F\left(\frac{x}{a}\right)dx = \frac{1}{a}\int_0^\infty e^{-pt}F\left(\frac{t}{a}\right)dt = L\left[\frac{1}{a}F\left(\frac{t}{a}\right)\right]$.

Taking inverse Laplace transform of both sides, we get

$$L^{-1}[f(ap)] = \frac{1}{a}F\left(\frac{t}{a}\right).$$

3.6 Inverse Transform of Derivatives

THEOREM

If $L^{-1}[f(p)] = F(t)$, then $L^{-1}[f^n(p)] = (-1)^n t^n F(t)$.

Proof: We have $L[t^n F(t)] = (-1)^n \frac{d^n}{dp^n} f(p) = (-1)^n f^n(p)$.

Taking inverse Laplace transform of both sides, we get

$$L^{-1}[f^n(p)] = (-1)^n t^n F(t).$$

3.7 Inverse Transform of Integrals

THEOREM

If $L^{-1}[f(p)] = F(t)$, then

$$L^{-1}\left[\int_p^\infty f(u)\,du\right] = \frac{F(t)}{t}.$$

Proof: We have $L[F(t)] = f(p)$, then we have

$$L\left[\frac{F(t)}{t}\right] = \int_p^\infty f(u)\,du, \text{ provided } \lim_{t\to 0}\frac{F(t)}{t} \text{ exists.}$$

Taking inverse Laplace transform of both sides, we get

$$L^{-1}\left[\int_p^\infty f(u)\,du\right] = \frac{F(t)}{t}.$$

3.8 Multiplication by *p*

THEOREM

If $L^{-1}[f(p)] = F(t)$ and $F(0) = 0$, then $L^{-1}[pf(p)] = F'(t)$.

Proof: Using the result $L[F'(t)] = pL[F(t)] - F(0)$, we get

$$L[F'(t)] = pL[F(t)] = pf(p).$$

Taking inverse Laplace transform, we get

$$L^{-1}[pf(p)] = F'(t).$$

3.9 Division by *p*

THEOREM

If $L^{-1}[f(p)] = F(t)$, then $L^{-1}\left[\dfrac{f(p)}{p}\right] = \int_0^t F(u)\,du.$

Proof: Let $G(t) = \int_0^t F(u)\,du$, then $G'(t) = F(t)$ and $G(0) = 0$. Using the result

$$L[G'(t)] = pL[G(t)] - G(0),$$

we get $$L[F(t)] = f(p) = pL[G(t)]$$

or, $$L[G(t)] = \frac{f(p)}{p}.$$

Taking inverse Laplace transform we get

$$L^{-1}\left[\frac{f(p)}{p}\right] = G(t) = \int_0^t F(u)\,du.$$

WORKED-OUT EXAMPLES

1. Find L^{-1} of the following functions:

(i) $\dfrac{1}{p^2 - 6p + 10}$, (ii) $\dfrac{1}{p^2 + 8p + 16}$, (iii) $\dfrac{3p-2}{p^2 - 4p + 20}$,

(iv) $\dfrac{3p+7}{p^2 - 2p - 3}$, (v) $\dfrac{1}{(p+a)^n}$, (vi) $\dfrac{p}{(p+a)^3}$,

(vii) $\dfrac{p-1}{p^2 + 6p + 25}$, (viii) $\dfrac{3p+5}{(p-1)^3}$, (ix) $\dfrac{1}{(p+a)^3}$,

(x) $\dfrac{p}{(p^2-a^2)^2}$, (xi) $\dfrac{p}{(p^2+a^2)^2}$,

2. Prove that $L^{-1}\left[\ln\left(\dfrac{p+2}{p+1}\right)\right]=\dfrac{e^{-t}-e^{-2t}}{t}$.

3. Show that $L^{-1}\left[\dfrac{1}{p}\sin\dfrac{1}{p}\right]=t-\dfrac{t^3}{(3!)^2}+\dfrac{t^5}{(5!)^2}-\dfrac{t^7}{(7!)^2}+\cdots$.

4. Prove that $L^{-1}\left[\ln\left(1-\dfrac{1}{p^2}\right)\right]=\dfrac{2[1-\cosh t]}{t}$.

5. For $a>0$, prove that $L^{-1}[f(p)]=F(t)$ implies

$$L^{-1}[f(ap+b)]=\frac{1}{a}e^{-\frac{bt}{a}}F\left(\frac{t}{a}\right).$$

6. Evaluate $L^{-1}\left\{\dfrac{1}{p^2(p^2+1)}\right\}$.

7. Evaluate $L^{-1}\left\{\dfrac{e^{-2p}}{(p-3)^2}\right\}$.

Solutions

1. (i) $L^{-1}\left(\dfrac{1}{p^2-6p+10}\right)$

$=L^{-1}\left(\dfrac{1}{(p-3)^2+1}\right)=e^{3t}L^{-1}\left(\dfrac{1}{p^2+1}\right)$, by first shifting theorem

$=e^{3t}\sin t$.

(ii) $L^{-1}\left(\dfrac{1}{p^2+8p+16}\right)$

$=L^{-1}\left[\dfrac{1}{(p+4)^2}\right]$, by first shifting theorem

$=e^{-4t}L^{-1}\left[\dfrac{1}{p^2}\right]=e^{-4t}t$.

(iii) $L^{-1}\left(\frac{3p-2}{p^2-4p+20}\right)$

$$= 3L^{-1}\left(\frac{p-2}{(p-2)^2+16}\right) + 4L^{-1}\left(\frac{1}{(p-2)^2+16}\right)$$

$$= 3e^{2t}L^{-1}\left(\frac{p}{p^2+16}\right) + e^{2t}L^{-1}\left(\frac{4}{p^2+16}\right), \text{ by first shifting theorem}$$

$$= 3e^{2t}\cos 4t + e^{2t}\sin 4t.$$

(iv) $L^{-1}\left(\frac{3p+7}{p^2-2p-3}\right)$

$$= 3L^{-1}\left(\frac{p-1}{(p-1)^2-4}\right) + 10L^{-1}\left(\frac{1}{(p-1)^2-4}\right)$$

$$= 3e^{t}L^{-1}\left(\frac{p}{p^2-4}\right) + 5e^{t}L^{-1}\left(\frac{2}{p^2-4}\right), \text{ by first shifting theorem}$$

$$= 3e^{t}\cosh 2t + 5e^{t}\sinh 2t.$$

(v) $L^{-1}\left[\frac{1}{(p+a)^n}\right]$

$$= e^{-at}L^{-1}\left[\frac{1}{p^n}\right], \text{ by first shifting theorem}$$

$$= \frac{e^{-at}t^{n-1}}{(n-1)!}.$$

(vi) $L^{-1}\left[\frac{p}{(p+a)^3}\right]$, by first shifting theorem

$$= e^{-at}L^{-1}\left[\frac{p-a}{p^3}\right]$$

$$= e^{-at}L^{-1}\left[\frac{1}{p^2}\right] - ae^{-at}L^{-1}\left[\frac{1}{p^3}\right]$$

$$= te^{-at} - \frac{a}{2}t^2e^{-at}.$$

(vii) $L^{-1}\left(\dfrac{p-1}{p^2+6p+25}\right)$

$$= L^{-1}\left(\frac{p+3-4}{(p+3)^2+16}\right), \text{ by first shifting theorem}$$

$$= e^{-3t}L^{-1}\left(\frac{p}{p^2+16}\right) - e^{-3t}L^{-1}\left(\frac{4}{p^2+16}\right)$$

$$= e^{-3t}[\cos 4t - \sin 4t].$$

(viii) $L^{-1}\left[\dfrac{3p+5}{(p-1)^3}\right]$

$$= L^{-1}\left[\frac{3p-3+8}{(p-1)^3}\right]$$

$$= 3L^{-1}\left[\frac{1}{(p-1)^2}\right] + 8L^{-1}\left[\frac{1}{(p-1)^3}\right], \text{ by first shifting theorem}$$

$$= 3e^t L^{-1}\left[\frac{1}{p^2}\right] + 8e^t L^{-1}\left[\frac{1}{p^3}\right]$$

$$= 3t\,e^t + 4t^2 e^t.$$

(ix) Since $\dfrac{d^2}{dp^2}\left(\dfrac{1}{p+a}\right) = \dfrac{2}{(p+a)^3}$,

$$L^{-1}\left[\frac{1}{(p+a)^3}\right] = \frac{1}{2}L^{-1}\left[\frac{d^2}{dp^2}\left(\frac{1}{p+a}\right)\right] = \frac{1}{2}(-1)^2 t^2 L^{-1}\left(\frac{1}{p+a}\right) = \frac{t^2 e^{-at}}{2}.$$

(x) Since $\dfrac{d}{dp}\left(\dfrac{1}{p^2-a^2}\right) = \dfrac{-2p}{(p^2-a^2)^2}$,

$$L^{-1}\left[\frac{p}{(p^2-a^2)^2}\right] = -\frac{1}{2}L^{-1}\left[\frac{d}{dp}\left(\frac{1}{p^2-a^2}\right)\right] = -\frac{1}{2}(-1)\,t\,L^{-1}\left[\left(\frac{1}{p^2-a^2}\right)\right] = \frac{t}{2a}\sinh at.$$

(xi) Since $\dfrac{d}{dp}\left(\dfrac{1}{p^2+a^2}\right) = \dfrac{-2p}{(p^2+a^2)^2}$,

$$L^{-1}\left[\frac{p}{(p^2+a^2)^2}\right] = -\frac{1}{2}L^{-1}\left[\frac{d}{dp}\left(\frac{1}{p^2+a^2}\right)\right] = -\frac{1}{2}(-1)\,t\,L^{-1}\left[\left(\frac{1}{p^2+a^2}\right)\right] = \frac{t}{2a}\sin at.$$

2. Let $f(p) = \log(p+2) - \log(p+1)$, then

$$f'(p) = \frac{1}{p+2} - \frac{1}{p+1}.$$

So, $\quad L^{-1}[f'(p)] = e^{-2t} - e^{-t}$

$$\Rightarrow (-1)t\,L^{-1}[f(p)] = e^{-2t} - e^{-t}$$

$$\Rightarrow L^{-1}[f(p)] = \frac{e^{-t} - e^{-2t}}{t}$$

$$\Rightarrow L^{-1}\left[\ln\left(\frac{p+2}{p+1}\right)\right] = \frac{e^{-t} - e^{-2t}}{t}.$$

3. We have $L^{-1}\left[\frac{1}{p}\sin\frac{1}{p}\right] = L^{-1}\left[\frac{1}{p}\left(\frac{1}{p} - \frac{1}{3!p^3} + \frac{1}{5!p^5} - \cdots\right)\right]$

$$= L^{-1}\left[\frac{1}{p^2} - \frac{1}{3!p^4} + \frac{1}{5!p^6} - \cdots\right]$$

$$= L^{-1}\left[\frac{1}{p^2}\right] - L^{-1}\frac{1}{3!p^4} + L^{-1}\left[\frac{1}{5!p^6}\right] - \cdots$$

$$= t - \frac{t^3}{(3!)^2} + \frac{t^5}{(5!)^2} - \cdots.$$

4. Let $f(p) = \ln\left(1 - \frac{1}{p^2}\right) = \ln(p^2 - 1) - 2\ln p$. Then

$$f'(p) = \frac{2p}{p^2 - 1} - \frac{2}{p}$$

$$\Rightarrow L^{-1}[f'(p)] = 2\cosh t - 2$$

$$\Rightarrow -t\,L^{-1}[f(p)] = 2\cosh t - 2$$

$$\Rightarrow L^{-1}[f(p)] = L^{-1}\left[\ln\left(1 - \frac{1}{p^2}\right)\right] = \frac{2[1 - \cosh t]}{t}.$$

5. Given that $f(p) = L[F(t)] = \int_0^\infty e^{-pt} F(t)\,dt$.

So, $$f(ap+b) = \int_0^\infty e^{-(ap+b)t} F(t)\,dt$$

$$\Rightarrow f(ap+b) = \int_0^\infty e^{-px} e^{-bx/a} F\left(\frac{x}{a}\right) d\left(\frac{x}{a}\right), \text{ on putting } t = \frac{x}{a}.$$

$$\Rightarrow f(ap+b) = \frac{1}{a}\int_0^\infty e^{-pt}\left[e^{-bt/a} F\left(\frac{t}{a}\right)\right] dt$$

$$\Rightarrow f(ap+b) = L\left[\frac{1}{a} e^{-bt/a} F\left(\frac{t}{a}\right)\right]$$

$$\Rightarrow L^{-1}[f(ap+b)] = \frac{1}{a} e^{-bt/a} F\left(\frac{t}{a}\right).$$

6. We have $L^{-1}\left\{\dfrac{1}{p^2+a}\right\} = \sin t$. So

$$L^{-1}\left\{\frac{1}{p(p^2+1)}\right\} = \int_0^t \sin x\,dx = 1-\cos t,$$

and $$L^{-1}\left\{\frac{1}{p^2(p^2+1)}\right\} = \int_0^t (1-\cos x)\,dx = t-\sin t.$$

7. Let $L[F(t)] = \dfrac{1}{(p-3)^2}$, then

$$F(t) = L^{-1}\left[\frac{1}{(p-3)^2}\right] = e^{3t}L^{-1}\left[\frac{1}{p^2}\right] = te^{3t}.$$

By second shifting theorem,

$$L^{-1}\left[\frac{e^{-2p}}{(p-3)^2}\right] = \begin{cases}(t-2)e^{3(t-2)} & t>2\\ 0, & t<2.\end{cases}$$

EXERCISE

- Find L^{-1} of the following functions (1-34):

1. $\dfrac{1}{(p-1)(p+2)}$, **2.** $\dfrac{1}{(p-1)^2(p+2)}$, **3.** $\dfrac{4p+3}{(p-2)^3}$,

4. $\dfrac{p}{(p+1)^5}$, **5.** $\dfrac{3p+2}{4p^2+12p+5}$, **6.** $\dfrac{p+1}{p^2+6p+25}$,

7. $\dfrac{1}{(p+1)^2(p^2+4)}$, **8.** $\dfrac{p}{(p-2)^2(p+3)}$, **9.** $\dfrac{4p+5}{(p-1)^2(p+2)}$,

10. $\dfrac{4p+5}{(p-4)^2(p+3)}$, **11.** $\dfrac{1}{\sqrt{2p+3}}$, **12.** $\dfrac{p+1}{p^2+p+1}$,

13. $\dfrac{1}{(p+2)^{5/2}}$, **14.** $\dfrac{1}{p^2(p^2+a^2)}$, **15.** $\dfrac{p^2-2}{(p+1)(p-2)(p-3)}$,

16. $\dfrac{p^2}{(p+2)^3}$ **17.** $\dfrac{2p-1}{p(p^2-1)}$, **18.** $\dfrac{p+1}{(p^2+2p+2)^2}$,

19. $\dfrac{p}{(p^2+4)^2}$, **20.** $\dfrac{3p+1}{(p-1)(p^2+1)}$, **21.** $\dfrac{p^2+2p+3}{(p^2+2p+2)(p^2+2p+5)}$,

22. $\dfrac{p}{(p^2+2p+2)(p^2-2p+2)}$, **23.** $\dfrac{6p^2+22p+18}{p^3+6p^2+11p+6}$,

24. $\dfrac{4p+5}{(p-1)^2(p+2)}$, **25.** $\dfrac{4p+5}{(p-4)^2(p+3)}$, **26.** $\dfrac{2p+1}{(p-1)^2(p+2)^2}$,

27. $\dfrac{p+1}{6p^2+7p+2}$, **28.** $\dfrac{9-4p}{(p+4)(p^2+9)}$, **29.** $\dfrac{p-1}{(p+3)(p^2+2p+2)}$,

30. $\dfrac{p^2-2p+3}{(p-1)^2(p+1)}$, **31.** $\dfrac{p}{p^4+p^2+1}$, **32.** $\dfrac{4p+5}{(p-1)^2(p+2)}$,

33. $\dfrac{1}{p(p^2+2p+2)}$, **34.** $\dfrac{p+1}{(p^2+2p+2)^2}$.

35. Prove that $L^{-1}\left[\ln\left(\dfrac{p+3}{p+2}\right)\right]=\dfrac{e^{-2t}-e^{-3t}}{t}$.

36. Prove that $L^{-1}\left[\ln\left(1+\dfrac{1}{p^2}\right)\right]=\dfrac{2(1-\cos t)}{t}$.

37. Show that $L^{-1}\left[\dfrac{1}{p}\cos\left(\dfrac{1}{p}\right)\right]=1-\dfrac{t^2}{(2!)^2}+\dfrac{t^4}{(4!)^2}-\dfrac{t^6}{(6!)^2}-\cdots$.

38. Show that $L^{-1}\left[\dfrac{1}{(p^2+a^2)^2}\right]=\dfrac{\sin at-at\cos at}{2a^3}$.

39. Find $L^{-1}\left\{\dfrac{1}{p^3(p^2+1)}\right\}$. **40.** Find $L^{-1}\left\{\dfrac{1}{p(p+1)^3}\right\}$.

41. Find $L^{-1}\left\{\dfrac{1}{p^2(p+1)^2}\right\}$. **42.** Find $L^{-1}\left\{\dfrac{1}{p^2(p^2-1)}\right\}$.

43. If $L[t\sin t]=\dfrac{2p}{(p^2+1)^2}$, then find $L^{-1}\left\{\dfrac{1}{(p^2+1)^2}\right\}$.

44. Find $L^{-1}\left\{\dfrac{pe^{-p\pi}}{p^2+1}\right\}$. **45.** Find $L^{-1}\left\{\dfrac{e^{-p}}{p^2-1}\right\}$.

46. Find $L^{-1}\left\{\dfrac{e^{4-3p}}{(p+1)^{5/2}}\right\}$.

ANSWERS

1. $\dfrac{e^t - e^{-2t}}{3}$,

2. $\frac{1}{9}[(3t-1)e^t + e^{-2t}]$,

3. $e^{2t}\left[4t + \frac{11}{2}t^2\right]$,

4. $\frac{e^{-t}}{24}(4t^3 - t^4)$,

5. $\frac{1}{8}e^{-\frac{3t}{2}}(6-5t)$,

6. $e^{-3t}\left[\cos 4t - \frac{1}{2}\sin 4t\right]$,

7. $\dfrac{5t+2}{25}e^{-t} - \dfrac{4\cos 2t + 3\sin 2t}{50}$,

8. $\frac{3}{25}(e^{2t} - e^{-3t}) + \frac{2}{5}te^{2t}$,

9. $\dfrac{e^t - e^{-2t}}{3} + 3te^t$,

10. $\dfrac{e^{4t} - e^{-3t}}{7} + 3te^{4t}$,

11. $\dfrac{e^{-\frac{3t}{2}}}{\sqrt{2\pi t}}$,

12. $e^{-t/2}\left(\cos\frac{\sqrt{3}t}{2} + \frac{1}{\sqrt{3}}\sin\frac{\sqrt{3}t}{2}\right)$,

13. $\dfrac{4t^{\frac{3}{2}}e^{-2t}}{3\sqrt{\pi}}$,

14. $\dfrac{at - \sin at}{a^3}$,

15. $\frac{7}{4}e^{3t} - \dfrac{e^{-t} + 8e^{2t}}{12}$,

16. $e^{-2t}[1 - 4t + 2t^2]$,

17. $1 + \dfrac{e^t - 3e^{-t}}{2}$,

18. $\dfrac{te^{-t}\sin t}{2}$,

19. $\dfrac{t\sin 2t}{4}$,

20. $2e^t - 2\cos t + \sin t$,

21. $\dfrac{e^{-t}(\sin t + \sin 2t)}{3}$,

22. $\frac{1}{2}\sin t \sinh t$,

23. $e^{-t} + 2e^{-2t} + 3e^{-3t}$,

24. $3te^t + \frac{1}{3}e^t - \frac{1}{3}e^{-2t}$,

25. $-\frac{1}{7}e^{-3t} + \frac{1}{7}e^{4t} + 3te^{4t}$,

26. $\frac{1}{3}t(e^t - e^{-2t})$,

27. $\frac{1}{2}e^{-\frac{t}{2}} - \frac{1}{3}e^{-\frac{2t}{3}}$,

28. $e^{-4t} - \cos 3t$,

29. $\frac{1}{5}e^{-t}(4\cos t - 3\sin t) - \frac{4}{5}e^{-3t}$,

30. $\left(t - \frac{1}{2}\right)e^t + \frac{3}{2}e^{-t}$,

31. $\frac{2}{\sqrt{3}}\sinh\frac{t}{2}\sin\frac{\sqrt{3}}{2}t$,

32. $3te^t + \frac{1}{3}e^t - \frac{1}{3}e^{-2t}$,

33. $\frac{1}{2}\left[1-(\sin t+\cos t)e^{-t}\right]$, **34.** $\frac{1}{2}te^{-t}\sin t$;

39. $\frac{t^2}{2}+\cos t-1$, **40.** $1-e^{-t}\left(\frac{t^2}{2}+t+1\right)$,

41. $(t+2)e^{-t}+t-2$, **42.** $\sinh t-1$,

43. $\frac{1}{2}(\sin t-t\cos t)$, **44.** $F(t)=\begin{cases}\cos(t-\pi), & t>\pi\\ 0, & t<\pi\end{cases}$

45. $F(t)=\begin{cases}\sinh(t-1), & t>1\\ 0, & t<1\end{cases}$

46. $F(t)=\begin{cases}\frac{4e^4}{3\sqrt{\pi}}e^{-4(t-3)}(t-3)^{3/2}, & t>3\\ 0, & t<3.\end{cases}$

3.10 Convolution

Let $F(t)$ and $G(t)$ be two functions of class A, then the convolution of the two functions $F(t)$ and $G(t)$ denoted by $F * G$ is defined by the relation

$$F*G=\int_0^t F(x)\,G(t-x)\,dx.$$

The convolution integral satisfies the following properties:

(1) *$F * G$ is commutative.*

Proof: $F*G=\int_0^t F(x)\,G(t-x)\,dx$

$$=-\int_t^0 F(t-y)G(y)\,dy \text{ on putting } x=t-y$$

$$=\int_0^t G(x)F(t-x)\,dx=G*F.$$

(2) *$F * G$ is associative.*

Proof: $(F*G)*H=\int_0^T (F*G)_t H(T-t)\,dt$

$$=\int_{t=0}^T\left(\int_{x=0}^t F(x)G(t-x)\,dx\right)H(T-t)\,dt.$$

Changing the order of integration, we get

$$(F*G)*H = \int_{x=0}^{T} F(x)\,dx \int_{t=x}^{T} G(t-x)H(T-t)\,dt$$

$$= \int_{x=0}^{T} F(x)\,dx \int_{y=0}^{T-x} G(y)H(T-x-y)\,dy, \text{ on putting } y = t-x$$

$$= \int_{x=0}^{T} F(x)(G*H)_{T-x}\,dx$$

$$= F*(G*H).$$

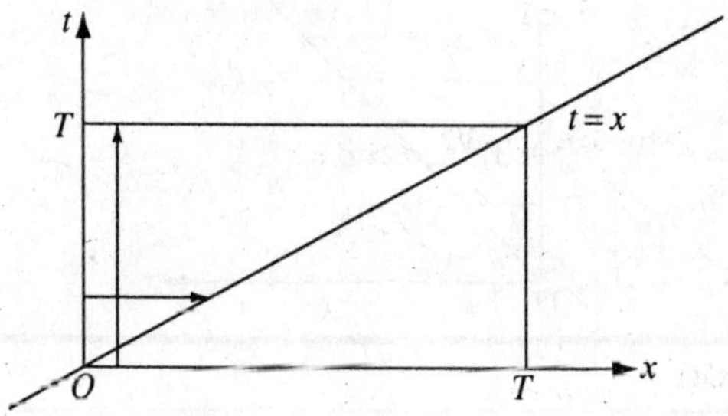

Fig.3.1 Change of order of integration

(3) *$F*G$ is distributive with respect to addition.*

Proof: $F*(G+H) = \int_0^t F(x)[G(t-x)+H(t-x)]\,dx$

$$= \int_0^t F(x)G(t-x)\,dx + \int_0^t F(x)H(t-x)\,dx = F*G + F*H.$$

3.11 Convolution Theorem

THEOREM

If $F(t)$ and $G(t)$ be two functions of class A and let $F(t) = L^{-1}[f(p)]$ and $G(t) = L^{-1}[g(p)]$, then

$$L^{-1}\{f(p)g(p)\} = \int_0^t F(x)G(t-x)\,dx = F*G.$$

Proof: We shall prove:

$$L\left[\int_0^t F(x)\,G(t-x)\,dx\right] = f(p)g(p),$$

which is the equivalent form of the required result. Now

$$L\left[\int_0^t F(x)G(t-x)\,dx\right] = \int_{t=0}^{\infty} e^{-pt}\left\{\int_0^t F(x)\,G(t-x)\,dx\right\}dt$$

$$= \lim_{T\to\infty} \int_{t=0}^{T}\int_{x=0}^{t} e^{-pt}F(x)G(t-x)\,dxdt$$

$$= \lim_{T\to\infty} I_T$$

where $I_T = \int_{t=0}^{T}\int_{x=0}^{t} e^{-pt}F(x)G(t-x)\,dxdt.$

Taking $x + y = t$ we see that the shaded area A_{tx} of the $t - x$ plane is transformed to the shaded area A_{xy} of $x - y$ plane.

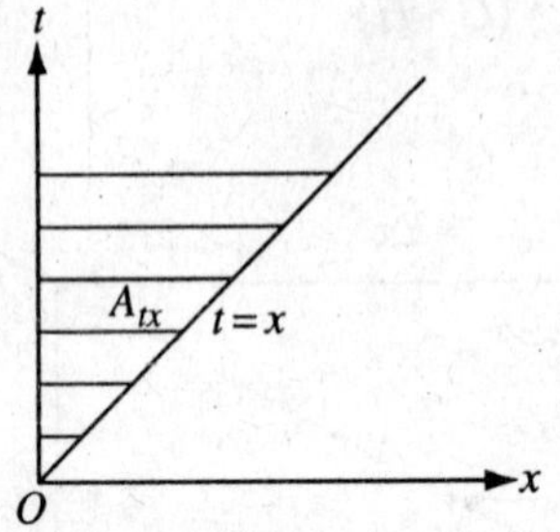

Fig.3.2 The *tx*–plane

Thus $I_T = \iint_{A_{tx}} e^{-pt}F(x)\,G(t-x)\,dxdt$

$$= \iint_{A_{xy}} e^{-p(x+y)}F(x)\,G(y)\left|\frac{\partial(x,t)}{\partial(x,y)}\right| dxdy,$$

where the Jacobian of the transformation

$$\left|\frac{\partial(x,t)}{\partial(x,y)}\right| = \begin{vmatrix} \frac{\partial x}{\partial x} & \frac{\partial x}{\partial y} \\ \frac{\partial t}{\partial x} & \frac{\partial t}{\partial y} \end{vmatrix} = \begin{vmatrix} 1 & 0 \\ 1 & 1 \end{vmatrix} = 1 \neq 0.$$

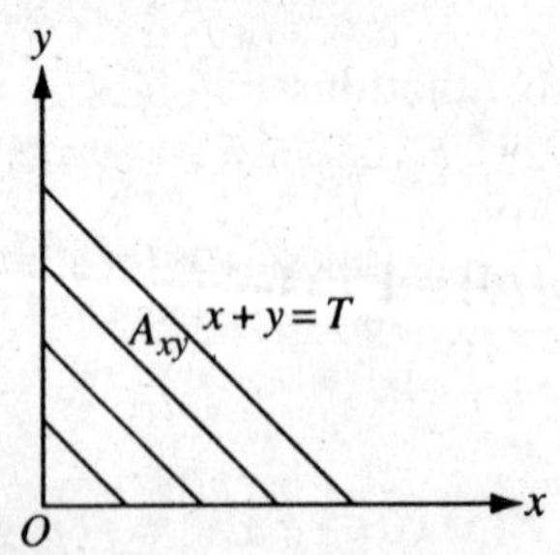

Fig.3.3 The *xy*–plane

So, $$I_T = \int_{y=0}^{T}\int_{x=0}^{T-y} e^{-p(x+y)}F(x)G(y)\,dxdy.$$

Now we define a function

$$K(x, y) = \begin{cases} e^{-p(x+y)}F(x)G(y), & x+y \le T \\ 0, & x+y > T. \end{cases}$$

which is defined over a square of side T such that the integral over the un-shaded area is zero.

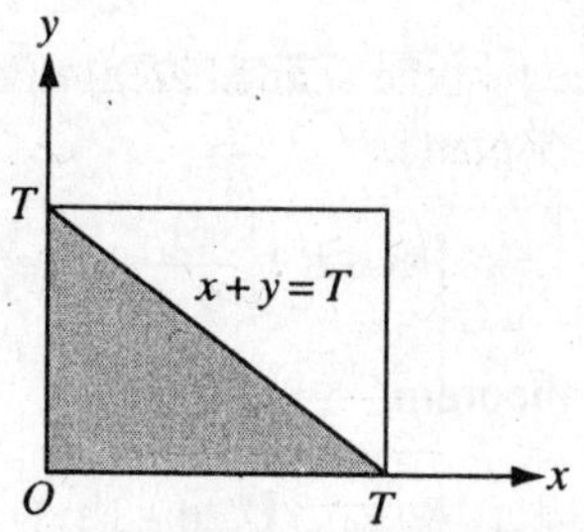

So, $$I_T = \int_{y=0}^{T}\int_{x=0}^{T} K(x, y)\,dxdy.$$

Now, $$L\left[\int_0^t F(x)G(t-x)\,dx\right] = \lim_{T\to\infty} I_T$$

$$= \lim_{T\to\infty}\int_{y=0}^{T}\int_{x=0}^{T} K(x, y)\,dxdy$$

$$= \left[\int_0^{\infty} e^{-px}F(x)\,dx\right]\left[\int_0^{\infty} e^{-py}G(y)\,dy\right]$$

$$= L[F(t)]\,L[G(t)]$$

$$= f(p)g(p).$$

Hence, $$F*G = \int_0^t F(x)G(t-x)\,dx = L^{-1}[f(p)g(p)].$$

Note: In particular, if $f = g$, then $L[F*F] = [f(p)]^2$.

WORKED-OUT EXAMPLES

- Use convolution theorem to find L^{-1} of the following functions (1-6):

1. $\dfrac{1}{(p+1)(p-2)}$, **2.** $\dfrac{1}{(p-a)^2}$,

3. $\dfrac{1}{(p^2+1)(p-2)}$, **4.** $\dfrac{p}{(p^2+a^2)^2}$,

5. $\dfrac{1}{p(p^2+4)^2}$, **6.** $\dfrac{1}{p^2(p+1)^2}$.

7. Using convolution theorem of Laplace transform deduce the formula

$$\int_0^1 t^{a-1}(1-t)^{b-1}dt = \frac{\Gamma(a)\Gamma(b)}{\Gamma(a+b)}, a>0, b>0.$$

8. Apply the convolution theorem to show that

$$\int_0^t \sin x \cos(t-x)dx = \frac{t}{2}\sin t.$$

Solutions

1. We have $L^{-1}\left[\frac{1}{p+1}\right] = e^{-t}$ and $L^{-1}\left[\frac{1}{p-2}\right] = e^{2t}$.

   So, by convolution theorem

$$L^{-1}\left[\frac{1}{(p+1)(p-2)}\right] = \int_0^t e^{-x}e^{2(t-x)}dx = \frac{1}{3}e^{2t}\left[-e^{-3x}\right]_0^t = \frac{1}{3}e^{2t}\left[1-e^{-3t}\right] = \frac{1}{3}\left[e^{2t}-e^{-t}\right].$$

2. We have $L^{-1}\left[\frac{1}{p-a}\right] = e^{at}$, so by convolution theorem

$$L^{-1}\left[\frac{1}{(p-a)^2}\right] = L^{-1}\left[\frac{1}{p-a}.\frac{1}{p-a}\right] = \int_0^t e^{ax}e^{a(t-x)}dx = e^{at}\int_0^t dx = te^{at}.$$

3. We have $L^{-1}\left[\frac{1}{p^2+1}\right] = \sin t$ and $L^{-1}\left[\frac{1}{p-2}\right] = e^{2t}$.

   So, by convolution theorem,

$$L^{-1}\left[\frac{1}{(p^2+1)(p-2)}\right] = \int_0^t e^{2(t-x)}\sin x\,dx = e^{2t}\int_0^t e^{-2x}\sin x\,dx =$$

$$= e^{2t}\left[\frac{e^{-2x}}{5}(-2\sin x - \cos x)\right]_0^t = \frac{e^{2t}-2\sin t-\cos t}{5}.$$

4. We have $L^{-1}\left[\frac{p}{p^2+a^2}\right] = \cos at$ and $L^{-1}\left[\frac{1}{p^2+a^2}\right] = \frac{1}{a}\sin at$. So by convolution theorem,

$$L^{-1}\left[\frac{p}{(p^2+a^2)^2}\right] = L^{-1}\left[\frac{p}{p^2+a^2}.\frac{1}{p^2+a^2}\right]$$

$$= \frac{1}{a}\int_0^t \cos ax \sin a(t-x)dx$$

$$= \frac{1}{a}\sin at\int_0^t \cos^2 ax\,dx - \frac{1}{2a}\cos at\int_0^t \sin 2ax\,dx$$

$$= \frac{1}{a}\sin at\left(\frac{t}{2} + \frac{\sin 2at}{4a}\right) - \frac{1}{2a}\cos at\left(\frac{1-\cos 2at}{2a}\right)$$

$$= \frac{t \sin at}{2a}.$$

5. We have $L^{-1}\left[\frac{p}{(p^2+4)^2}\right] = \frac{t\sin 2t}{4}$ and $L^{-1}\left[\frac{1}{p^2}\right] = t$.

So, by convolution theorem,

$$L^{-1}\left[\frac{1}{p(p^2+4)^2}\right] = \frac{1}{4}\int_0^t (t-x)\,x\sin 2x\,dx$$

$$= \frac{t}{4}\int_0^t x\sin 2x\,dx - \frac{1}{4}\int_0^t x^2 \sin 2x\,dx$$

$$= \frac{t}{4}\left[-\frac{1}{2}x\cos 2x - \frac{1}{4}\sin 2x\right]_0^t - \frac{1}{4}\left[-\frac{1}{2}x^2\cos 2x + \frac{1}{2}x\sin 2x + \frac{1}{4}\cos 2x\right]_0^t$$

$$= \frac{1 - t\sin 2t - \cos 2t}{16}$$

6. We have $L^{-1}\left[\frac{1}{p^2}\right] = t$ and $L^{-1}\left[\frac{1}{(p+1)^2}\right] = te^{-t}$. So by convolution theorem,

$$L^{-1}\left[\frac{1}{p^2(p+1)^2}\right] = \int_0^t xe^{-x}(t-x)\,dx$$

$$= t\int_0^t xe^{-x}dx - \int_0^t x^2e^{-x}dx$$

$$= t\left[-xe^{-x} - e^{-x}\right]_0^t - \left[x^2e^{-x}\right]_0^t + 2\left[-xe^{-x} - e^{-x}\right]_0^t$$

$$= t\left[-te^{-t} - e^{-t} + 1\right] - t^2e^{-t} + 2\left(-te^{-t} - e^{-t} + 1\right)$$

$$= te^{-t} + 2e^{-t} + t - 2.$$

7. Let $f(t) = t^{a-1}, (a>0)$ and $g(t) = t^{b-1}, (b>0)$, then

$$L[f(t)] = \frac{\Gamma(a)}{p^a} \quad \text{and} \quad L[g(t)] = \frac{\Gamma(b)}{p^b}.$$

Using convolution theorem of Laplace transform, we get

$$L^{-1}\left[\frac{\Gamma(a)}{p^a}\cdot\frac{\Gamma(b)}{p^b}\right] = \int_0^t f(x)g(t-x)\,dx$$

or, $$\Gamma(a)\Gamma(b)L^{-1}\left[p^{-(a+b)}\right]=\int_0^t x^{a-1}(t-x)^{b-1}dx$$

or, $$\int_0^t x^{a-1}(t-x)^{b-1}dx=\frac{\Gamma(a)\Gamma(b)}{\Gamma(a+b)}t^{a+b-1}.$$

Putting $t = 1$, we get

$$\int_0^1 t^{a-1}(1-t)^{b-1}dt=\frac{\Gamma(a)\Gamma(b)}{\Gamma(a+b)},\ a>0, b>0.$$

8. Let $f(t)=\int_0^t \sin x\cos(t-x)dx$. Then by convolution theorem, we have

$$L[f(t)]=L[\sin t]L[\cos t]$$

$$=\frac{1}{p^2+1}\cdot\frac{p}{p^2+1}$$

$$=\frac{p}{(p^2+1)^2}.$$

So, $$f(t)=L^{-1}\left[\frac{p}{(p^2+1)^2}\right]=-\frac{1}{2}L^{-1}\left[\frac{d}{dp}\left\{\frac{1}{p^2+1}\right\}\right]$$

$$=-\frac{1}{2}t(-1)L^{-1}\left[\frac{1}{p^2+1}\right]=\frac{t}{2}\sin t.$$

EXERCISE

Use convolution theorem to find L^{-1} of the following functions:

1. $\dfrac{1}{(p+1)(p-2)}$,
2. $\dfrac{1}{(p+3)(p-1)}$,
3. $\dfrac{1}{(p+1)p^3}$,
4. $\dfrac{1}{(p+2)^2(p-2)}$,
5. $\dfrac{1}{(p+1)(p^2+1)}$,
6. $\dfrac{p^2}{(p^2+4)^2}$,
7. $\dfrac{1}{(p+1)^2p^2}$,
8. $\dfrac{1}{(p+2)(p^2+4)}$,
9. $\dfrac{1}{(p^2+1)^3}$,
10. $\dfrac{p}{(p^2+4)^3}$,

11. $\dfrac{1}{p^3(p^2+1)}$, 12. $\dfrac{1}{p(p+1)^3}$,

13. $\dfrac{1}{p(p-a)^2}$ 14. $\dfrac{2p+3}{p^2(p+3)^2}$.

ANSWERS

1. $\dfrac{e^{2t}-e^{-t}}{3}$, 2. $\dfrac{e^{t}-e^{-3t}}{4}$,

3. $1-t+\dfrac{t^2}{2}-e^{-t}$, 4. $\dfrac{e^{2t}-(4t+1)e^{-2t}}{16}$,

5. $\dfrac{1}{2}(\sin t-\cos t+e^{-t})$, 6. $\dfrac{t}{2}\cos 2t+\dfrac{1}{4}\sin 2t$,

7. $(t+2)e^{-t}+t-2$, 8. $[\sin 2t-\cos 2t+e^{-2t}]/8$,

9. $\dfrac{1}{8}[(3-t^2)\sin t-3t\cos t]$, 10. $\dfrac{1}{64}t(\sin 2t-2t\cos 2t)$,

11. $\dfrac{t^2}{2}+\cos t-1$, 12. $1-e^{-t}\left(\dfrac{t^2}{2}+t+1\right)$,

13. $[1-e^{at}(at+1)]/a^2$ 14. $\dfrac{t}{3}(1-e^{-3t})$.

3.12 Application of Laplace Transform to Solutions of Differential Equations

3.12.1 Differential equation with constant coefficients

The Laplace transform is a powerful technique to solve ordinary differential equation. In this section we will consider the differential equations with constant coefficients and with given initial or boundary conditions. Using Laplace transform on both sides of ordinary differential equation having constant coefficients and using the given conditions we get an algebraic equation. Next taking inverse Laplace transform we get the required solutions.

Consider the differential equation

$$ay''+by'+cy=f(t) \tag{3.1}$$

with the initial conditions

$$y(0)=y_0 \text{ and } y'(0)=y_1. \tag{3.2}$$

Taking Laplace transform on both sides of the given equation, we get

$$aL[y'']+bL[y']+cL[y]=L[f(t)]$$

$$\Rightarrow a[p^2L[y]-py(0)-y'(0)]+b[pL(y)-y(0)]+cL[y]=F(p)$$

$$\Rightarrow (ap^2+bp+c)L[y]=(ap+b)y_0+ay_1+F(p)$$

$$\Rightarrow L[y]=\frac{(ap+b)y_0+ay_1+F(p)}{ap^2+bp+c}$$

$$\Rightarrow y=L^{-1}\left[\frac{(ap+b)y_0+ay_1+F(p)}{ap^2+bp+c}\right].$$

This is the solution of the differential equation (3.1) subject to the initial conditions given by (3.2).

3.12.2 Differential equation with variable coefficients

Laplace transform can also be used to solve ordinary differential equations with variable coefficients of the form $t^m y^n(t)$ whose Laplace transform is $(-1)^m \frac{d^m}{dp^m} L[y^n(t)]$. Consider the differential equation

$$t^2y''+ty'+y=f(t)$$

with the initial conditions $y(0)=a,\ y'(0)=b$. Taking Laplace transform of both sides, we get

$$L[t^2y'']+L[ty']+L[y]=L[f(t)]$$

$$\Rightarrow \frac{d^2L[y'']}{dp^2}-\frac{dL[y']}{dp}+L[y]=L[f(t)]$$

$$\Rightarrow \frac{d^2}{dp^2}[p^2L[y]-py(0)-y'(0)]-\frac{d}{dp}[pL[y]-y(0)]+L[y]=L[f(t)]$$

$$\Rightarrow \frac{d^2}{dp^2}[p^2L[y]-ap-b]-\frac{d}{dp}[pL[y]-a]+L[y]=L[f(t)]$$

$$\Rightarrow p^2\frac{d^2z}{dp^2}+3p\frac{dz}{dp}+2z=L[f(t)],\ \text{where } z=L[y].$$

Solving this equation, we get $z=L[y]$, which tends to zero as p tends to infinity. Now taking inverse Laplace transform we get the value of y.

3.12.3 Simultaneous differential equation

Laplace transform is also useful in solving a system of differential equations. Let a system of differential equations be given in the form

$$f(D)x+F(D)y=T;\ g(D)x+G(D)y=T',$$

where $f(D), g(D), F(D), G(D)$ are all rational function of D with constant coefficients and T, T' are functions of the independent variable t.

Taking Laplace transform of both the equations and using the given conditions we get two system of equations in $L[x(t)], L[y(t)]$. Solving these we get the values of $L[x(t)], L[y(t)]$. Next taking inverse Laplace transform, we get the values of $x(t), y(t)$.

WORKED-OUT EXAMPLES

(*In the following problems prime denotes differentiation w.r.t.t*)

1. Solve: $y'' + y = 0$, given $y = 1, y' = 0$, when $t = 0$.

Solution

Taking Laplace transform of both sides of the given equation, we get

$$L[y''] + L[y] = 0$$

$$\Rightarrow p^2 L[y] - py(0) - y'(0) + L[y] = 0$$

$$\Rightarrow (p^2 + 1) L[y] = p$$

$$\Rightarrow L[y] = \frac{p}{p^2 + 1}.$$

Taking inverse Laplace transform, $y = \cos t$. This is the required solution.

2. Solve: $y'' + y = t$, given $y = 1, y' = -2$, when $t = 0$.

Solution

Taking Laplace transform of both sides of the given equation, we get

$$L[y''] + L[y] = L[t]$$

$$\Rightarrow p^2 L[y] - py(0) - y'(0) + L[y] = \frac{1}{p^2}$$

$$\Rightarrow (p^2 + 1) L[y] - p + 2 = \frac{1}{p^2}$$

$$\Rightarrow L[y] = \frac{1}{p^2(p^2 + 1)} + \frac{p - 2}{p^2 + 1}$$

$$\Rightarrow L[y] = \frac{1}{p^2} - \frac{3}{p^2 + 1} + \frac{p}{p^2 + 1}.$$

Taking inverse Laplace transform,

$$y = t + \cos t - 3\sin t.$$

This is the required solution.

3. Solve: $y'' + y = 2e^t$, given that $y(0) = 2 = y'(0)$.

Solution

Taking Laplace transform of both sides of the given equation, we get

$$L[y''] + L[y] = 2L[e^t]$$

$$\Rightarrow p^2L[y] - py(0) - y'(0) + L[y] = \frac{2}{p-1}$$

$$\Rightarrow (p^2+1)\,L[y] - 2p - 2 = \frac{2}{p-1}$$

$$\Rightarrow L[y] = \frac{2}{(p-1)(p^2+1)} + \frac{2(p+1)}{p^2+1} = \frac{2p^2}{(p-1)(p^2+1)} = \frac{1}{p-1} + \frac{p+1}{p^2+1}.$$

Taking inverse Laplace transform,

$$y = e^t + \cos t - \sin t.$$

This is the required solution.

4. Solve: $y'' + 25y = 10\cos 5t$, given that $y(0) = 2 = y'(0) = 0$.

Solution

Taking Laplace transform of both sides of the given equation, we get

$$L[y''] + 25L[y] = 10L[\cos 5t]$$

$$\Rightarrow p^2L[y] - py(0) - y'(0) + 25L[y] = \frac{10p}{p^2+25}$$

$$\Rightarrow (p^2+25)\,L[y] = \frac{10p}{p^2+25} + 2p$$

$$\Rightarrow L[y] = \frac{2p}{p^2+25} + \frac{10p}{(p^2+25)^2}.$$

Taking inverse Laplace transform,

$$y = 2L^{-1}\left[\frac{p}{p^2+25}\right] + 10L^{-1}\left[\frac{p}{(p^2+25)^2}\right]$$

$$= 2\cos 5t + 5t\sin 5t.$$

5. Solve: $y'' - 3y' + 2y = 4e^{2t}$, given that $y = -3$, $y' = 5$, when $t = 0$.

Solution

Taking Laplace transform of both sides of the given equation, we get

$$L[y''] + 3L[y'] + 2L[y] = 4L[e^{2t}]$$

$$\Rightarrow p^2L[y] - py(0) - y'(0) - 3\{pL[y] - y(0)\} + 2L[y] = \frac{4}{p-2}$$

$$\Rightarrow (p^2 - 3p + 2)\,L[y] + 3p - 14 = \frac{4}{p-2}$$

$$\Rightarrow L[y] = \frac{4}{(p-2)(p^2-3p+2)} + \frac{14-3p}{p^2-3p+2} = \frac{-3p^2+20p-24}{(p-1)(p-2)^2}$$

$$= -\frac{7}{p-1} + \frac{4}{p-2} + \frac{4}{(p-2)^2}.$$

Taking inverse Laplace transform,

$$y = -7e^t + 4e^{2t} + 4te^{2t}.$$

This is the required solution.

6. Solve: $y'' - 2y' + 2y = 0$, given that $y = y' = 1$, when $t = 0$.

Solution

Taking Laplace transform of both sides of the given equation, we get

$$L[y''] + 2L[y'] + 2L[y] = 0$$

$$\Rightarrow p^2 L[y] - py(0) - y'(0) - 2\{pL[y] - y(0)\} + 2L[y] = 0$$

$$\Rightarrow (p^2 - 2p + 2)\, L[y] = p - 1$$

$$\Rightarrow L[y] = \frac{p-1}{p^2 - 2p + 2} = \frac{p-1}{(p-1)^2 + 1}.$$

Taking inverse Laplace transform,

$$y = L^{-1}\left[\frac{p-1}{(p-1)^2 + 1}\right] = e^t L^{-1}\left[\frac{p}{p^2 + 1}\right] = e^t \cos t,$$

which is the required solution.

7. Solve: $y'' + 9y = \cos 2t$, given that $y(0) = 1$, $y'\left(\frac{\pi}{2}\right) = -1$.

Solution

Applying Laplace transform of both sides of the given equation, we get

$$L[y''] + 9L[y] = \frac{p}{p^2 + 4}$$

$$\Rightarrow p^2 L[y] - py(0) - y'(0) + 9L[y] = \frac{p}{p^2 + 4}$$

$$\Rightarrow (p^2 + 9)\, L[y] = p + C + \frac{p}{p^2 + 4} \quad (\text{where } y'(0) = C)$$

$$\Rightarrow L[y] = \frac{p + C}{p^2 + 9} + \frac{p}{(p^2 + 9)(p^2 + 4)}$$

$$= \frac{p}{p^2 + 9} + \frac{C}{p^2 + 9} + \frac{p}{5(p^2 + 4)} - \frac{p}{5(p^2 + 9)}.$$

Taking inverse Laplace transform, we get $y = \cos 3t + \frac{C}{3}\sin 3t + \frac{1}{5}[\cos 2t - \cos 3t]$.

Now $y'\left(\frac{\pi}{2}\right) = -1$ given that $C = \frac{12}{5}$. So, the required solution is

$$y = \frac{4}{5}[\cos 3t + \sin 3t] + \frac{1}{5}\cos 2t.$$

8. Solve: $y'' + y = t$, given that $y'(0) = 1$, $y(\pi) = 0$.

Solution

Taking Laplace transform of both sides of the given equation we get,

$$L[y''] + L[y] = L[t]$$

$$\Rightarrow p^2 L[y] - py(0) - y'(0) + L[y] = \frac{1}{p^2}$$

$$\Rightarrow (p^2+1)\, L[y] = \frac{1}{p^2} + 1 + Cp, \text{ where } y(0) = C$$

$$\Rightarrow L[y] = \frac{1}{p^2(p^2+1)} + \frac{Cp+1}{p^2+1}.$$

Now taking inverse Laplace transform and using the following result

$$L^{-1}\left[\frac{1}{p^2+1}\right] = \sin t$$

$$\Rightarrow L^{-1}\left[\frac{1}{p(p^2+1)}\right] = \int_0^t \sin t\, dt = 1 - \cos t$$

$$\Rightarrow L^{-1}\left[\frac{1}{p^2(p^2+1)}\right] = \int_0^t (1-\cos t)\, dt = t - \sin t,$$

we get
$$y = L^{-1}\left[\frac{1}{p^2(p^2+1)}\right] + CL^{-1}\left[\frac{p}{p^2+1}\right] + L^{-1}\left[\frac{1}{p^2+1}\right]$$

$$= t - \sin t + C\cos t + \sin t$$

$$= t + C\cos t.$$

Using $y(\pi) = 0$, we get $C = \pi$. Hence the solution of the given problem is

$$y = t + \pi\cos t.$$

9. Solve: $y'' + 2y' + y = 0$, given that $y(0) = 0$ and $y(1) = 2$.

Solution

Taking Laplace transform of both sides of the given equation we get,

$$L[y''] + 2L[y'] + L[y] = 0$$

$$\Rightarrow p^2 L[y] - py(0) - y'(0) + 2\{pL[y] - y(0)\} + L[y] = 0$$

$$\Rightarrow (p+1)^2 L[y] = C \text{ where } y'(0) = C,$$

$$\Rightarrow L[y] = \frac{C}{(p+1)^2}.$$

Taking inverse Laplace transform, we get $y = Cte^{-t}$.

Using the condition $y(1) = 2$, we get $C = 2e$. Hence the solution of the given problem is

$$y = 2te^{1-t}.$$

10. Solve: $ty'' + 2y' + ty = 0$, given that $y(0) = 1$, $y(\pi) = 0$.

Solution

Taking Laplace transform of both sides, we get

$$L[ty''] + 2L[y'] + L[ty] = 0$$

$$\Rightarrow -\frac{d}{dp}[p^2L[y] - py(0) - y'(0)] + 2[pL[y] - y(0) - \frac{d}{dp}L[y] = 0$$

$$\Rightarrow -2pz - p^2\frac{dz}{dp} + 1 + 2pz - 2 - \frac{dz}{dp} = 0, \text{ where } L[y] = z$$

$$\Rightarrow \frac{dz}{dp} = -\frac{1}{p^2+1}.$$

Integrating, we get

$$L[y] = z = -\tan^{-1}p + K.$$

Since $L[y] \to 0$ as $p \to \infty$, we must have $K = \frac{\pi}{2}$. Hence,

$$L[y] = -\tan^{-1}p + \frac{\pi}{2} = \cot^{-1}p = \tan^{-1}\frac{1}{p}.$$

Taking inverse Laplace transform, we get

$$y = L^{-1}\left[\tan^{-1}\frac{1}{p}\right] = \frac{\sin t}{t}.$$

11. Solve: $y'' + ty' - y = 0$, given that $y(0) = 0$, $y'(0) = 1$.

Solution

Taking Laplace transform of both sides, we get

$$L[y''] + L[ty'] - L[y] = 0$$

$$\Rightarrow [p^2L[y] - py(0) - y'(0)] - \frac{d}{dp}[pL[y]] - y(0)] - L[y] = 0$$

$$\Rightarrow p\frac{dz}{dp} - (p^2 - 2)z = -1$$

$$\Rightarrow \frac{dz}{dp} - \left(p - \frac{2}{p}\right)z = -1/p.$$

This is a linear differential equation, whose integrating factor is $e^{-\int\left(p-\frac{2}{p}\right)dp} = p^2e^{-\frac{p^2}{2}}$. Multiplying both sides by this I.F., we get

$$\frac{d}{dp}\left[zp^2e^{-\frac{p^2}{2}}\right] = -pe^{\frac{p^2}{2}}$$

$$\Rightarrow zp^2e^{-\frac{p^2}{2}} = -\int pe^{-\frac{p^2}{2}}\,dp = e^{-\frac{p^2}{2}} + C$$

$$\Rightarrow z = \frac{1}{p^2} + Ce^{\frac{p^2}{2}}.$$

Since $L[y] \to 0$ as $p \to \infty$, we must have $C = 0$. Hence,

$$L[y] = \frac{1}{p^2}.$$

Taking inverse Laplace transform, we get $y = t$.

12. Solve by Laplace transform, when x, y are both functions of t:

$$(D-2)x + 3y = 0;\ 2x + (D-1)y = 0, \text{ if } x(0) = 9,\ y(0) = 3.$$

Solution

Taking Laplace transform of both sides of $(D-2)x + 3y = 0;\ 2x + (D-1)y = 0$, and using the given conditions, we get

$$(p-1)L[x] + 2L[y] = 8 \text{ and } 2L[x] + (p-1)L[y] = 3.$$

Solving the above equation for $L[x]$, $L[y]$, we get

$$L[x] = \frac{8p-17}{(p-4)(p+1)} = \frac{3}{p-4} + \frac{5}{p+1} \text{ and } L[y] = \frac{3p-22}{(p-4)(p+1)} = -\frac{2}{p-4} + \frac{5}{p+1}.$$

Taking inverse Laplace transform, we get

$$x = 3e^{4t} + 5e^{-t};\ y = -2e^{4t} + 5e^{-t}.$$

13. Solve by Laplace transform, when x, y are both functions of t:

$$(D^2+2)x - Dy = 1;\ Dx + (D^2+2)y = 0, \text{ if } x = 0,\ y = 0,\ Dx = 0 \text{ at } t = 0.$$

Solution

Taking Laplace transform of both sides of $(D^2+2)x - Dy = 1;\ Dx + (D^2+2)y = 0$, and using the given conditions, we get

$$(p^2+2)L[x] - pL[y] = \frac{1}{p} \text{ and } pL[x] + (p^2+2)L[y] = 0.$$

Solving the above equation for $L[x]$, $L[y]$, we get

$$L[x] = \frac{p^2+2}{p(p^2+1)(p^2+4)} = \frac{1}{3p(p^2+1)} + \frac{2}{3p(p^2+4)}$$

and $$L[y] = \frac{-1}{(p^2+1)(p^2+4)} = \frac{-1}{3(p^2+1)} + \frac{1}{3(p^2+4)}.$$

Taking inverse Laplace transform, we get

$$x = \frac{1}{3}\int_0^t \sin t\, dt + \frac{1}{3}\int_0^t \sin 2t\, dt = \frac{1}{2} - \frac{1}{3}\cos t - \frac{1}{6}\cos 2t, \text{ and } y = \frac{1}{6}\sin 2t - \frac{1}{3}\sin t.$$

EXERCISE

- **Solve the following differential equations with the help of Laplace transform:**

1. $y'' + 4y = 4t$, given that $y(0) = 1, y'(0) = 5$.

2. $y'' + m^2 y = a\cos nt, t > 0$, given that $y = y_0, y' = y_1$ when $t = 0$ and $n \neq m$.

3. $y'' - 2y' - 8y = 0$, given that $y(0) = 3, y'(0) = 6$.

4. $y'' - y = t$, given that $y(0) = 0, y'(0) = 1$.

5. $y'' + 4y' + 4y = 4e^{-2t}$, given that $y = -1, y' = 4$, when $t = 0$.

6. $y'' - y' - 2y = 20\sin 2t$, given that $y = -1, y' = 2$, when $t = 0$.

7. $y'' - 3y' + 2y = 1 - e^{2t}$, given that $y = 1, y' = 0$, when $t = 0$.

8. $y'' + y = 2t + t^2$, given that $y = 4, y' = -2$, when $t = 0$.

9. $y'' + y' = 2t + t^2$, given that $y = 4, y' = -2$, when $t = 0$.

10. $y'' + 9y = 18t$, given that $y(0) = 0$ and $y\left(\frac{\pi}{2}\right) = 0$.

11. $y'' + y = 0$, given that $y(0) = 1$ and $y\left(\frac{\pi}{2}\right) = 0$.

12. $y''' - 3y'' + 3y' - y = t^2 e^t$, given that $y = 1, y' = 0$ and $y'' = -2$, when $t = 0$.

13. $y'' + 2y' + 5y = e^{-t}\sin t$, given that $y = 0, y' = 1$, when $t = 0$.

14. $y'' - 3y' + 2y = 4e^{2t}$, given that $y = -3, y' = 5$, when $t = 0$.

15. $y''' - y'' - y' + y = 8te^{-t}$, given that $y = 0, y' = 1$ and $y'' = 0$, when $t = 0$.

16. $y'' + y = 2e^{-2t}\sin t$, given that $y(0) = 0 = y'(0)$.

17. $y'' - ty' + y = 1$, given that $y(0) = 1, y'(0) = 2$.

18. $y'' + aty' - 2ay = 1$, given that $y(0) = 0 = y'(0), a > 0$.

19. $ty'' + (t-1)y' = 0$, given that $y(0) = 5, y(\infty) = 0$.

20. $ty'' + (1-2t)y' - 2y = 0$, given that $y(0) = 1, y'(0) = 2$.

21. $(D^2 - 3)x - 4y = 0;\ x + (D^2 + 1)y = 0$, given that

$$x = y = Dy = 0,\ Dx = 2 \text{ at } t = 0.$$

22. $(D^2 - 1)x + 5Dy = t;\ -2Dx + (D^2 - 4)y = -2$, given that

$$x = y = Dy = Dx = 0, \text{ at } t = 0.$$

23. $Dx + Dy = t,\ D^2x - y = e^{-t}$, given that $x(0) = 3,\ x'(0) = -2,\ y(0) = 0$.

24. $(D-2)x - (D+1)y = 6e^{3t};\ (2D-3)x + (D-3)y = 6e^{3t}$, given that

$$x(0) = 3,\ y(0) = 0.$$

ANSWERS

1. $y = t + \cos 2t + 2\sin 2t$,

2. $y = y_0 \cos mt + \dfrac{y_1}{m}\sin mt + \dfrac{a}{m^2 - n^2}[\cos nt - \cos mt]$,

3. $y = 2e^{4t} + e^{-2t}$,

4. $y = -t + 2\sinh t$,

5. $y = e^{-2t}[2t^2 + 2t - 1]$,

6. $y = 2e^{2t} - 4e^{-t} + \cos 2t - 3\sin 2t$,

7. $y = \dfrac{1}{2}[1 + e^{2t} - 2te^{2t}]$,

8. $y = t^2 + 2t - 2 + 6\cos t - 4\sin t$,

9. $y = \dfrac{t^3}{3} + 2(1 + e^{-t})$,

10. $y = 2t + \pi \sin 3t$,

11. $y = \cos t$,

12. $y = e^t\left[1 - t - \dfrac{t^2}{2} + \dfrac{t^5}{60}\right]$,

13. $y = \dfrac{1}{3}e^{-t}[\sin t + \sin 2t]$,

14. $y = -7e^t + 4e^{2t}(t+1)$,

15. $y = (t+1)^2 e^{-t} - (1-t)e^t$,

16. $y = \dfrac{1}{8}(\sin t - \cos t) + \dfrac{e^{-2t}}{8}(\sin t + \cos t)$,

17. $y = 2t + 1$,

18. $y = \dfrac{1}{2}t^2$,

19. $y = 5e^{-t}$,

20. $y = e^{2t}$,

21. $x = t(e^t + e^{-t});\ y = \dfrac{1}{2}(1-t)e^t - \dfrac{1}{2}(1+t)e^{-t}$,

22. $x = -t + 5\sin t - 2\sin 2t;\ y = 1 - 2\cos t + \cos 2t$,

23. $x = 2 + \dfrac{1}{2}t^2 + \dfrac{1}{2}e^{-t} + \dfrac{1}{2}\cos t - \dfrac{3}{2}\sin t;\ y = 1 - \dfrac{1}{2}e^{-t} - \dfrac{1}{2}\cos t + \dfrac{3}{2}\sin t$,

24. $x = (1 + 2t)e^t + 2e^{3t};\ y = (1-t)e^t - e^{3t}$.

3.13 Application of Laplace Transform to Solutions of Difference Equations

A difference equation is an equation involving an independent variable x, the dependent variable u_x and the successive differences Δu_x, $\Delta^2 u_x$. Thus $\varphi\{x, u_x, \Delta u_x, \Delta^2 u_x, \cdots\} = 0$, i.e.,

$$u_{x+n} + A_1 u_{x+n-1} + A_2 u_{x+n-2} + \cdots + A_n u_x = \psi(x),$$

where A_i's are known constants and $\psi(x)$ is a known function, is a difference equation. In various applications it is possible to formulate a difference equation from which we seek the unknown function subject to specified conditions. The Laplace transform is also a powerful technique to solve difference equation with constant coefficients. Using Laplace transform on both sides of difference equation and using the given conditions we get an algebraic equation. Next taking inverse Laplace transform we get the required solutions. Before going to solve difference equation with the aid of Laplace transform we shall establish a few basic results.

1. If $F(t) = r^n, n \le t < n+1, n = 0, 1, 2, 3, \cdots$, find $L[F(t)]$.

Solution

$$\begin{aligned}
L[F(t)] &= \int_0^0 e^{-pt} F(t)\,dt = \int_0^1 e^{-pt} r^0 dt + \int_1^2 e^{-pt} r^1 dt + \int_2^3 e^{-pt} r^2 dt \cdots \\
&= 1\left(\frac{1-e^{-p}}{p}\right) + r\left(\frac{e^{-p} - e^{-2p}}{p}\right) + r^2\left(\frac{e^{-2p} - e^{-3p}}{p}\right) + r^3\left(\frac{e^{-3p} - e^{-4p}}{p}\right) + \cdots \\
&= \left(\frac{1-e^{-p}}{p}\right)(1 + re^{-p} + r^2 e^{-2p} + \cdots) \\
&= \frac{1-e^{-p}}{p(1-re^{-p})}.
\end{aligned}$$

So, $L^{-1}\left[\dfrac{1-e^{-p}}{p(1-re^{-p})}\right] = r^2$ for $n \le t < n+1, n = 0, 1, 2, 3, \cdots$.

2. Let $Y(t) = a_n, n \le t < n+1, n = 0, 1, 2, 3, \ldots$ Find $L[Y(t+1)]$, $L[Y(t+2)]$ in terms of $L[Y(t)] = y(p)$.

Solution

$$\begin{aligned}
L[Y(t+1)] &= \int_0^\infty e^{-pt} Y(t+1)\,dt = e^p \int_1^\infty e^{-pu} Y(u)\,du \\
&= e^p \int_0^\infty e^{-pu} Y(u)\,du - e^p \int_0^1 e^{-pu} Y(u)\,du \\
&= e^p y(p) - e^p \int_0^1 e^{-pu} a_0\,du \\
&= e^p y(p) - \frac{a_0 e^p (1-e^{-p})}{p}.
\end{aligned}$$

$$L[Y(t+2)] = \int_0^\infty e^{-pt}Y(t+2)\,dt = e^{2p}\int_2^\infty e^{-pu}Y(u)\,du$$

$$= e^{2p}\int_0^\infty e^{-pu}Y(u)\,du - e^{2p}\int_0^1 e^{-pu}Y(u)\,du - e^{2p}\int_1^2 e^{-pu}Y(u)\,du$$

$$= e^{2p}y(p) - e^{2p}\int_0^1 e^{-pu}a_0\,du - e^{2p}\int_1^2 e^{-pu}a_1\,du$$

$$= e^{2p}y(p) - \frac{a_0 e^{2p}(1-e^{-p})}{p} - \frac{a_1 e^{2p}(e^{-p}-e^{-2p})}{p}$$

$$= e^{2p}y(p) - \frac{e^{p}(a_0 e^{p} + a_1)(1-e^{-p})}{p}.$$

3. Find $L[F(t)] = n,\ n \le t < n+1,\ n = 0,1,2,3.$

Solution

We have $L[F(t)] = \int_0^\infty e^{-pt}F(t)\,dt = \int_0^1 e^{-pt}(0)\,dt + \int_1^2 e^{-pt}(1)\,dt + \int_2^3 e^{-pt}(2)\,dt \ldots$

$$= 1\left(\frac{e^{-p}-e^{-2p}}{p}\right) + 2\left(\frac{e^{-2p}-e^{-3p}}{p}\right) + 3\left(\frac{e^{-3p}-e^{-4p}}{p}\right) + \cdots$$

$$= e^{-p}\left(\frac{1-e^{-p}}{p}\right)(1 + 2e^{-p} + 3e^{-2p} + \cdots)$$

$$= \frac{e^{-p}(1-e^{-p})}{p(1-e^{-p})^2}$$

$$= \frac{e^{-p}}{p(1-e^{-p})}.$$

So, $L^{-1}\left[\frac{e^{-p}}{p(1-e^{-p})}\right] = n,$ for $n \le t < n+1,\ n = 0, 1, 2, 3, \ldots$

Note: $L^{-1}\left[\frac{e^{-p}}{p(1-re^{-p})}\right] = \frac{r(r^n-1)}{r-1},\ r \ne 1.$

WORKED-OUT EXAMPLES

1. Solve the difference equation: $a_{n+2} - 5a_{n+1} + 6a_n = 0,\ a_0 = 0,\ a_1 = 1.$

Solution

We define the function $Y(t) = a_n,\ n \le t < n+1,\ n = 0,1,2,3,\cdots$. Then the given recursion formula becomes

$$Y(t+2) - 5Y(t+1) + 6y(t) = 0.$$

Taking the Laplace transform of this equation and using $a_0 = 0, a_1 = 1$, we get

$$e^{2p}y(p) - \frac{e^p(1-e^{-p})}{p} - 5e^p y(p) + 6y(p) = 0$$

$$\Rightarrow [e^{2p} - 5e^p + 6]\, y(p) = \frac{e^p(1-e^{-p})}{p}$$

$$\Rightarrow y(p) = \frac{e^p(1-e^{-p})}{p(e^{2p} - 5e^p + 6)} = \frac{e^p(1-e^{-p})}{p(e^p - 3)(e^p - 2)}$$

$$\Rightarrow y(p) = \frac{e^p(1-e^{-p})}{p}\left[\frac{1}{e^p - 3} - \frac{1}{e^p - 2}\right] = \frac{1-e^{-p}}{p}\left[\frac{1}{1-3e^{-p}} - \frac{1}{1-2e^{-p}}\right].$$

Taking inverse Laplace transform, we get $a_n = 3^n - 2^n, n = 0,1,2,\cdots$.

2. Solve the difference equation. $a_{n+2} - 5a_{n+1} + 6a_n = 4^n, a_0 = 0, a_1 = 1$.

Solution

We define the function $Y(t) = a_n, n \le t < n+1, n = 0,1,2,3, \cdots$. Then the given recursion formula becomes

$$Y(t+2) - 5Y(t+1) + 6Y(t) = F(t), \text{ where } F(t) = 4^n.$$

Taking the Laplace transform of this equation and using $a_0 = 0, a_1 = 1$, we get

$$e^{2p}y(p) - \frac{e^p(1-e^{-p})}{p} - 5e^p y(p) + 6y(p) = \frac{1-e^{-p}}{p(1-4e^{-p})}$$

$$\Rightarrow y(p) = \frac{1-e^{-p}}{p}\left[\frac{1}{1-3e^{-p}} - \frac{1}{1-2e^{-p}}\right] + \frac{1-e^{-p}}{p}\left[\frac{1}{2(1-2e^{-p})} - \frac{1}{1-3e^{-p}} + \frac{1}{2(1-4e^{-p})}\right]$$

Taking inverse Laplace transform, we get

$$a_n = 3^n - 2^n + \frac{1}{2}2^n - 3^n + \frac{1}{2}4^n = \frac{1}{2}(4^n - 2^n), \quad n = 0,1,2,\cdots.$$

3. Solve the difference equation $a_{n+2} - 7a_{n+1} + 10a_n = 16n, a_0 = 0, a_1 = 0$.

Solution

We define the function $Y(t) = a_n, n \le t < n+1, n = 0,1,2,3,\cdots$. Then the given recursion formula becomes

$$Y(t+2) - 7Y(t+1) + 10Y(t) = F(t), \text{ where } F(t) = 16n.$$

Taking the Laplace transform of this equation and using $a_0 = 0, a_1 = 0$, we get

$$e^{2p}y(p) - 7e^p y(p) + 10y(p) = \frac{16e^{-p}}{p(1-e^{-p})}$$

$$\Rightarrow y(p) = \frac{16e^{-p}}{p(1-e^{-p})(e^p - 5)(e^p - 2)} = \frac{1}{p}\left[\frac{4}{e^p - 1} + \frac{4}{3(e^p - 5)} - \frac{16}{3(e^p - 2)}\right]$$

$$= \frac{1}{p}\left[\frac{4e^{-p}}{1-e^{-p}} + \frac{4e^{-p}}{3(1-5e^{-p})} - \frac{16}{3(1-2e^{-p})}\right].$$

Taking inverse Laplace transform, we get

$$a_n = 4n + \frac{5}{3}(5^n - 1) - \frac{32}{3}(2^n - 1), n = 0, 1, 2, 3, \cdots.$$

EXERCISE

- Solve the following difference equation using Laplace transform:

1. $a_{n+2} - 3a_{n+1} + 2a_n = 0, a_0 = 0, a_1 = 1,$
2. $a_{n+2} - 5a_{n+1} + 6a_n = 0, a_0 = 1, a_1 = 0,$
3. $a_{n+2} - 7a_{n+1} + 10a_n = 0, a_0 = 6, a_1 = 2,$
4. $a_{n+2} - (b+c)a_{n+1} + bca_n = d^n, a_0 = 0, a_1 = 0,$
5. $a_{n+2} - (b+c)a_{n+1} + bca_n = b^n, a_0 = 0, a_1 = 0,$
6. $a_{n+2} - 2ca_{n+1} + c^2 a_n = 0,$
7. $a_{n+2} + 5a_{n+1} - 6a_n = 2n + 1, a_0 = 0, a_1 = 1,$
8. $a_{n+2} + 4a_{n+1} - 5a_n = 24n - 8, a_0 = 3, a_1 = -5,$
9. $a_{n+2} - 2a_{n+1} + a_n = n + 2, a_0 = 0, a_1 = 1,$
10. $a_{n+3} - 2a_{n+2} - a_{n+1} + 2a_n = 0, a_0 = 0, a_1 = a_2 = 1,$
11. $a_{n+2} - 2ca_{n+1} + c^2 a_n = c^n, a_0 = a_1 = 0.$

ANSWERS

1. $a_n = \frac{1}{4}[1 - (-3)^n],$
2. $a_n = 3\left(\frac{2}{3}\right)^n - 2,$
3. $a_n = \frac{28}{3}2^n - \frac{10}{3}5^n,$
4. $a_n = \frac{(c-d)b^n + (d-b)c^n + (b-c)d^n}{(b-c)(b-d)(c-d)},$
5. $a_n = \frac{(n-1)b^n + c^n - ncb^{n-1}}{(b-c)^2},$
6. $a_n = a_0 c^n + (a_1 - ca_0)\, nc^{n-1},$
7. $a_n = \frac{5}{2}(3^n + 1) - 5 \cdot 2^n + n + \frac{5}{2},$
8. $a_n = 2n^2 - 4n + 2 + (-5)^n,$
9. $a_n = \frac{1}{4}(3n-1)(-1)^n + \frac{1}{4}(n+1),$
10. $a_n = \frac{1}{3}[2^n - (-1)^n],$
11. $a_n = n(n-1)\, c^{n-2}.$

INDEX